Chemistry and Physics of Carbon

VOLUME 30

Chemistry and Physics of Carbon

Edited by
LJUBISA R. RADOVIC
Department of Energy and Geo-Environmental Engineering
The Pennsylvania State University
University Park, Pennsylvania

VOLUME 30

CRC Press is an imprint of the
Taylor & Francis Group, an **informa** business

CRC Press
Taylor & Francis Group
6000 Broken Sound Parkway NW, Suite 300
Boca Raton, FL 33487-2742

First issued in paperback 2020

ISBN-13: 978-0-367-57756-8 (pbk)
ISBN-13: 978-1-4200-4298-6 (hbk)

Visit the Taylor & Francis Web site at
http://www.taylorandfrancis.com

and the CRC Press Web site at
http://www.crcpress.com

Contents

Preface

This anniversary volume contains four chapters on a wide range of chemical and physical aspects of carbon science and technology. With it, as with its 29 predecessors, we want to illustrate the breadth and depth of knowledge required from carbon researchers for mastery of their subject. In today's age of emphasis on teamwork and networking, there is no question that carbon research is the ultimate example of both *multi*disciplinarity and *inter*disciplinarity.

No one knows more about the physics *and* chemistry of activated carbon *surfaces* than Prof. Angel Linares and his collaborators in Alicante. Following up on their important review of microporosity and adsorption phenomena in Vol. 21, here we can learn from the most authoritative source about important details and some general trends in a specific (and rather remarkable!) example of careful preparation and thorough characterization of activated carbons. No one knows more about the physics *and* chemistry of carbon *materials* than Prof. Michio Inagaki, and we are privileged to include in this volume another valuable contribution from his research group. Clean-up of oil spills is a fascinating, demanding and, alas, an increasingly important application of porous carbon materials. The authors inject an insight into the fundamentals of heavy oil sorption while attempting to resolve an eminently practical problem using carbons ranging from graphite to charcoal.

The remaining contributions are from relative 'newcomers' in carbon research, who have nevertheless taken head on the ambitious task of summarizing the state of the art of an old and a new issue in carbon (nano)science and (nano)engineering. Dr. Burg and Prof. Cagniant present their view, and an update, regarding the methodology of characterizing carbon surfaces, a topic that has been reviewed in Vol. 8 and more recently in Vols. 24, 25 and 27; they emphasize that a thorough and reliable knowledge of the essential details of these chameleonic surfaces requires the use of a battery of complementary techniques. Dr. Zhao and his collaborators discuss the fascinating (and still evolving!) field of molecular-level design of the porosity and pore size distribution of carbons using template-based synthesis; they illustrate the fact that opportunities for developing tailor-made carbon materials for wide-ranging applications – in adsorption, catalysis, electricity storage and medicine – are today more exciting than ever.

Our previous 29 volumes were published over the past four decades by Marcel Dekker. We hope that the next four decades with our new publisher, Taylor & Francis (CRC Press), will be an equally fruitful, pleasant and rewarding experience.

Ljubisa R. Radovic
LRR3@psu.edu
University Park, PA, and Concepción, Chile, September 2007

Contributors to Volume 30

Philippe Burg *Laboratoire de Chimie et Applications, Université Paul Verlaine, Metz, France.*

Denise Cagniant *Laboratoire de Chimie et Applications, Université Paul Verlaine, Metz, France.*

D. Cazorla-Amorós *Inorganic Chemistry Department, Carbon Materials and Environment, University of Alicante, Spain.*

Wanping Guo *Department of Chemical and Biomolecular Engineering, National University of Singapore, Singapore.*

Michio Inagaki *Faculty of Engineering, Aichi Institute of Technology, Yakusa, Japan.*

Norio Iwashita *National Institute of Advanced Industrial Science and Technology, Ibaraki, Japan.*

M. A. Lillo-Ródenas *Inorganic Chemistry Department, Carbon Materials and Environment, University of Alicante, Spain.*

A. Linares-Solano *Inorganic Chemistry Deparatment, Carbon Materials and Environment, University of Alicante, Spain.*

Jiajia Liu *Department of Chemical and Biomolecular Engineering, National University of Singapore, Singapore.*

D. Lozano-Castelló *Inorganic Chemistry Department, Carbon Materials and Environment, University of Alicante, Spain.*

Fabing Su *Department of Chemical and Biomolecular Engineering, National University of Singapore, Singapore.*

Xiao Ning Tian *Department of Chemical and Biomolecular Engineering, National University of Singapore, Singapore.*

Masahiro Toyoda *Department of Applied Chemistry, Faculty of Engineering, Oita University, Oita, Japan.*

X. S. Zhao *Department of Chemical and Biomolecular Engineering, National University of Singapore, Singapore.*

Zuocheng Zhou *Department of Chemical and Biomolecular Engineering, National University of Singapore, Singapore.*

Contents of Other Volumes

Internal Friction Studies on Graphite, *T. Tsuzuku and M. H. Saito*
The Formation of Some Graphitizing Carbons, *J. D. Brooks and G. H. Taylor*
Catalysis of Carbon Gasification, *P. L. Walker, Jr., M. Shelef, and R. A. Anderson*

VOLUME 5
Deposition, Structure, and Properties of Pyrolytic Carbon, *J. C. Bokros*
The Thermal Conductivity of Graphite, *B. T. Kelly*
The Study of Defects in Graphite by Transmission Electron Microscopy, *P. A. Thrower*
Intercalation Isotherms on Natural and Pyrolytic Graphite, *J. G. Hooley*

VOLUME 6
Physical Adsorption of Gases and Vapors on Graphitized Carbon Blacks, *N. N. Avgul and A. V. Kiselev*
Graphitization of Soft Carbons, *Jacques Maire and Jacques Méring*
Surface Complexes on Carbons, *B. R. Puri*
Effects of Reactor Irradiation on the Dynamic Mechanical Behavior of Graphites and Carbons, *R. E. Taylor and D. E. Kline*

VOLUME 7
The Kinetics and Mechanism of Graphitization, *D. B. Fischbach*
The Kinetics of Graphitization, *A. Pacault*
Electronic Properties of Doped Carbons, *André Marchand*
Positive and Negative Magnetoresistances in Carbons, *P. Delhaes*
The Chemistry of the Pyrolytic Conversion of Organic Compounds to Carbon, *E. Fitzer, K Mueller, and W. Schaefer*

VOLUME 8
The Electronic Properties of Graphite, *I. L. Spain*
Surface Properties of Carbon Fibers, *D. W. McKee and V. J. Mimeault*
The Behavior of Fission Products Captured in Graphite by Nuclear Recoil, *Seishi Yajima*

VOLUME 9
Carbon Fibers from Rayon Precursors, *Roger Bacon*
Control of Structure of Carbon for Use in Bioengineering, *J. C. Bokros, L. D. LaGrange, and F. J. Schoen*
Deposition of Pyrolytic Carbon in Porous Solids, *W. V. Kotlensky*

VOLUME 10
The Thermal Properties of Graphite, *B. T. Kelly and R. Taylor*
Lamellar Reactions in Graphitizable Carbons, *M. C. Robert, M. Oberlin, and J. Méring*

1 Carbon Activation by Alkaline Hydroxides

*Preparation and Reactions, Porosity and Performance**

A. Linares-Solano, D. Lozano-Castelló, M. A. Lillo-Ródenas, and D. Cazorla-Amorós

CONTENTS

* This chapter is based on a plenary lecture presented at "Carbon2004," Brown University.

I. INTRODUCTION

Activated carbon (AC) is the collective name for a group of porous materials, consisting mostly of carbon, that exhibit appreciable apparent surface area and micropore volume (MPV) [1–4]. They are solids with a wide variety of pore size distributions (PSDs) and micropore size distributions (MPSDs), which can be prepared in different forms, such as powders, granules, pellets, fibers, cloths, and others. Owing to these features and their special chemical characteristics, they can be used for very different applications, for example, liquid- and gas-phase treatments and energy storage [1–7]. Considering the variety of fields in which AC is being used, it is extremely important to develop a suitable characterization for it. This will enable determination of the effect that its porous structure has on a given application, which will permit control and performance optimization, which in turn will facilitate the discovery of new applications. For this purpose, the three steps depicted in Figure 1.1 are needed. This figure emphasizes that characterization is one of the three essential steps that should not be omitted to optimize both the preparation of ACs and their applications.

In regard to the characterization of porous carbons, it should be pointed out, for a better understanding of the results presented in this chapter and of their importance, that all our ACs have been thoroughly characterized. Always, both N_2 at 77 K and CO_2 at 273 K adsorption isotherms have been determined for each sample studied. The combined use of both adsorptives improves their characterization considerably [8–13], allowing conclusions to be drawn that would not be possible without adsorption of CO_2 at 273 K. Nevertheless, for those readers interested in characterization of porous carbons, there is an extensive literature available, and the following general books, and their corresponding references, are recommended [14–17].

The development of ACs with tailored porosity is necessary to improve their performance in classical applications and to prepare better adsorbents to satisfy new and emerging applications [4–7,17,18]. The preparation of such ACs can be carried out by two different methods: the so-called physical and chemical activations [1,19–28]. The differences between them lie mainly in the procedure and the activating agents used.

Physical activation has traditionally included a controlled gasification of the carbonaceous material that has previously been carbonized, although occasionally the activation of the precursor can be done directly. Many different carbonaceous precursors have been employed for physical activation: lignocellulosic materials, coals, woods, and materials of polymeric origin. The samples are typically treated to 800–1100°C with an oxidant gas, mainly CO_2 or steam, so that carbon atoms are removed selectively. Although this process obviously involves a chemical reaction (and is not merely a physical process), it is known as *physical activation*.

Careful control of the carbon atom removal process by gasification in CO_2 or steam, usually termed *burn-off (BO) degree*, allows selection of the adsorption characteristics of an AC. Thus, highly activated carbons can be prepared

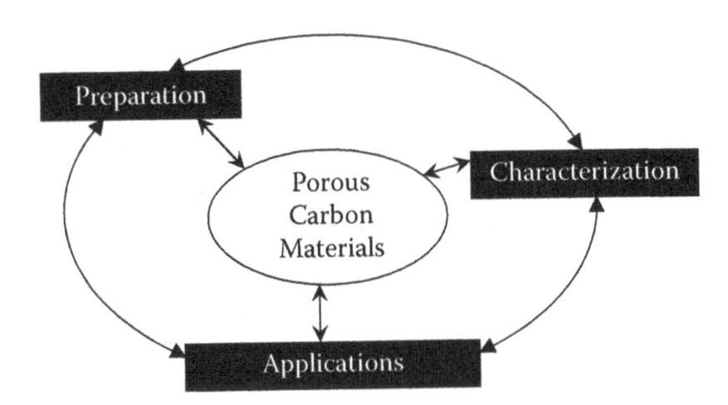

FIGURE 1.1 Relationship between the preparation, characterization, and applications of porous carbon materials that should always hold.

by physical activation, reaching high BO degrees. However, high BO degree is directly related not only to the development of the micropore volume, but also to the widening of porosity. Therefore, an AC with high adsorption capacity and narrow MPSD cannot be prepared by this process. This is one of the main disadvantages of physical activation: narrow MPSDs are needed for some applications (e.g., methane or hydrogen storage). The chemical activation process consists of contacting a carbonaceous precursor with a chemical activating agent, followed by a heat treatment stage, and finally by a washing step to remove the chemical agent and the inorganic reaction products [1,3,6]. In the literature, the use of several activating agents, such as phosphoric acid [29–32], zinc chloride [19,20,33–35], alkaline carbonates [36,37], KOH [23,38–43], and more recently NaOH [23,35,39,43,44], has been reported. In the case of chemical activation, we prefer not to use the term *BO degree*, as for physical activation; we prefer to talk about the *extent carbon reacted* (or *degree of activation*).

Chemical activation offers well-known advantages [3,19,21,42,43] over physical activation, which can be summarized as follows: (1) it uses lower temperatures and heat treatment times, (2) it usually consists of one stage, and (3) the carbon yields obtained are typically higher. On the other hand, chemical activation has some disadvantages [3,6,21,42,43], such as the need for a washing stage after heat treatment and the more corrosive behavior of the chemical agents used in comparison with CO_2 or steam. Traditionally, chemical activation has been carried out using one of two activating agents: phosphoric acid or zinc chloride.

In the case of chemical activation with phosphoric acid, lignocellulosic materials are preferred as precursors [6,30,45]. At low degrees of activation, the ACs do not have a highly developed area and they are essentially microporous, whereas at higher activation degrees, the surface area and the MPV increase, but there is also a remarkable increase in the mesopore volume and a widening of the MPSD [46,47]. Therefore, as with physical activation, in the case of activation with phosphoric acid both high adsorption capacity and narrow MPSD cannot be achieved. However, for ACs that need a well-developed mesoporosity, for example, for gasoline removal [5], phosphoric acid activation is a very suitable activation method [45].

The ACs prepared by chemical activation with $ZnCl_2$ are essentially microporous [19,20,33]. The loading of zinc has an important effect on porosity: samples activated at high loadings exhibit high porosity development and MPV, but also a more heterogeneous MPSD [33]. Although higher micropore volumes can be obtained by $ZnCl_2$ activation than by physical activation or by phosphoric acid activation, the increase in porosity development is also accompanied by a widening of the microporosity. The main disadvantage of this activating agent is that the emission of metallic zinc may cause serious environmental problems, which strongly limits its present use.

In summary, as is the case with physical activation, highly activated carbons with narrow MPSDs can hardly be prepared by chemical activation with either phosphoric acid or zinc chloride. Therefore, they can neither satisfy the demand for improving some AC applications nor for finding new ones, as high-pressure gas storage (e.g., CO_2, H_2, natural gas) requires both high MPV and narrow MPSD.

The purpose of developing ACs with tailored porosity in the whole range of microporosity has motivated continued research toward the use of other activating agents, such as alkaline hydroxides. After the pioneering patent of Wennerberg and O'Grady [48], the production of very-high-surface-area (superactive) ACs using alkaline hydroxide activation began on a commercial basis, first by Amoco and then by Kansai Coke and Chemical Company (Japan) [41,48]. In addition, a considerable number of studies have been carried out over the years that focus on chemical activation with hydroxides [20,21,23,35,38–45,48–97].

An example of the great attention that alkaline hydroxides have attracted recently can be seen in Figure 1.2, which divides the papers published in the journal *Carbon* in 2004 according to their AC preparation methods.

In this large number of papers, we can see that most of the variables affecting the activation process have been analyzed, especially the precursor, the hydroxide/carbon ratio, and the reaction temperature. From a cursory analysis of all these papers, it can be concluded that carbon activation by alkaline hydroxides is a promising process. However, very often a comparison between the results obtained by different authors is difficult and almost impossible to make, because of the different variables used (e.g., the activation process of a given lignocellulosic precursor is very different from that of its corresponding char). Therefore, general conclusions about this interesting activation process have not been compiled and are missing from the otherwise abundant literature.

During the last 15 years our group has been researching carbon activation by alkaline hydroxides, analyzing its different aspects. Thus, we have compared both NaOH and KOH with other chemical agents [23], we have analyzed several variables of this activation process [21,23,42,43,70–72,77,88,98], we have studied the reactions taking place [99–101], we have worked on their porosity characterization [12,13,70,102–106], and, finally, we have studied these ACs in several applications [71,107–114]. From all these studies, which cover most of the aspects of carbon activation by alkaline hydroxides, we have been able to identify additional advantages to those mentioned earlier: (1) the precursor ash content has no

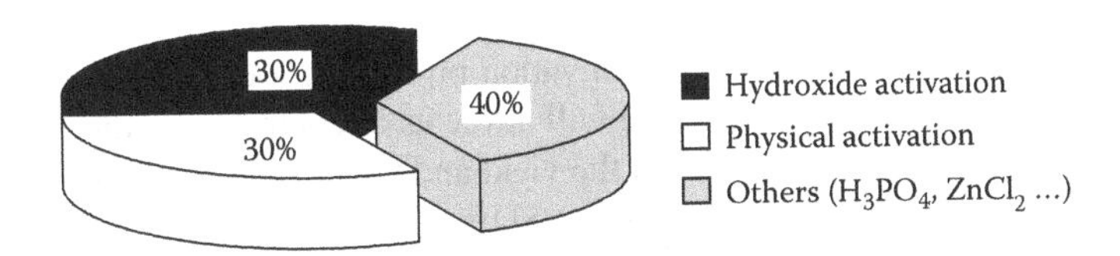

FIGURE 1.2 Distribution of the number of papers published in the year 2004 in the journal *Carbon* related to preparation of activated carbons by different methods.

effect on the activation, which is especially interesting in the case of coals; (2) the resulting ACs have low ash content (usually less than 1–2 wt% even with simple water washing); (3) the activation process is highly reproducible; (4) the control of experimental variables allows production of highly microporous ACs; and (5) precursors such as nanotubes that cannot be activated by physical activation can be activated with hydroxides [115].

Considering the increasing interest in the use of alkaline hydroxides as activating agents and the large number of published papers, the present review is timely because it provides an update on the most important aspects of this activation process. As our studies of carbon activation by alkaline hydroxides do not significantly differ from most of those available in the literature, the results obtained over the last 15 years will be used as the main source for review of this interesting process. The advantage of using our own results is that much better comparisons can be made, and the main objective of the chapter can be fulfilled more readily: to analyze and summarize the fundamentals of carbon activation by alkaline hydroxides (variables, reactions, types of porosity produced, and applications). The general concepts discussed are thought to be useful tools that will help the reader understand previously published results, even if they were obtained using different experimental conditions.

To achieve the general objective of this chapter, the following aspects of activation by sodium and potassium hydroxides will be analyzed:

1. Variables of the alkaline hydroxide activation method and suitability of the process
2. Singular characteristics of the AC prepared by alkaline hydroxide activation that can have both high MPV and narrow MPSD
3. Main reactions occurring during activation by hydroxides
4. Importance of controlling MPV and MPSD to improve performance applications of these ACs

To limit the length of the chapter, the discussion of these issues will be restricted to carbon materials (any type, independent of their origin, e.g., coals, chars, nanotubes). Hence, other carbonaceous materials (such as polyaromatic hydrocarbons, resins, polymers, lignocellulosic materials, pitches) that can also be activated when heated in the presence of hydroxides will not be considered.

II. PREPARATION AND VARIABLES

Many variables are involved in the activation process of a carbon by a hydroxide. A knowledge of their importance will allow us to control them and hence to improve the process in terms of both the yield and the characteristics of the ACs prepared (i.e., adsorption capacity and MPSD).

Among these variables, the two crucial ones are the chemical nature of the activating agent and the precursor used. KOH and more recently NaOH are the two most-used hydroxides for the preparation of ACs. In principle, any carbonaceous material can be used to prepare ACs; however, among the precursors most often used there are mainly two types: lignocellulosic materials and coals. Once the precursor and the activating agent are chosen, other experimental variables also affect the overall activation process in any of the main steps of the process: (1) hydroxide–carbon contacting, (2) heat treatment, and (3) washing. A schematic representation of these preparation steps is shown in Figure 1.3.

To acquire a good knowledge of the effect of the different preparation variables on the resulting porous texture, a thorough analysis of the literature must be performed. The main variables affecting the final porosity of the AC are the nature of the precursor, the nature of the hydroxide, the hydroxide/precursor ratio, the contacting method, the heating rate, the final temperature, the reaction time, and the flow rate used during the heat treatment. An inspection of the literature shows that most of the papers related to chemical activation with hydroxides focus on preparation variables (several or a few of them) [20,23,35,38–44,49–97]. However, there are some parameters that have not been analyzed by most investigators. This is true of the hydroxides used and their comparison (not only KOH but also NaOH), the use of new precursors (such as carbon nanotubes), the nitrogen flow rate, the method of mixing of carbonaceous precursor and activating agent (e.g., impregnation or physical mixing), and the washing stage (e.g., washing with water or with hydrochloric acid). Over the last 15 years, we have paid attention not only to the variables most frequently studied but also to the less frequently analyzed ones. Considering that in a given paper not all the variables can be analyzed and that, from our experience, all of them may be very important, in this section their effect will be analyzed in some detail.

A. Effect of Precursor

We begin by emphasizing that only carbon precursors will be analyzed (and not carbonaceous precursors) and reiterate that the selection of the precursor is crucial for the final porous texture of the AC. In this sense, Figure 1.4 presents the

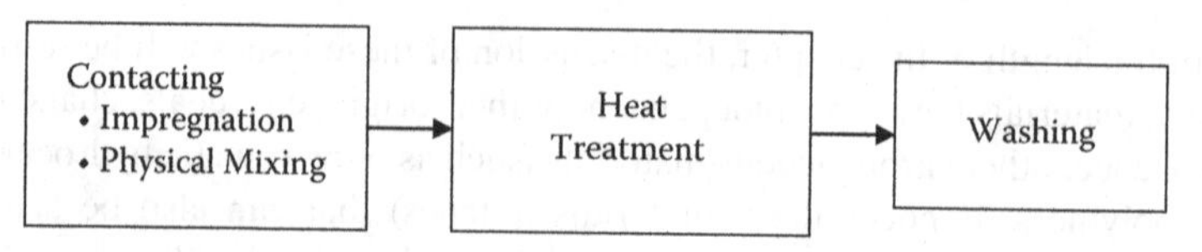

FIGURE 1.3 Scheme of the chemical activation procedure.

N_2 adsorption isotherms at 77 K corresponding to samples prepared from different precursors. All the samples were prepared by chemical activation with either NaOH or KOH. The selected precursors include materials with very different structures, from coals to nanotubes. The results of Figure 1.4a, corresponding to NaOH activation, show that all the precursors, except graphite, can be successfully activated, allowing us to get ACs with very different adsorption capacities regardless of their ash content. Accordingly, these isotherms clearly show that the precursor itself has a remarkable influence on adsorption capacity and MPV, as well as on the isotherm shape, which itself is indicative of different MPSDs. Most of the isotherms in Figures 1.4a and 1.4b are of Type I, corresponding to essentially microporous adsorbents, except for the activated multiwalled carbon nanotubes (MWCNTs).

The adsorption capacity sequence of Figure 1.4a (increasing from graphite to subbituminous coal) can be different when changing the nature of the activating agent (e.g., KOH to NaOH) even if most other variables are kept constant. Figure 1.4b presents the results of activation by KOH. We observe better porosity development for the anthracite than for the subbituminous coal, and the activation of the MWCNT is much more effective with KOH than when NaOH is used. A remarkable observation is that, although graphite reacts with these compounds, no development of porosity occurs.

B. Effect of Activating Agent: NaOH versus KOH

As illustrated earlier, the activation results for KOH and NaOH are not necessarily the same. In the following text, the comparison between the performance of NaOH and KOH will be made using four different precursors and keeping the preparation method constant: (1) anthracite [43]; (2) subbituminous coal [116]; (3) carbon fibers (CF) [88,98]; and (4) MWCNTs [72,101].

Figure 1.5a includes the N_2 adsorption isotherms of samples prepared from anthracite with either NaOH or KOH at a constant hydroxide/anthracite ratio, 3/1 by weight [43]. These isotherms show that both activating agents are effective for the preparation of ACs from anthracite, producing samples with high micropore volume and apparent BET (Brunauer Emmett Teller) surface area (up to 1 cm^3/g and 2750 m^2/g, respectively) easily. These results show that KOH produces a higher porosity development than NaOH. This observation and the conclusion derived from it, that KOH is a better activating agent than NaOH, cannot be generalized, as is usually done in the literature. In fact, the relative effectiveness of the hydroxides depends very much on the nature of the AC precursor. An example is shown in Figure 1.5b, where the results obtained with a lower-rank coal (subbituminous) using KOH and NaOH are presented. Contrary to what happened when activating the anthracite, in the case of subbituminous coal NaOH produces an AC with a higher adsorption capacity than KOH.

Figures 1.6a and 1.6b present the N_2 isotherms for ACs prepared using two types of carbon fiber (CF) precursors [88,98]: (1) anisotropic PAN-based carbon fibers from Hexcel, and (2) isotropic coal tar-pitch-based fibers from Donac. It

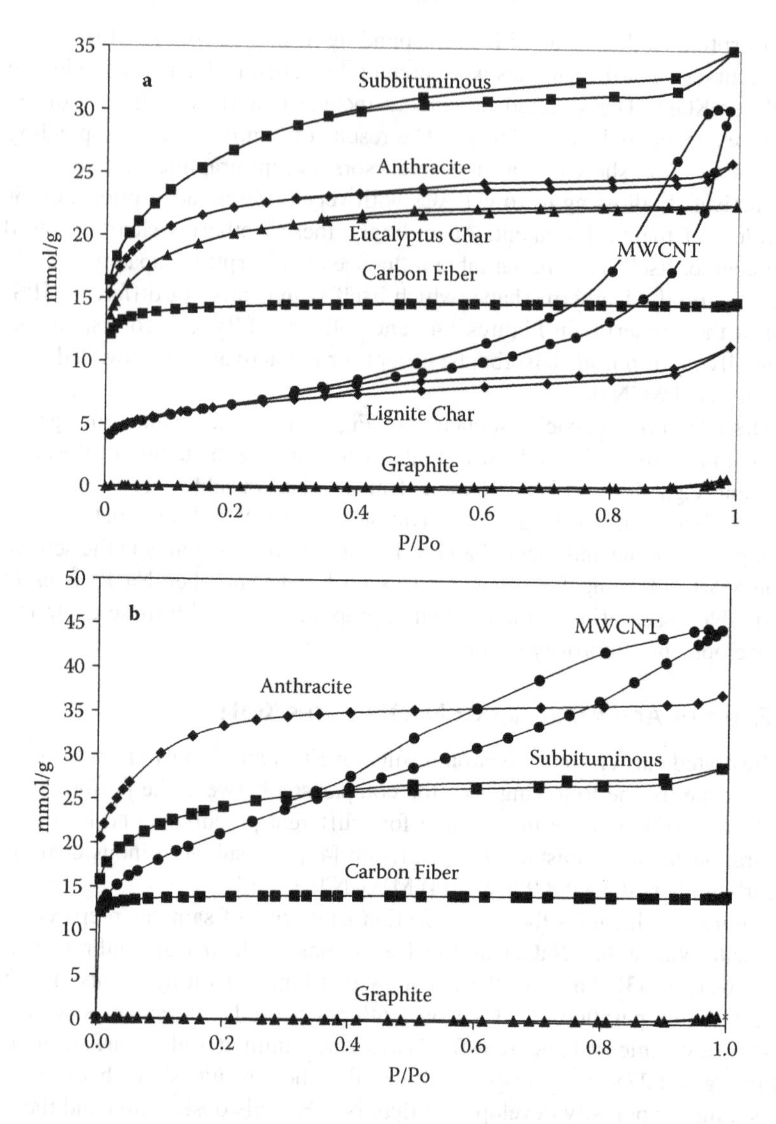

FIGURE 1.4 N_2 adsorption isotherms corresponding to samples prepared by chemical activation of different precursors with (a) NaOH and (b) KOH. Preparation conditions: physical mixing; hydroxide/precursor ratio 3/1 (weight); heating rate 5°C/min to 750°C; soaking time 1 h; N_2 flow rate during heat treatment 500 mL/min, except for the fibers and carbon nanotubes (carbon fibers: hydroxide/precursor 2/1 (weight); carbon nanotubes: hydroxide/precursor 4/1 (weight), heating to 800°C; N_2 flow rate during heat treatment 250 mL/min).

was observed that for the carbon fibers from Hexcel KOH is a better activating agent, whereas NaOH is a better activating agent for the carbon fibers from Donac. The main difference between these two precursors, and also between those in

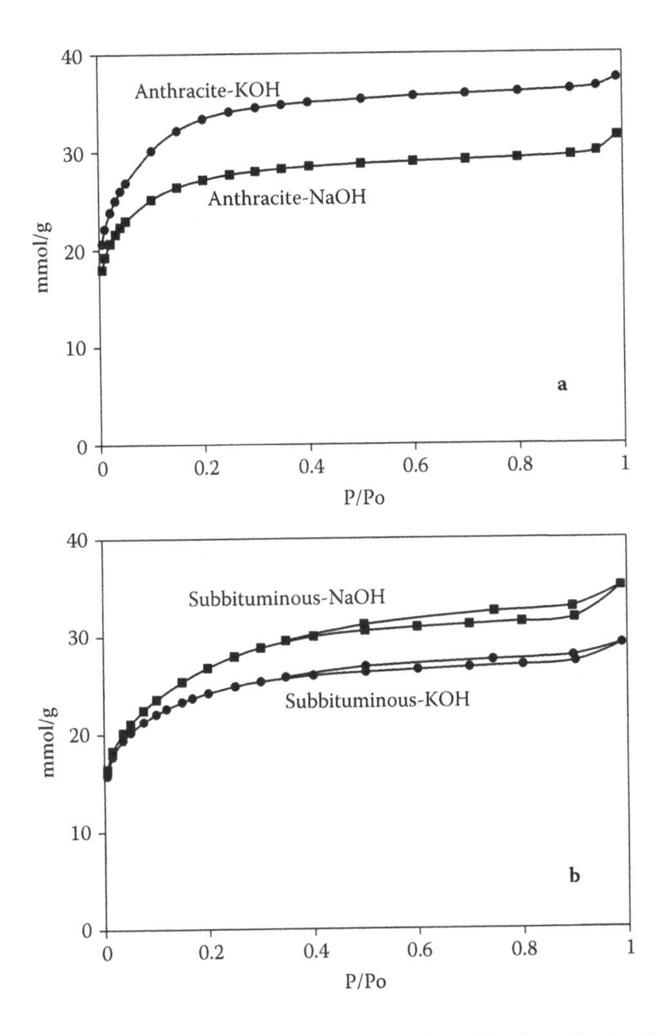

FIGURE 1.5 KOH versus NaOH activation comparison. N_2 adsorption isotherms at 77 K of (a) two activated carbons prepared from an anthracite and (b) two activated carbons prepared from a subbituminous coal. In both figures the same preparation procedure was employed (physical mixing, hydroxide/anthracite ratio 3/1 weight, heating rate 5°C/min to 750°C; soaking time 1 h; N_2 flow rate during heat treatment 500 mL/min).

Figure 1.5, seems to be their structural organization: both the anthracite and the Hexcel precursor (a high-performance anisotropic carbon fiber) have, relatively speaking, a much more ordered structure. These data show that, depending on the precursor, the best activating agent can be either KOH (e.g., for anthracite or Hexcel CF) or NaOH (e.g., subbituminous coal or Donac CF).

The results summarized in these figures have puzzled us for many years because there was no explanation for them in the literature. Analyzing many series of ACs prepared in our laboratory using both NaOH and KOH, and putting on one

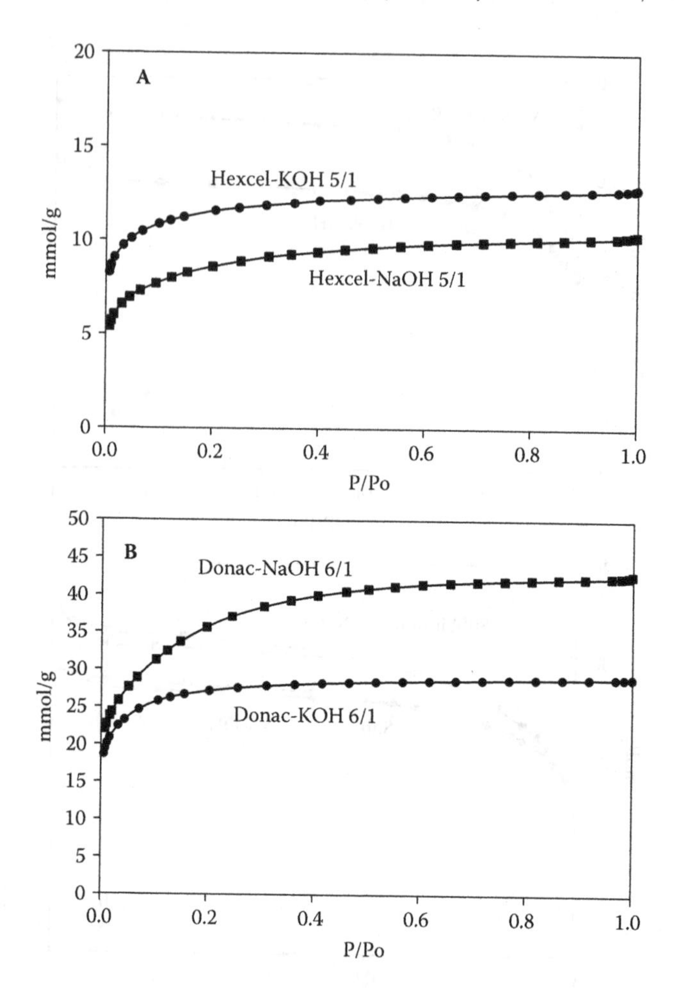

FIGURE 1.6 KOH versus NaOH activation comparison. N_2 adsorption isotherms at 77 K of (a) ACF from anisotropic carbon fibers from Hexcel and (b) ACF from isotropic carbon fibers from Donac (redrawn from Maciá-Agulló, J.A., Moore, B.C., Cazorla-Amorós, D., and Linares-Solano, A. *Carbon* 42(7): 1367–1370, 2004. With permission). In both cases the preparation procedure was by physical mixing (heating rate 5°C/min to 750°C; soaking time 1 h; N_2 flow rate during heat treatment 500 mL/min). The hydroxide/anthracite ratio was 5/1 (weight) for those prepared by Hexcel and 6/1 (weight) for the ACF prepared from Donac.

side those in which KOH gives better results and on the other those in which NaOH gives better results, we realized the importance of the structural order of the precursor and its effect in the comparative performance of both hydroxides. The main conclusion is that NaOH is better for carbons without structural order, whereas KOH produces better results for carbons having some structural order. The reasons for such behavior will be discussed on in more detail in Section IV.D.

This correlation can be corroborated by keeping constant the nature of the precursor (e.g., MWCNT) and changing its degree of structural order, which can be obtained by modifying the synthesis temperature [101]. Thus, KOH and NaOH activation have been compared in Figures 1.7a and 1.7b. The results confirm that for this ordered material the best activating agent is KOH, as with the more ordered precursors (Figures 1.4–1.6). Going from Figure 1.7a to 1.7b, the preparation temperature of the MWCNT has increased from 450 to 600°C and with it the structural order observed by x-ray diffraction (XRD). The comparison of these two figures allows us to observe again that the higher the order, the more difficult the hydroxide activation becomes.

C. Effect of Contacting Method

Two different ways of putting in contact the hydroxide and the AC precursor have been used: physical mixing and impregnation [42,43]. In the physical mixing method, different amounts of dry pelletized hydroxide are typically mixed at room temperature with a given amount of the carbonaceous precursor, also in a dry (powdered) state. In the impregnation method, which is the most common, the first stage consists of mixing by stirring, at a given temperature (e.g., 60°C), the precursor with different volumes of a hydroxide solution. After the impregnation process, the sample is often dried (usually by heating to 110°C) before being heat-treated for activation. This drying step can be an additional variable. Figure 1.8 compares the N_2 and CO_2 micropore volumes of samples prepared by impregnation using different times of drying and keeping constant the other experimental variables. Impregnation is seen to provide better porosity development when shorter drying times are used, the zero drying time being the best option, because formation of the alkaline carbonate is minimized. Based on these results, a drying step prior to heat treatment is not recommended.

The most popular method for chemical activation with hydroxides uses KOH and impregnation, probably because better porosity development results have been reported. An example is presented in Figure 1.9a, which compares the two methods. However, its implication (better impregnation than physical mixing) should not be mistaken for a general conclusion because the effect of the preparation method is strongly dependent on the activating agent used. In this sense, taking NaOH as the activating agent (rather scarcely studied in the literature [23,35,39,43,44]), a comparison of the two preparation methods shows that the sample prepared by physical mixing has a much better porosity development than that obtained by impregnation (Figure 1.9b). The apparent BET surface area is 1872 m^2/g for the AC prepared by physical mixing and 1248 m^2/g for that prepared by impregnation. As will be discussed next, this trend is confirmed for other NaOH/precursor ratios. This is an important observation because simple physical mixing of NaOH pellets with the carbonaceous precursor powder allows us to prepare high-adsorption-capacity ACs in less time and with less work than by impregnation.

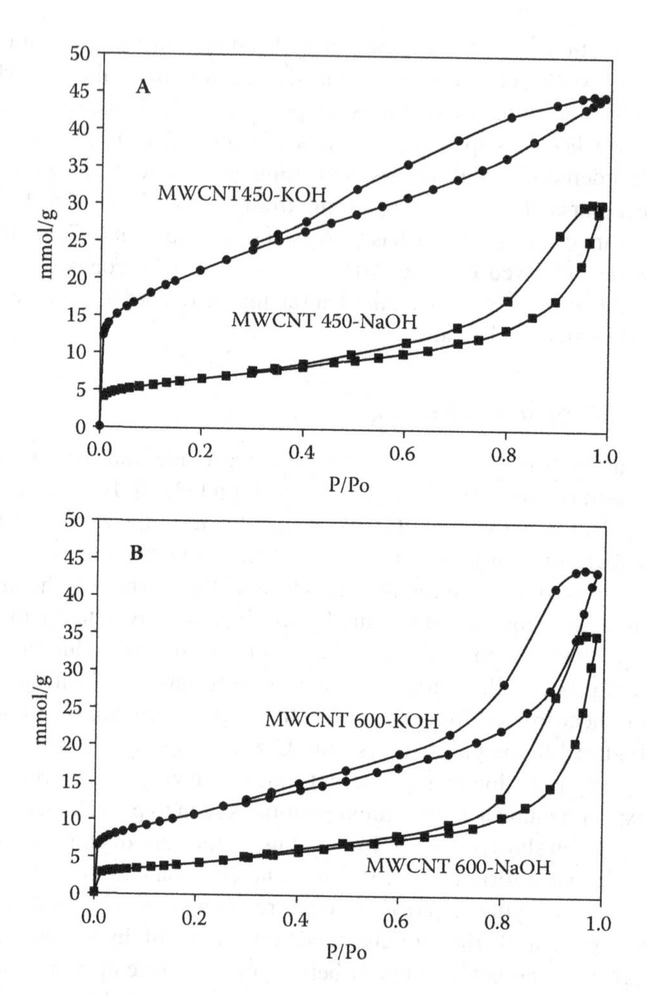

FIGURE 1.7 KOH versus NaOH activation comparison. N_2 adsorption isotherms at 77 K for chemically activated MWCNT (redrawn from Raymundo-Piñero, E., Azaïs, P., Cacciaguerra, T., Cazorla-Amorós, D., Linares-Solano, A., and Béguin, F. *Carbon* 43(4): 786–795, 2005): (a) MWCNT prepared at 450°C and (b) MWCNT prepared at 600°C.

The importance of the size of precursor and activating agent for porosity development of samples prepared by physical mixing has been analyzed. The decrease in the carbonaceous precursor particle size produces some increase in the final porosity. As an example, the decrease in the particle size from 2 to 0.6 mm leads to an increase in porosity that could be up to 10% for a bituminous coal. Regarding the activating agent size, comparison between an anthracite activated with hydroxide lentils or powder hydroxide shows no meaningful porosity difference.

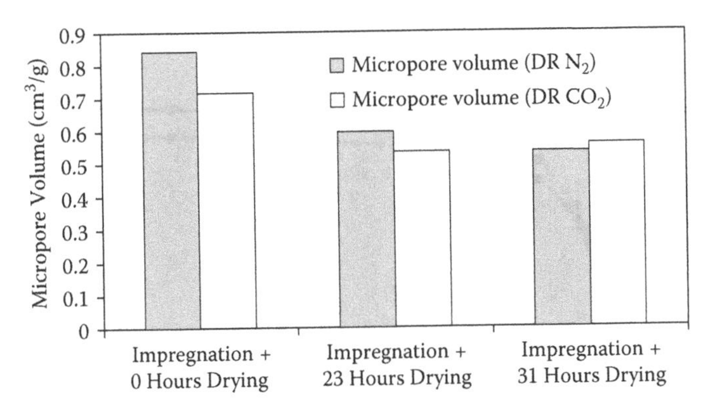

FIGURE 1.8 Comparison of the microporosity of samples prepared by impregnation using different drying times. Activating agent: NaOH (redrawn from Lillo-Ródenas, M.A., Lozano-Castelló, D., Cazorla-Amorós, D., and Linares-Solano, A. *Carbon* 39(5): 751–759, 2001. With permission).

D. Effect of Hydroxide/Precursor Ratio

The hydroxide/carbon ratio has been reported to be the most important parameter in a chemical activation process and, hence, it is the most frequently studied and used variable to develop the adsorption capacity of AC. Figure 1.10 summarizes the effect of KOH/anthracite ratio on the N_2 adsorption isotherms, as an example of the importance of this variable. An increase in this ratio produces a great enhancement of adsorption capacity at 77 K. A closer look at this figure shows that the shape of the adsorption isotherm also changes. An increase in the ratio produces a widening of the knee of the isotherm, indicating a change in the MPSD. Table 1.1 contains the porous texture characterization results obtained by applying the BET equation to N_2 adsorption at 77 K, and the DR equation to N_2 adsorption at 77 K and CO_2 adsorption at 273 K. The apparent BET surface area and the micropore volume both increase continuously with increasing KOH/anthracite ratio, reaching a maximum for a value of 4/1. These results also show that chemical activation of the anthracite produces adsorbents with a very high adsorption capacity, with apparent BET surface areas in some cases higher than 3000 m^2/g.

According to these results, and many others published in the literature [20,23,35,38–44,49–97], to increase the adsorption capacity of an AC high hydroxide/carbon ratios need to be used. However, in addition to the increase in surface area and micropore volume, it is also important to analyze the effect on the MPSD. Figure 1.11 presents the MPSD calculated by applying the Dubinin–Stoeckli (DS) equation [10,11] to the N_2 adsorption data. The higher the KOH/anthracite ratio, the wider the pore size distribution and the higher the mean pore size. These MPSD curves agree with what can be deduced from the difference in the micropore volumes calculated from N_2 and CO_2 adsorption data,

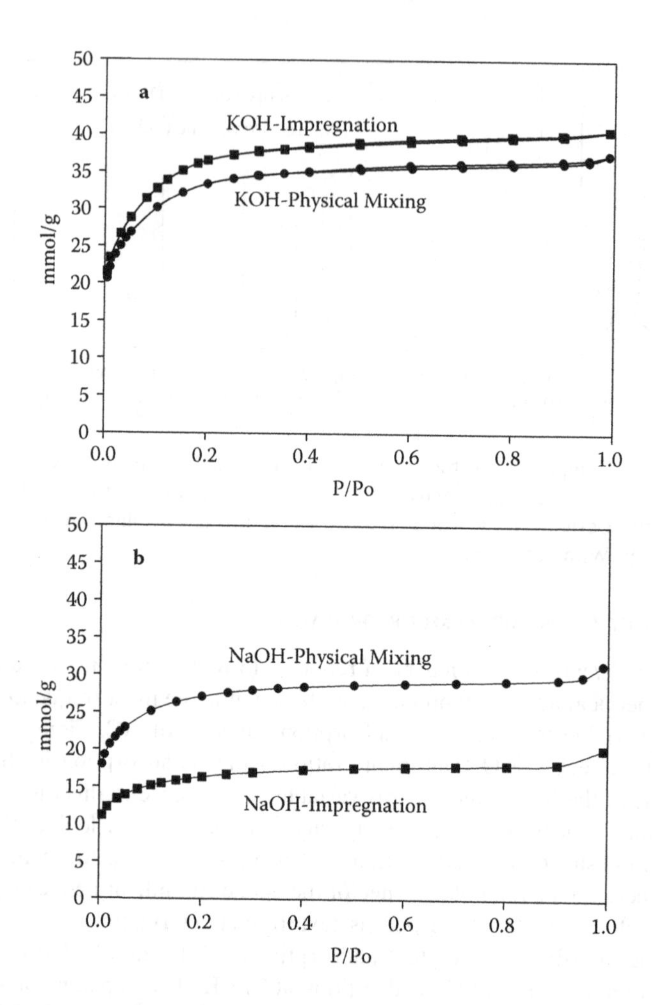

FIGURE 1.9 N_2 adsorption isotherms of samples prepared by chemical activation using either physical mixing or impregnation with a hydroxide/anthracite ratio of 3/1 (weight): (a) KOH and (b) NaOH.

which increases with the ratio (Table 1.1). The larger this difference, the wider the MPSD.

Thus, the KOH/anthracite ratio not only affects the micropore volume but also the micropore size distribution, which should be taken into account for the final use of the AC. If for a given application (e.g., for gas storage; see Section V.D) both high adsorption capacity and narrow MPSD are required, the hydroxide/carbon ratio has important limitations as a variable, and it cannot be useful for optimizing the preparation protocol of this type of AC.

It should be mentioned that, in the case of NaOH activation, independent of the preparation method (physical mixing or impregnation) and the precursor used, the effect of the hydroxide/carbon ratio shows similar tendencies in the evolution

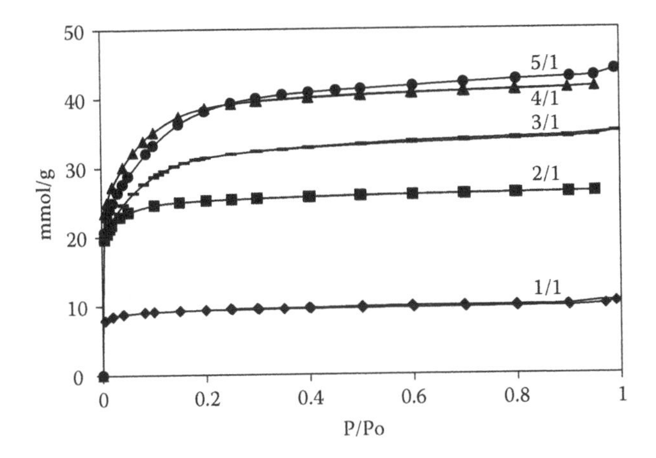

FIGURE 1.10 N_2 adsorption isotherms corresponding to samples prepared by chemical activation using different KOH/anthracite ratios. Preparation conditions: impregnation method; heating rate 5°C/min to 700°C; soaking time 1 h; N_2 flow rate during heat treatment 800 mL/min (redrawn from Lozano-Castelló, D., Lillo-Ródenas, M.A., Cazorla-Amorós, D., and Linares-Solano, A. *Carbon* 39(5): 741–749, 2001. With permission).

TABLE 1.1
Effect of KOH/anthracite ratio on porous texture (5°C/min, 800 mL/min, 700°C, 1 h)

Ratio	BET (m^2/g)	$V_{DR\ N_2}$ (cm^3/g)	$V_{DR\ CO_2}$ (cm^3/g)	$V_{DR\ N_2} - V_{DR\ CO_2}$ (cm^3/g)
1/1	726	0.33	0.37	<0
2/1	2021	0.89	0.86	0.03
3/1	2758	1.35	0.72	0.63
4/1	3290	1.45	0.81	0.64
5/1	3350	1.48	0.67	0.81

of surface area, MPV, and MPSD, as in the case of KOH activation. As an example, in Figure 1.12a we have compiled the N_2 adsorption isotherms of samples prepared from anthracite by physical mixing at different NaOH/carbon ratios, whereas Figure 1.12b shows the mean pore size (*L* calculated from the Dubinin equation) of the AC versus the ratio used. The conclusion is the same: as the ratio increases, the adsorption capacity increases and the MPSD becomes wider.

The influence of both variables, the ratio and the preparation method, can be seen in Table 1.2 for the case of an anthracite activated by NaOH. The table allows a comparison for samples prepared by impregnation and by physical mixing at different ratios. In both methods, the increase in the hydroxide/precursor ratio

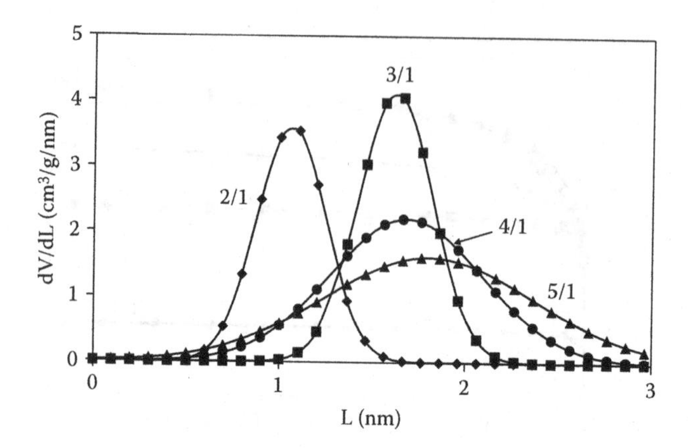

FIGURE 1.11 MPSD calculated by applying the Dubinin–Stoeckli (DS) equation to the N_2 adsorption data for samples prepared with different KOH/anthracite ratios (redrawn from Lozano-Castelló, D., Cazorla-Amorós, D., Linares-Solano, A., and Quinn, D.F. *Carbon* 40(7): 989–1002, 2002. With permission).

produces more extensive porosity development. This table also shows that, as previously mentioned, the physical mixing method provides higher adsorption capacities for all the NaOH/coal ratios, the differences between both methods being reduced as the activating agent/coal ratio increases. This behavior can be clearly observed in Figure 1.13, where the micropore volumes (calculated from N_2 adsorption at 77 K) obtained by impregnation and physical mixing are compared.

It should be pointed out that, using a very simple method—i.e., physical mixing—and a low activating agent/coal ratio (i.e., 1/1 and 2/1), ACs with well-developed surfaces can be obtained (e.g., 1000 and 1500 m^2/g, respectively). Higher ratios allow us to reach much higher values. Table 1.2 shows that, independent of the NaOH/anthracite ratio used, physical mixing gives better results than impregnation.

E. Effect of Heat Treatment

Under this category, three different variables are considered: heating rate, as well as time and temperature of heat treatment. These parameters affect the activation process independently of the activating agent (KOH or NaOH) and the precursor used.

In Table 1.3 we compiled an example of the heating rate effect on the activation process for a coconut shell char activated with NaOH. As a general rule, an increase in the heating rate produces a lowering of the adsorption capacity. During the heating process the hydroxide melts (melting points of KOH and NaOH are 360 and 318°C, respectively [117]); therefore, it is reasonable that a lower heating rate allows longer contact between the carbon and the molten hydroxide before the final reaction temperature is reached.

The effect of soaking time can be seen in Table 1.4, corresponding to the activation process of an anthracite using KOH (experimental conditions: KOH/anthracite, 2/1 (weight); activation temperature, 700°C; flow rate, 800 mL/min;

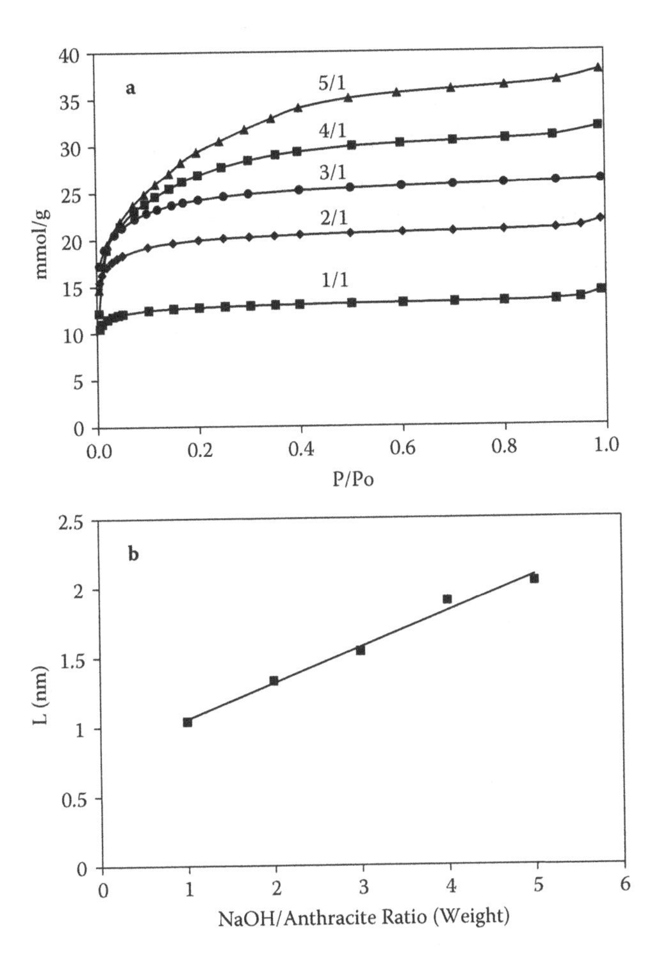

FIGURE 1.12 (a) N_2 adsorption isotherms corresponding to samples prepared from anthracite by physical mixing at different NaOH/carbon ratios and (b) mean pore size (*L* calculated from Dubinin equation) of the AC versus the ratio used.

heating rate, 5°C/min). At the maximum reaction temperature it can be observed that the micropore volume increases with soaking time, and that the higher the soaking time, the higher becomes the difference between micropore volume deduced from N_2 and CO_2 isotherms, which implies a wider MPSD [9,11].

The effect of the maximum heat treatment temperature (HTT) will be discussed next using two precursors: a lignocellulosic char and an anthracite. Figure 1.14a shows the N_2 adsorption isotherms of oak char heat-treated in the presence of NaOH at 550, 750, and 900°C. Figure 1.14b presents the results corresponding to four samples prepared at HTT from 700 to 850°C using KOH activation. In addition, results corresponding to a sample prepared by combining a higher heating rate and a higher HTT are also included.

TABLE 1.2
Porosity characterization for chemically activated carbons prepared from anthracite with NaOH activation by impregnation and physical mixing (5°C/min, 500 mL/min, 700°C, 1 h)

Ratio	Mixing Method	BET (m^2/g)	$V_{DR\ N_2}$ (cm^3/g)	$V_{DR\ CO_2}$ (cm^3/g)
1/1	Impregnated	101	0.04	0.07
1/1	Physical mix	1017	0.45	0.46
2/1	Impregnated	759	0.33	0.36
2/1	Physical mix	1594	0.67	0.67
3/1	Impregnated	1248	0.53	0.56
3/1	Physical mix	1872	0.80	0.73

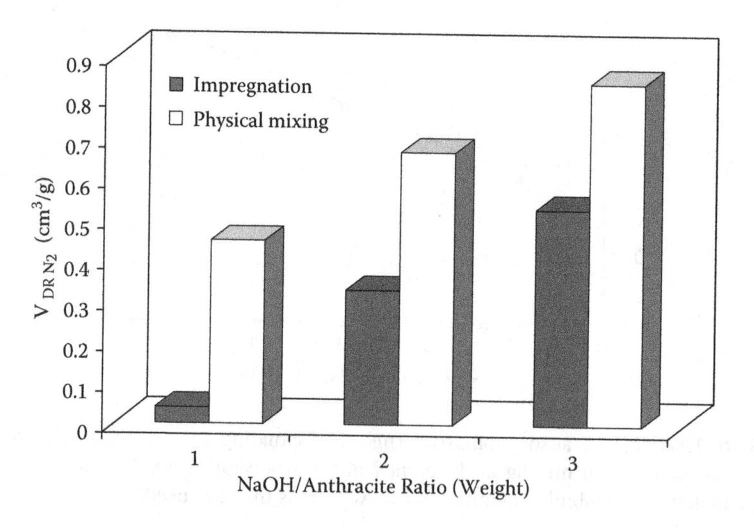

FIGURE 1.13 Comparison of the micropore volumes calculated from N_2 adsorption at 77 K for different ACs prepared by NaOH, using either impregnation or physical mixing method and different NaOH/anthracite ratios (redrawn from Lillo-Ródenas, M.A., Lozano-Castelló, D., Cazorla-Amorós, D., and Linares-Solano, A. *Carbon* 39(5): 751–759, 2001. With permission).

The N_2 isotherms of Figure 1.14a and 1.14b are, for all the samples, Type I isotherms, corresponding to samples that are essentially microporous. It can be seen that the maximum HTT has a great effect; as a general rule, an increase in the HTT produces an important enhancement of the adsorption capacity, concurrent with a widening of the MPSD.

As an example of porosity changes caused by different maximum HTT values, Table 1.5 summarizes the porous texture characterization results corresponding

TABLE 1.3
Effect of heating rate on porous texture (experimental conditions: physical mixing; NaOH/coconut shell char 3/1[weight]; activation temperature 750°C; flow rate 500 mL/min; soaking time 1 h)

Heating Rate (°C/min)	BET (m^2/g)	$V_{DR\,N_2}$ (cm^3/g)	$V_{DR\,CO_2}$ (cm^3/g)
5	2196	0.92	0.65
20	1704	0.75	0.59
50	1318	0.57	0.46

TABLE 1.4
Effect of soaking time on porous texture (KOH/anthracite 2/1 [weight], 800 mL/min, 700°C, 5°C/min)

Soaking Time (h)	BET (m^2/g)	$V_{DR\,N_2}$ (cm^3/g)	$V_{DR\,CO_2}$ (cm^3/g)	$V_{DR\,N_2}$ - $V_{DR\,CO_2}$ (cm^3/g)
1/2	1784	0.80	0.80	0
1	2021	0.89	0.86	0.03
2	2111	0.91	0.86	0.05

to ACs prepared from anthracite (Figure 1.14b). First, it should be pointed out that a remarkable adsorption capacity can be obtained from an anthracite precursor activated by KOH. Second, it can be seen that, from 700 to 850°C, the higher the HTT, the higher the apparent BET surface area and micropore volume. A closer look at the N_2 isotherms (Figure 1.14b) and at the N_2 and CO_2 micropore volumes (Table 1.5) allows us to observe that by increasing the HTT, the micropore size distribution becomes wider (the knee of the N_2 isotherms becomes wider and the difference between the micropore volumes determined by N_2 and CO_2 adsorption becomes larger). Figure 1.15 plots the mean pore size (L, calculated from the Dubinin equation) as a function of the HTT used, and confirms that L increases almost linearly with temperature, which is indicative of MPSD widening.

As seen in Tables 1.3 and 1.5 and Figures 1.14 and 1.15, both the heating rate and the maximum HTT affect the activation process. A drastic combination of these two variables is analyzed next. The sample prepared by changing in a remarkable way the HTT (950°C) and the heating rate (higher than 50°C/min) shows a very special behavior. In Figure 1.14b it can be observed that the shape of the isotherm for this sample changes drastically, indicating the presence of a large number of mesopores and a much smaller number of micropores (see V_{N_2} in Table 1.5). It should be pointed out that, for this sample, $V_{DR\,N_2} - V_{DR\,CO_2}$ is

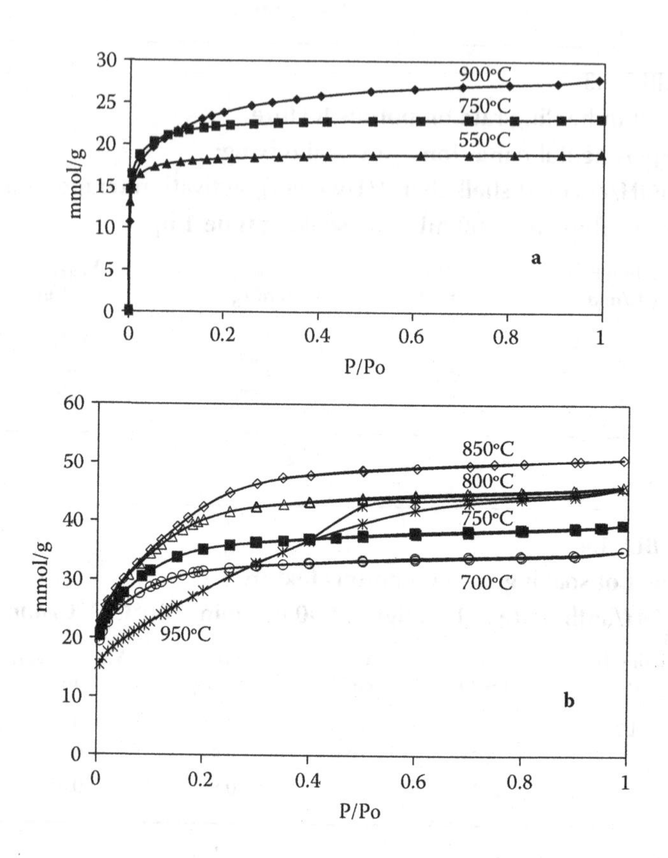

FIGURE 1.14 N_2 adsorption isotherms corresponding to samples prepared at different activation temperatures: (a) an oak char by chemical activation with NaOH. Preparation conditions: physical mixing; NaOH/char 2/1 (weight); heating rate 20°C/min; soaking time 1 h; N_2 flow rate during heat treatment 500 mL/min, (b) anthracite, KOH activation. Preparation conditions: impregnation; KOH/anthracite ratio 3/1 (weight); heating rate 20°C/min; soaking time 1 h; N_2 flow rate during heat treatment 500 mL/min (redrawn from Lozano-Castellô, D., Cazorla-Amorós, D., Linares-Solano, A., and Quinn, D.F. *Carbon* 40(7): 989–1002, 2002. With permission).

smaller than expected from its wide pore size distribution. This is because the mesopore volume contribution was not included in the linear region of the DR equation we selected to calculate V_{N_2}. If a different region of the DR plot were taken, the value of the micropore volume would be much higher and so would $V_{DR\,N_2} - V_{DR\,CO_2}$ (e.g., 1.02 cm^3/g).

These results show that heat treatment affects the activation process and that a high-adsorption-capacity AC with a wide MPSD, which can be useful for some applications, can be prepared by increasing (in addition to the hydroxide/carbon ratio, as discussed in Section II.D) the HTT and the reaction time. However, if a narrow MPSD is required, all these variables should be carefully controlled, especially the maximum HTT, because, as will be discussed in Section IV, a reaction

TABLE 1.5
Effect of maximum heat treatment temperature (HTT) on porous texture (KOH/anthracite 3/1 [weight], 5°C/min, 1 h, 800 mL/min)

HTT (°C)	BET (m^2/g)	$V_{DR\,N_2}$ (cm^3/g)	$V_{DR\,CO_2}$ (cm^3/g)	$V_{DR\,N_2}$ - $V_{DR\,CO_2}$ (cm^3/g)
700	2647	1.19	0.64	0.55
750	2758	1.35	0.72	0.63
800	3357	1.50	0.72	0.78
850	3506	1.51	0.68	0.83
950	2179	1.01	0.38	0.63 (1.02)[a]

[a] Higher linear DR range.

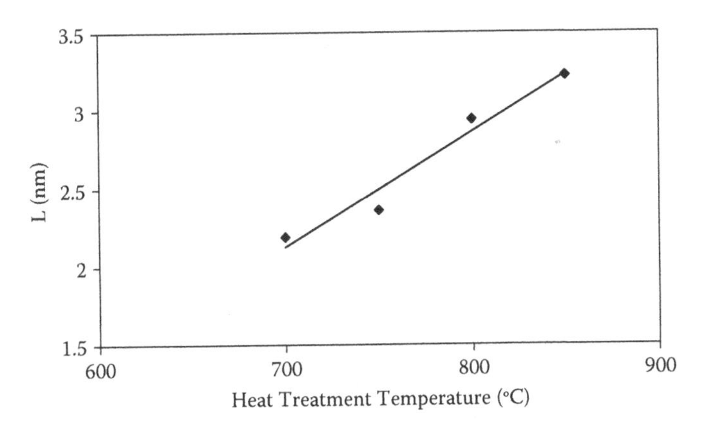

FIGURE 1.15 Mean pore size (*L* calculated from Dubinin–Stoeckli [DS] equation) of AC versus the maximum HTT.

temperature higher than 750°C will produce activation by a different mechanism (by a physical activation process) and will cause a significant decrease in the carbon yield of the activated product.

F. Effect of Nitrogen Flow Rate

In most chemical activation studies, nitrogen is chosen as the activation atmosphere (although, as will be shown in Section IV.B, we have analyzed the use of other inert gases). However, the effect of the nitrogen flow rate during the heat treatment process was not studied, or the modifications of the nitrogen flow rate were very small. In our previous studies [42,43], the results corresponding to a large variation of nitrogen flow rate were presented, showing its importance for the final porous texture of the samples. As a summary of these results, Figure 1.16

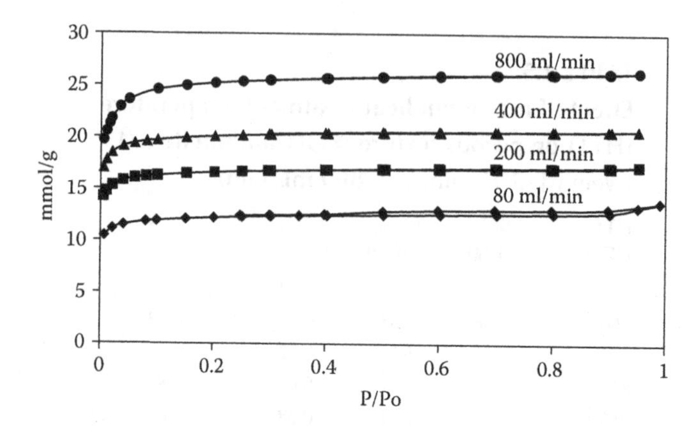

FIGURE 1.16 N_2 adsorption isotherms for activated carbons prepared from an anthracite by chemical activation with KOH using different N_2 flow rates during the heat treatment. Preparation conditions: impregnation method; KOH/anthracite 2/1 (weight); heating rate 5°C/min to 700°C; soaking time 1 h (redrawn from Lozano-Castelló, D., Lillo-Ródenas, M.A., Cazorla-Amorós, D., and Linares-Solano, A. *Carbon* 39(5): 741–749, 2001. With permission).

shows the N_2 isotherms for samples prepared using nitrogen flow rates from 80 to 800 mL/min. All these samples have been prepared using a KOH/anthracite ratio of 2/1 (by weight) and a HTT of 700°C for 1 h. It can be observed that, although all the preparation parameters are the same, the different nitrogen flow rates modify the porous texture of the samples drastically.

ACs with much higher adsorption capacities and much larger micropore volumes (see Figure 1.16 and Table 1.6) are obtained using higher nitrogen flow rates. It should be emphasized that for a 2/1 KOH/anthracite ratio, an increase of the nitrogen flow rate from 80 to 800 mL/min produces an AC with more than double the micropore volume and surface area. This increase is similar to that produced by changing the KOH/anthracite ratio from 1/1 to 2/1. Further comments, explaining the effect of the N_2 flow rate, will be analyzed in Section IV.

From a practical point of view, of course, not only are the surface area and MPV of porous materials important but so is their pore size distribution. The value of the difference $V_{DR\,N_2} - V_{DR\,CO_2}$ (Table 1.6) suggests that the flow rate does not produce an important change in the MPSD. To verify this effect, Figure 1.17 contains the MPSD obtained by the density functional theory (DFT) method [118] for samples presented in Table 1.6 and prepared at different N_2 flow rates.

The important finding here is that, contrary to the effect of hydroxide/anthracite ratio, an increase in N_2 flow rate does not appreciably change the MPSD of the samples. Thus, a flow characteristic (N_2 flow rate during the heat treatment process) opens a new way to prepare AC with both high adsorption capacity (high values of micropore volume and surface area) and narrow MPSD, which cannot be obtained by adjusting the other variables (i.e., hydroxide/carbon ratio, HTT, and soaking time).

TABLE 1.6
Effect of nitrogen flow rate on porous texture (KOH/anthracite 2/1, 5°C/min, 700°C, 1 h)

N_2 Flow Rate (mL/min)	BET (m^2/g)	$V_{DR\ N_2}$ (cm^3/g)	$V_{DR\ CO_2}$ (cm^3/g)	$V_{DR\ N_2}$ - $V_{DR\ CO_2}$ (cm^3/g)
80	945	0.43	0.43	0
200	1305	0.61	0.58	0.03
400	1580	0.73	0.71	0.02
800	2021	0.89	0.86	0.03

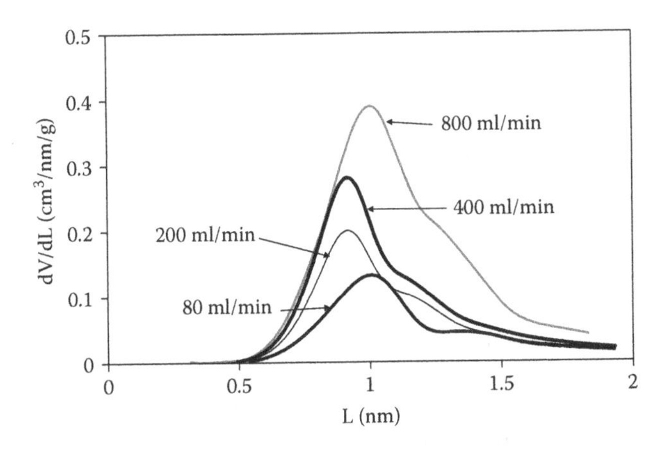

FIGURE 1.17 MPSD calculated by applying the DFT method to the N_2 adsorption data for samples prepared using different N_2 flow rates.

Although further work is in progress to understand the N_2 flow effect and its possible dependence on the type of furnace used, there is no doubt that the flow rate of inert gas (and thus mass transport) is a very important parameter. Hence, it has to be taken into account during KOH activation and also during NaOH activation, in which this effect has been also observed [43].

G. Additional Issues

In Section II we have seen that hydroxide activation is a complex process that is noticeably influenced by numerous variables. If these variables are known, and are controlled, this activation process is very reproducible, giving higher carbon yields than other activation processes (usually in the range of 50–80 wt%). In addition, the control of these variables will allow us to select the desired AC features for a given application. Thus, we can develop AC with different mechanical properties, packing densities, ash contents, micropore volumes, apparent surface

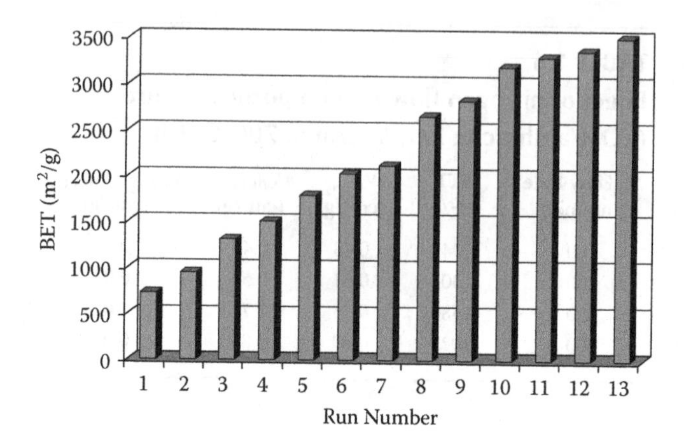

FIGURE 1.18 Apparent BET surface area corresponding to selected ACs prepared by KOH activation of an anthracite under different experimental conditions.

areas, and, additionally, it is possible to tailor the MPSD, even for ACs with high micropore volume.

To show an example of how these characteristics can be selected, Figure 1.18 presents the apparent BET surface areas of samples prepared by controlling selected variables of the activation process. The samples have been obtained by keeping constant the nature of the precursor (anthracite) and the nature of the hydroxide (KOH). It can be seen that a wide range of apparent BET surface areas can be easily obtained just by changing the preparation variables.

As will be discussed in Section V, ACs with a high packing density and good mechanical properties are required in several applications. A way to increase the packing density and to improve the mechanical properties is to agglomerate powdered AC using suitable binders. Figure 1.19 shows, as an example, a picture of two activated carbon monoliths (ACMs) prepared from a KOH AC using a commercial binder, as described in detail elsewhere [113]. These ACs are thermally stable and can be well agglomerated without significant loss of adsorption capacity. The fact that ACs can be conformed in different shapes and sizes broadens the scope of possible applications of these materials.

III. SINGULAR CHARACTERISTICS OF POROSITY DEVELOPED BY ALKALINE HYDROXIDE ACTIVATION

As discussed in Section II, ACs with well-developed porosity can be prepared by activation with alkaline hydroxides by changing the experimental conditions. However, from an application point of view, the precise tailoring of porosity (not only pore volume and surface area but also pore size distribution) is the most important aspect of the carbon activation process. Thus, in the present section we will emphasize the type of microporosity that can be developed during hydroxide

FIGURE 1.19 Photograph of activated carbon monoliths with different dimensions prepared from a KOH-activated carbon using a commercial binder.

activation, comparing it with that obtained by other activation processes (e.g., physical activation).

The singularity of the porosity developed during hydroxide treatment can be highlighted by analyzing the following simple experiment, which is nevertheless not well known [119]. If a given porous carbon (e.g., an activated carbon) is heat-treated to 700–900°C with a hydroxide, its starting porosity changes dramatically, exhibiting a noticeable narrowing of its MPSD. This narrowing is independent of the type and porosity of the AC used and of its preparation method. This behavior occurs even with low hydroxide/AC ratio (e.g., 0.5/1). Under such experimental conditions, the usual activation process does not take place; there is no increase in adsorption capacity and the yield of the product is almost 100%, indicating that carbon gasification by the hydroxide is negligible. However, an important change in the porosity of the starting porous material does occur. As an example, Figure 1.20 contains the N_2 isotherms at 77 K (Figure 1.20a) and the MPSD (Figure 1.20b) corresponding to the starting AC and the resulting KOH post-heat treatment AC at 750°C, using a KOH/AC ratio of 0.5/1 (by weight). It can be observed that after KOH post-heat treatment, the AC has a lower adsorption capacity and a narrower MPSD. Considering that when a carbon precursor is activated with KOH heat treatment at around 750°C is also carried out, it would be expected that, together with the porosity development, a narrowing of the porosity could also take place. Thus, chemical activation would give ACs with narrower MPSDs than physical activation.

To emphasize the advantage of chemical activation in producing ACs with narrow MPSDs, results obtained with chemical and physical activation are compared next. A comparison of the pore size distribution (PSD) developed during hydroxide activation (KOH and NaOH) with physical activation of carbon

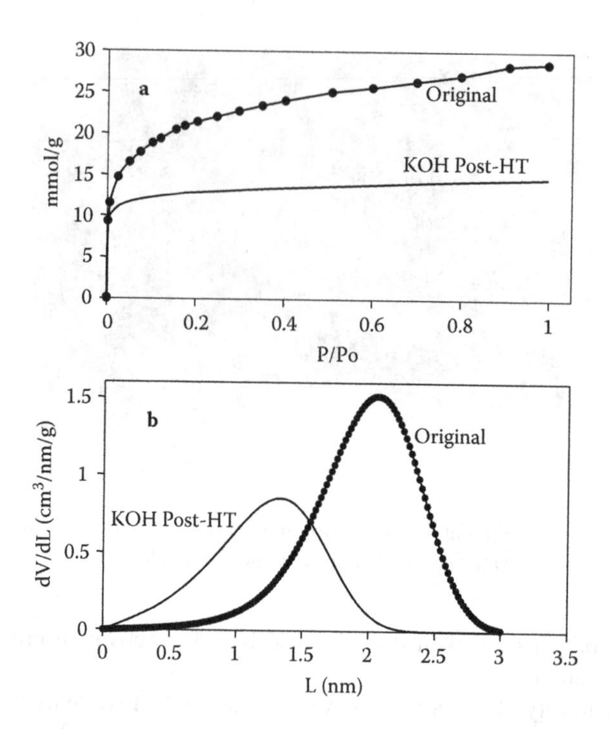

FIGURE 1.20 (a) N_2 adsorption isotherms at 77 K and (b) MPSD by applying the Dubinin–Stoeckli (DS) method corresponding to an AC, and the same material after KOH post-heat treatment (redrawn from Martin-Gullón, I., Marco-Lozar, J.P., Cazorla-Amorós, D., and Linares-Solano, A. *Carbon* 42(7): 1339–1343, 2004. With permission).

fibers (Donac) is shown in Figure 1.21 [88], using different hydroxide/carbon ratios (Figures 1.21a and 1.21b) and different BOs in CO_2 (Figure 1.21c). In both cases, an increase in hydroxide/carbon ratio up to 6/1 produces a widening in the MPSD. Interestingly, the change in PSD is different depending on the activating agent used: widening of porosity is more pronounced for NaOH than for KOH. In the case of physical activation with CO_2, an increase in the BO degree produces a larger widening of the PSD (Figure 1.21c), much larger than with any of the hydroxide activations.

For a more meaningful comparison between these activation processes, three ACFs with similar apparent surface area, prepared using KOH, NaOH, and CO_2, were selected. Table 1.7 summarizes some of their characteristics, including the product yields. It should be noted that the yield during hydroxide activation is much higher than during physical activation.

Because the porosity development of these three samples is very high (about 2500 m^2/g), SEM images have been obtained to compare the morphology changes as a function of the activating agent used. It can be observed in Figure 1.22 that fiber shape is not damaged after hydroxide activations whereas, for a very high activation degree with gaseous CO_2, the fiber is damaged and its diameter is considerably reduced.

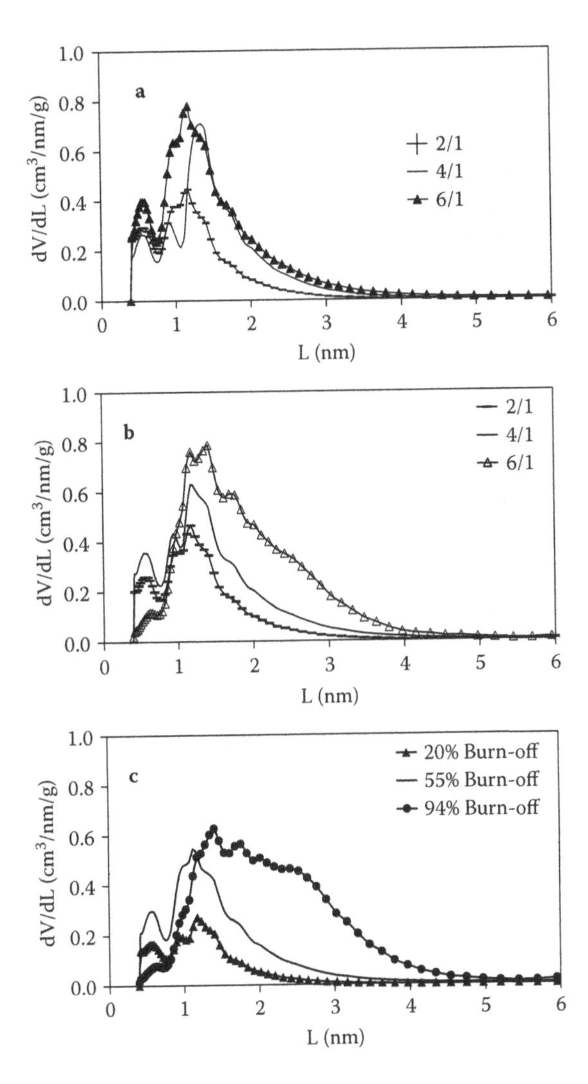

FIGURE 1.21 Pore size distribution by DFT method of ACF prepared from carbon fibers (Donac) by (a) KOH activation, (b) NaOH activation using different hydroxide/carbon fiber ratio, and (c) by physical activation with CO_2 at 890°C to different burn-off degrees (redrawn from Maciá-Agulló, J.A., Moore, B.C., Cazorla-Amorós, D., and Linares-Solano, A. *Carbon* 42(7): 1367–1370, 2004. With permission).

Figure 1.23 contains the PSD for the same three samples, showing again that it depends on the activation method. The narrowest MPSD is obtained with KOH, whereas the PSD becomes wider in the case of NaOH activation and much wider in the case of physical activation with CO_2. Thus, the widening of the MPSD during activation follows the order KOH < NaOH < CO_2.

TABLE 1.7
Apparent BET surface area, micropore volume, and yield for ACF prepared with different activating agents

Activating Agent	BET (m^2/g)	$V_{DR\,N2}$ (cm^3/g)	Yield (wt%)
CO_2	2487	0.86	6
NaOH	2541	0.96	40
KOH	2420	0.94	47

FIGURE 1.22 SEM images corresponding to ACF prepared with different activating agents: (a) KOH, (b) NaOH, and (c) CO_2 up to 94% burn-off.

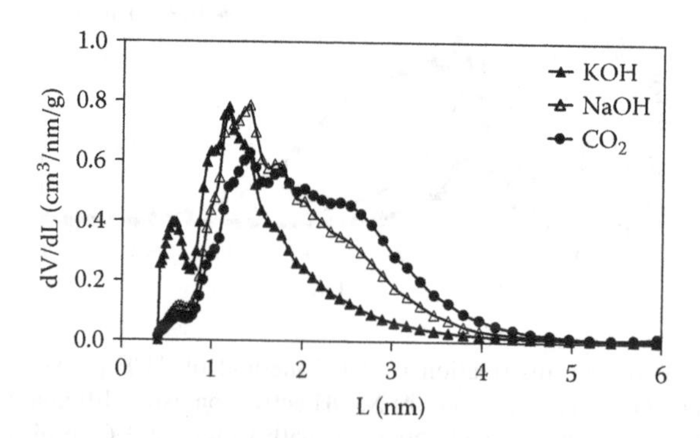

FIGURE 1.23 PSD by DFT method corresponding to ACF with similar apparent BET surface areas prepared with different activating agents.

These three independent results (product yield, SEM images, and PSD) highlight some of the advantages of carbon activation by hydroxides in relation to physical activation. Additional arguments and comparisons are presented next.

As has been shown for the case of carbon fibers, both a high adsorption capacity and a narrow MPSD cannot be obtained by physical activation. Is this a general feature of physical activation? To analyze in more detail the effect of the

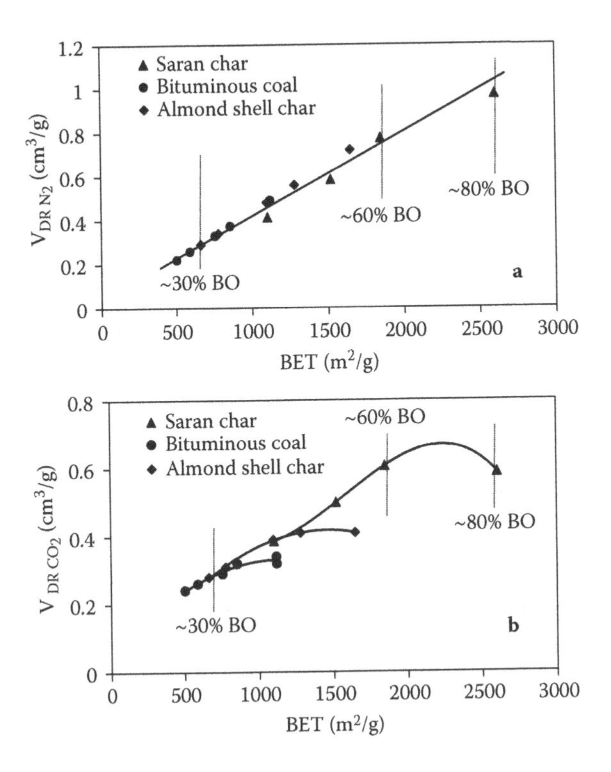

FIGURE 1.24 Relationship between (a) total micropore volume ($V_{DR\ N_2}$) and apparent BET surface area, and (b) narrow micropore volume ($V_{DR\ CO_2}$) and apparent BET surface area, for physically activated carbons with different burn-off degrees (redrawn from Lozano-Castelló, D., Cazorla-Amorós, D., and Linares-Solano, A. *Fuel Process Technol.* 77–78: 325–330, 2002. With permission).

activation method on the evolution of the MPSD, different precursors (e.g., carbon fibers, coal, Saran char, lignocellulosic materials) have been physically activated with CO_2 or with steam. Figure 1.24a plots the micropore volume (calculated from N_2 isotherms at 77 K) against the apparent BET surface area for samples with different BO degrees. It is seen that, as expected, microporosity develops linearly with BO, and there is a known and well-defined correlation between micropore volume and BET surface area [8]. This is the general behavior found for most ACs prepared by physical activation from different precursors [8,21,26,27,120,121]. However, if the evolution of the narrow micropore volume (calculated from CO_2 adsorption data) is analyzed in the same way (see Figure 1.24b), it can be seen that, although for low BO degrees the narrow micropore volume also increases linearly with BO, it reaches a maximum (for each precursor) whose value is a function of the precursor. These results indicate that $V_{DR\ N_2}$ and $V_{DR\ CO_2}$ develop differently with BO degree; for low BO, the activation process develops narrow microporosity, but for higher BOs (required for high adsorption capacity), narrow microporosity reaches a maximum and wider microporosity (supermicropores) is

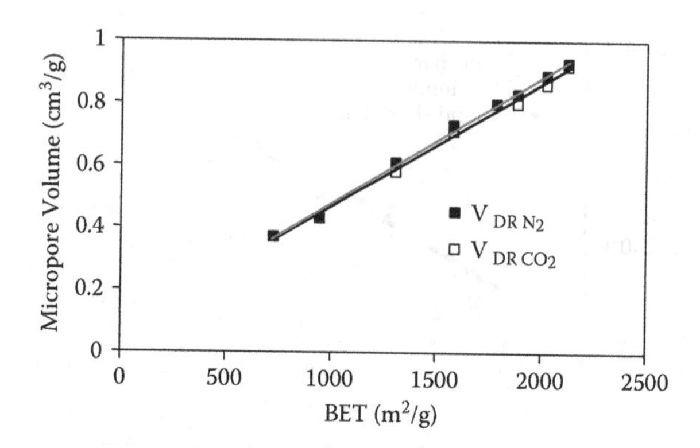

FIGURE 1.25 Relationship between the total micropore volume ($V_{DR\ N_2}$) and narrow micropore volume ($V_{DR\ CO_2}$) and the apparent BET surface area for KOH-activated carbons (redrawn from Lozano-Castelló, D., Cazorla-Amorós, D., and Linares-Solano, A. *Fuel Process Technol.* 77–78: 325–330, 2002. With permission).

also developed. Thus, there is no way to prepare an AC with high apparent surface area (BET > 2000 m^2/g) by physical activation without producing a widening of the microporosity.

From these results, it seems that an AC possessing simultaneously a high adsorption capacity (high micropore volume, $V_{DR\ N_2}$) and narrow MPSD ($V_{DR\ N_2} \approx V_{DR\ CO_2}$) cannot be prepared by physical activation. This could be an important drawback for applications such as methane and hydrogen storage (see Section V.D).

Considering the importance of preparing adsorbents with high adsorption capacity and a homogeneous MPSD, alkaline hydroxide activation with a selection of the most suitable variables has been carried out. Variable selection has been based on the results discussed in Section II.

Figure 1.25 summarizes the information analogous to that of Figures 1.24a and 1.24b, but this case corresponds to selected KOH-activated carbons prepared from different precursors (anthracite and bituminous coal). Contrary to what occurs in physical activation (Figures 1.24a and 1.24b), the plot of narrow micropore volume ($V_{DR\ CO_2}$) against BET surface area is linear, even for highly activated carbons ($V_{DR\ N_2}$ = 0.93 cm^3/g; apparent BET surface area = 2123 m^2/g; $V_{DR\ CO_2}$ = 0.92 cm^3/g). These results indicate that both activation processes develop microporosity but in different ways. A thorough analysis of the preparation variables of the chemical activation process has allowed us to select them carefully and to prepare ACs with very narrow micropore size distribution ($V_{DR\ N_2} \approx V_{DR\ CO_2}$), which cover a wide range of surface areas (from 700 to 2000 m^2/g). It should be noted that reaching an apparent BET surface area as high as 2000 m^2/g presents some difficulties, and it is even more difficult if this surface area corresponds mainly to narrow microporosity ($V_{DR\ CO_2}$ = 0.92 cm^3/g) [70]. This is an important advantage of hydroxide activation that merits special emphasis.

To illustrate the difficulty of obtaining a high narrow micropore volume, Figure 1.26 compares the narrow micropore volumes ($V_{DR\ CO_2}$) of a well-known

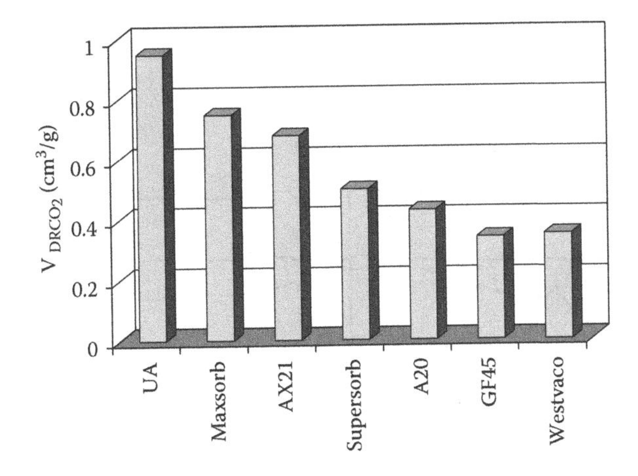

FIGURE 1.26 Narrow micropore volume ($V_{DR\,CO_2}$) corresponding to different commercial activated carbons and to a KOH-activated carbon prepared in our laboratory (UA).

(and high-quality) commercial AC, prepared using different activation processes and precursors: an ACF prepared by physical activation from Osaka Gas (A20); two phosphoric-acid-activated carbons from Norit (GF45) and from Westvaco (A1100); and three KOH-activated carbons (Maxsorb, AX21, and Supersorb). In addition, for comparison purposes, the figure also includes an AC prepared by KOH activation in our laboratory (UA). The results confirm that KOH activation gives the highest narrow micropore volumes ($V_{DR\,CO_2}$): the four samples prepared by KOH activation have much higher MPVs than those obtained by phosphoric acid and by physical activation. Also, it can be seen that sample UA has the highest narrow micropore volume, even higher than "Maxsorb" (whose apparent BET surface area is about 3100 m^2/g). The good performance of this material in various applications will be shown in Section V.

IV. REACTIONS OCCURRING DURING ACTIVATION

A full understanding of the variables that affect the hydroxide activation process requires a thorough knowledge of the main reactions occurring during heat treatment. Although there are published results dealing with the mechanism of chemical activation, most of them are related to phosphoric acid and zinc chloride activations, and only a few deal with hydroxide activation. However, in these studies, we often observe controversial results, perhaps because they have been obtained using very different carbonaceous materials (polyaromatic hydrocarbons, chars from lignocellulosic materials, polymers, coals, cokes, microbeads, and others [53,84,122–129]). Some carbonaceous precursors (e.g., polyaromatic hydrocarbons, lignocellulosic materials and polymers have, from a chemistry point of view, significant compositional and structural differences relative to other carbon-richer precursors, such as coals, cokes, and chars. Therefore, when

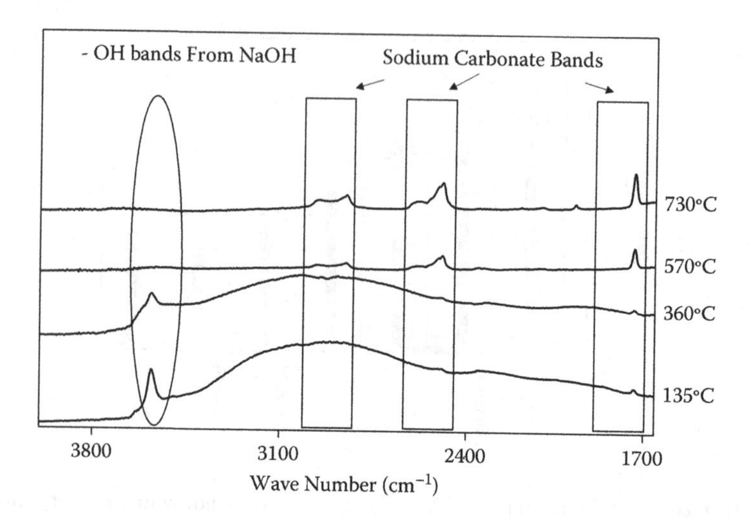

FIGURE 1.27 FTIR spectra of samples prepared by chemical activation with NaOH and anthracite up to different maximum heat treatment temperatures indicated in the figure (redrawn from Lillo-Ródenas, M.A., Cazorla-Amorós, D., and Linares-Solano, A. *Carbon* 41(2): 267–275, 2003. With permission).

they are heat-treated in the presence of a hydroxide, their complex reaction pathways may differ significantly from one another.

As discussed in Section I, in this chapter we strive to simplify the study of reactions occurring during chemical activation with alkaline hydroxides by limiting our analysis to the activation of carbon materials. In other words, carbonaceous precursors that are not carbons are excluded. To identify the reaction products, to understand the types of reactions occurring, and to propose a main global reaction for hydroxide activation, a combination of different techniques has been used: FTIR, in situ XRD, analysis of activation products followed by mass spectrometry, as well as thermodynamic data [99–101].

A. Reaction Products

Figure 1.27 presents the FTIR spectra of a carbon–hydroxide mixture as a function of the heat treatment temperature [99]. This example corresponds to an anthracite activated by NaOH using a physical mixing ratio of 3/1 by weight. Similar results have been obtained for other ratios, for other precursors, and for KOH activation. The figure shows that, as activation proceeds, sodium hydroxide is converted to sodium carbonate and that this reaction starts, for the system studied, between 360 and 570°C. We observe that NaOH is only identified up to about 400°C, whereas above 600°C, sodium carbonate is observed [99].

In situ XRD results [101] are consistent with the FTIR experiments. Figure 1.28 shows, as an example, the in situ XRD patterns obtained when MWCNTs are heat-treated in the presence of hydroxide (a 3/1 physical mixture of NaOH and MWCNT was used). We observe NaOH peaks up to about 400°C. At higher

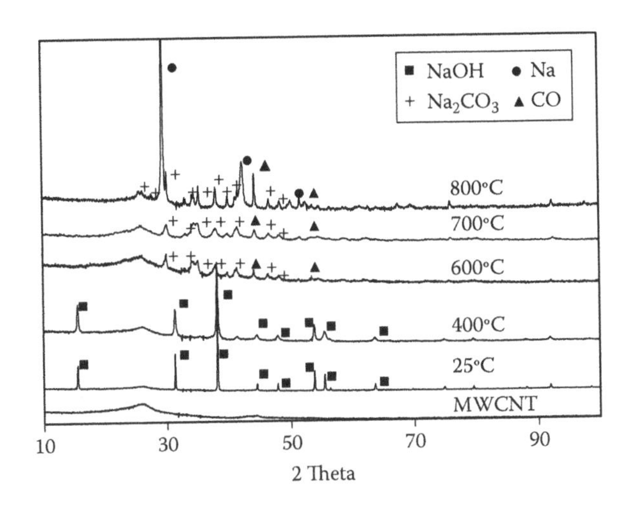

FIGURE 1.28 In situ XRD spectra of MWCNT activated with NaOH at the temperatures indicated in the figure (redrawn from Raymundo-Piñero, E., Azaïs, P., Cacciaguerra, T., Cazorla-Amorós, D., Linares-Solano, A., and Béguin, F. *Carbon* 43(4): 786–795, 2005. With permission).

temperatures (600–800°C), the peaks correspond to sodium carbonate. The XRD technique confirms and complements the FTIR results, showing metallic sodium at 800°C, which can be observed in all activations processes by NaOH (also by KOH) in any type of furnace used (e.g., horizontal or vertical). The presence of Co, not relevant for the discussion in this chapter, originated from the preparatory process of the MWCNT [72,101].

Thus, both FTIR and in situ XRD results lead to the same conclusion: carbon activation with alkaline hydroxides involves a reaction in which the hydroxide (NaOH or KOH) is converted to a carbonate. In addition, XRD shows that the metal (Na or K) is also formed at the highest temperature.

Activation experiments followed by mass spectrometry have been carried out to detect and quantify the gases produced during the chemical activation of an anthracite with hydroxide [99,100]. The evolution of three gases, H_2, CO, and CO_2, has been followed as shown in Figure 1.29. Hydrogen evolution starts at about 550°C and increases at higher temperatures. From FTIR and XRD it was seen that sodium carbonate starts to appear also at about this temperature (550°C). Thus, these three sets of results suggest that, if the temperature is maintained below 800°C (most of our activation results have been obtained at 750°C), activation takes place when the solid (carbon) reacts with the liquid (hydroxide) producing H_2 and carbonate as the main reaction products [99]. It is important to note that, at our activation temperatures, a small evolution of CO and CO_2 occurs. However, as can be seen in Figure 1.29, above this temperature CO and CO_2 evolution becomes important; this is due to the decomposition of the carbonate [99]. Identification of these reaction products (metal, carbonate, and hydrogen) agrees with the results of other authors, for example, those of

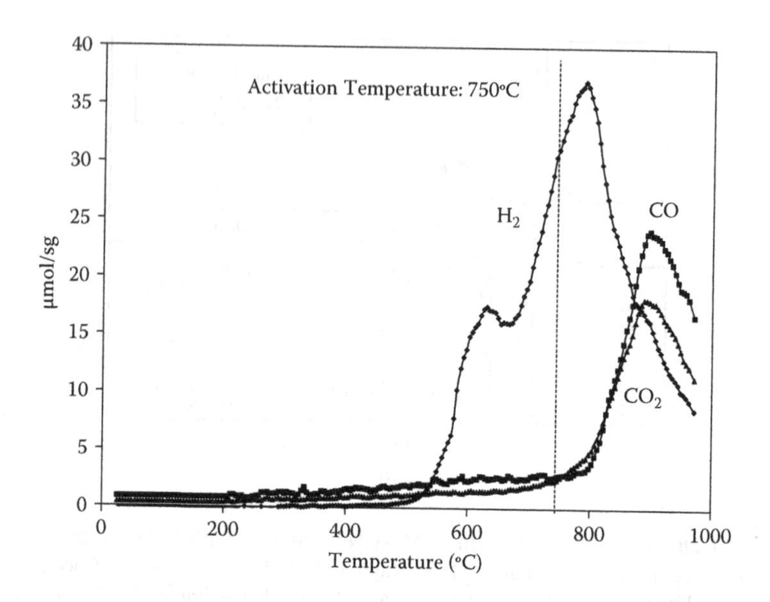

FIGURE 1.29 Hydrogen, carbon monoxide, and carbon dioxide evolution for the activation of an anthracite with NaOH (3/1 weight ratio, 20°C/min to 900°C).

Otowa et al. [53], who mixed petroleum cokes with KOH, as well as those of Béguin and Setton [122], Mochida et al. [123,124], and Yamashita and Ouchi [125–127] related to reactions between aromatic hydrocarbons and other precursors with hydroxides, and those of Diez-Teran et al. [84], who mixed KOH with a lignocellulosic material.

The results of Figure 1.29 are important because they illustrate the great significance of the heat treatment temperature discussed in Section II.E. In addition, a peculiarity of the hydroxide activation process is the narrow microporosity that can be obtained. However, if the activation temperature reaches 800°C, in addition to the carbon–hydroxide reaction, CO_2 is produced from carbonate decomposition. This CO_2 can also react with carbon by a different process (i.e., gas-phase catalytic carbon gasification), producing CO and thus causing a lowering of the product yield as well as a concomitant widening of the MPSD. In the literature a wide range of activation temperatures can be found, which are very often much higher than 750°C. Activation at around 750°C or lower can explain the singularity of the AC porosity development by hydroxides, an aspect that we strongly emphasize in this chapter, and which has not been observed or reported previously.

In addition, the results shown in Figure 1.29 are also important to explain the finding [36,37] that carbonate is a good chemical activating agent. If the temperature is above 800°C, activation will occur by CO_2 from carbonate decomposition, that is, not by a chemical activation process but by a catalyzed carbon gasification process. On the other hand, if HTT is below 750°C, carbonates are not good activating agents [99]. This explains our results showing that CO_2 (instead of nitrogen) cannot be used during activation up to 750°C, because this temperature is not sufficient to decompose the carbonate into CO or CO_2 and, therefore, no gas-phase activation is possible [99].

From the reaction products observed below 750°C and by calculation of Gibbs free energies [99], the main reaction occurring during the heat treatment of a carbon material with hydroxide has been proposed. For the case of NaOH activation, this reaction is:

$$6NaOH + 2C \leftrightarrow 2\,Na_2CO_3 + 2\,Na + 3H_2 \text{ (reaction 1)}$$

Regarding the global reaction for KOH activation, analogous results [99,100,101] have been obtained: potassium carbonate and potassium metal have been identified by XRD, and the evolution of hydrogen has been detected and quantified by mass spectrometry [99,100]. The global reaction proposed for chemical activation with KOH is therefore:

$$6KOH + 2C \leftrightarrow 2\,K_2CO_3 + 2\,K + 3H_2 \text{ (reaction 2)}$$

Both reactions (1 and 2) are redox reactions in which the carbon is oxidized to carbonate and the hydroxide is reduced, giving metallic sodium or potassium and hydrogen. Of course, in both reactions not all the carbon is converted to carbonate; the residual carbon is the activated carbon product.

It is important to note, according to the stoichiometry of these two reactions, that most of the published studies (including all our results) use reasonably lower hydroxide amounts than those needed for 100% carbon conversion (hydroxide/carbon ratios as high as 10/1 by weight (for NaOH) or 14/1 (for KOH) would be needed for 100% carbon conversion). This means that, under the experimental conditions employed, the extent of the carbon–hydroxide reaction is controlled by the hydroxide amount, which is the limiting reactant. However, the carbon reacted does not always develop porosity and, hence, the remaining carbon is not necessarily a good AC. An extreme example is the case of graphite (see Figures 1.4a and 1.4b). In these figures the graphite was reacted to about 40 to 50% weight with NaOH, but no activation can be observed. Therefore, the key point is to know how and why this residual carbon can be suitably activated.

B. Variables Affecting Reaction Kinetics

As shown in Section II, many variables affect the carbon–hydroxide activation process. If we reduce the number of variables by keeping constant the precursor, the hydroxide and the preparation method, we observe that the remaining variables can affect the kinetics of the reaction and, hence, the extent of the activation. These variables are, for example, the hydroxide/carbon ratio, the reaction temperature, and the N_2 flow rate. An increase in their values shifts the reaction toward products, thus affecting the development of porosity. The importance of

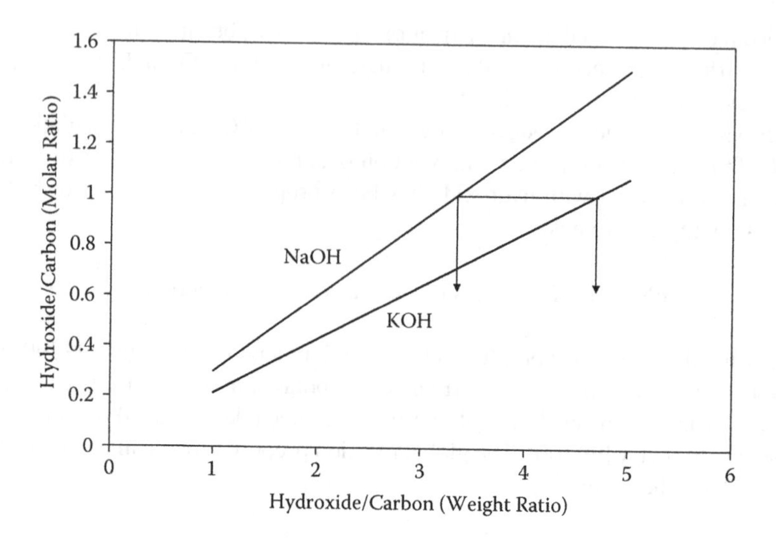

FIGURE 1.30 Relationship between hydroxide/carbon weight and molar ratios.

these variables, based on the stoichiometry of reactions 1 and 2, can now be discussed in greater detail.

a. Influence of Hydroxide/Carbon Ratio

In most of our studies the hydroxide/carbon ratio has been expressed on a weight basis, because the literature most often reports this variable in this way. However, and particularly when the behavior of NaOH and KOH is compared, this ratio should be more reasonably expressed in molar terms. As an example, Figure 1.30 shows the relationship between these ratios in weight and molar terms. Thus, if we want to use the same molar ratio for both activating agents (e.g., 1/1), the hydroxide/carbon weight ratio should be close to 3/1 for NaOH and close to 5/1 for KOH.

In addition to displacing the reaction equilibrium toward the reaction products, an increase in the hydroxide/carbon ratio is expected to enhance the carbon–hydroxide contact. Consequently, for a given reaction time (1 h in most of our experiments), the amount of reacted carbon is expected to increase with the hydroxide/carbon ratio. Table 1.8 contains the apparent BET surface areas, the MPVs ($V_{DR\ N_2}$ and $V_{DR\ CO_2}$), and the yields of AC prepared by NaOH at increasing NaOH/carbon weight ratios, while maintaining the remaining experimental variables constant. It can be readily concluded that, as the activation ratio increases, there is an increase in the apparent BET surface area and in the MPV ($V_{DR\ N_2}$), together with a decrease in the AC yield.

b. Influence of Heat Treatment Temperature (HTT)

The effect of the maximum HTT as an important variable has been discussed in some detail in Section II.E. Here we only need to point out that most often

TABLE 1.8
Apparent BET surface area, MPV, and yields of activated carbons prepared with NaOH at increasing NaOH/carbon ratios (weight), keeping the remaining experimental variables constant (physical mixing method, heating rate 5°C/min to 750°C, holding time 1 h, N_2 flow 500 mL/min)

Ratio	BET (m^2/g)	$V_{DR\ N_2}$ (cm^3/g)	$V_{DR\ CO_2}$ (cm^3/g)	Yield (wt%)
1/1	1017	0.45	0.46	74
2/1	1594	0.67	0.67	66
3/1	1872	0.80	0.73	52
4/1	2125	0.92	0.70	37
5/1	2370	0.96	0.65	35

this effect is analyzed for a given reaction time, keeping the remaining variables constant. In our studies, the reaction time was usually 1 h. When the maximum temperature is raised, two different phenomena take place: (1) the reaction rate increases considerably, of course, and so does the extent of carbon activation (i.e., more carbon atoms react with hydroxide for a given time), and (2) an additional catalytic carbon reaction, gas-phase CO_2 gasification, can occur for HTT ≥ 800°C. Consequently, the heat treatment temperature is a variable that must be controlled carefully to obtain a reproducible activation process, and it has to be considered when analyzing and comparing results from different studies.

c. Influence of N_2 Flow Rate

As shown in Section II, the N_2 flow rate considerably affects the degree of carbon activation by hydroxides. However, according to reactions 1 and 2, N_2 is not chemically involved in these reactions. Therefore, its influence has to be only as a purging gas, which enhances the rate of mass (or heat) transfer. An increase in the flow rate during heat treatment can enhance the adsorption capacity of the resulting AC as a consequence of a more efficient removal of the gaseous reaction products (H_2 and alkaline metals) and the shift of the activation equilibrium toward the reaction products. More important, perhaps, a higher N_2 flow rate will help increase the hydroxide diffusion rate within and between the carbon particles, thus reducing the diffusional control of the activation process.

To verify that the beneficial effect of the increase in the nitrogen flow rate is due to its purging effect, hydroxide activation has been studied using other inert atmospheres (argon and helium). The ACs have been prepared from an anthracite, keeping constant all the experimental variables. Table 1.9 summarizes porosity characterization of the samples prepared and shows that porosity development and yield are quite similar. From the isotherm shapes (not shown) and from the results in Table 1.9, it must be noted that porosity development follows the order He $<$ N_2

TABLE 1.9
Apparent BET surface area, MPV, and yields of activated carbons prepared by NaOH using a constant flow of 500 mL/min of different gases, keeping the remaining experimental variables constant (3/1 NaOH/anthracite ratio [weight], physical mixing method, heating rate 5°C/min to 750°C, holding time 1 h)

Atmosphere	BET (m^2/g)	$V_{DR\ N_2}$ (cm^3/g)	$V_{DR\ CO_2}$ (cm^3/g)	Yield (wt%)
Nitrogen	1872	0.80	0.73	52
Helium	1627	0.74	0.65	60
Argon	1939	0.86	0.61	57

< Ar. This order confirms the purging effect of nitrogen because it agrees, at least intuitively, with the efficiency of these three gases in removing the reaction products due to their molecular weight, helium, as expected, being the least effective.

C. Influence of Precursor Reactivity

It was shown in Section IV.A that chemical activation seems to be a consequence of a redox reaction between hydroxide and carbon, which is influenced by parameters that affect any solid-state reaction. In this context, the starting carbon reactivity will be analyzed next [100].

For this purpose, different precursors were used; their nomenclature is L for lignite, SB for subbituminous coal, and A for anthracite. Moreover, thermal treatments that affect reactivity have been applied for some of the precursors. Thus, the lignite has been carbonized to 850°C and the anthracite to 1000°C. The nomenclature for the carbonized lignite is LC, whereas for the anthracite it is an "A" followed by the treatment temperature (1000°C).

To assess the influence of the reactivity of different precursors, the onset of hydrogen appearance (To) has been determined from activation experiments followed by mass spectrometry, as presented in Figure 1.31. These results are compiled in Table 1.10. It can be seen that the chemical activation process depends on the rank and reactivity of the coal. As expected, the lower the rank of the coal, and hence the higher its reactivity [130], the lower the temperature for the beginning of chemical activation with both NaOH and KOH. In all the cases we observe that reaction starts for KOH at lower temperatures than for NaOH, in agreement with the higher chemical reactivity of KOH in relation to NaOH [117].

The hydrogen evolved during heating of the NaOH–precursor and KOH–precursor mixtures has been followed and quantified by mass spectrometry. Some of these data are included in Table 1.11. We can observe that H_2 evolution decreases with increasing coal rank or decreasing precursor reactivity. By assuming that the hydrogen evolved originates from the conversion of the hydroxides, according to

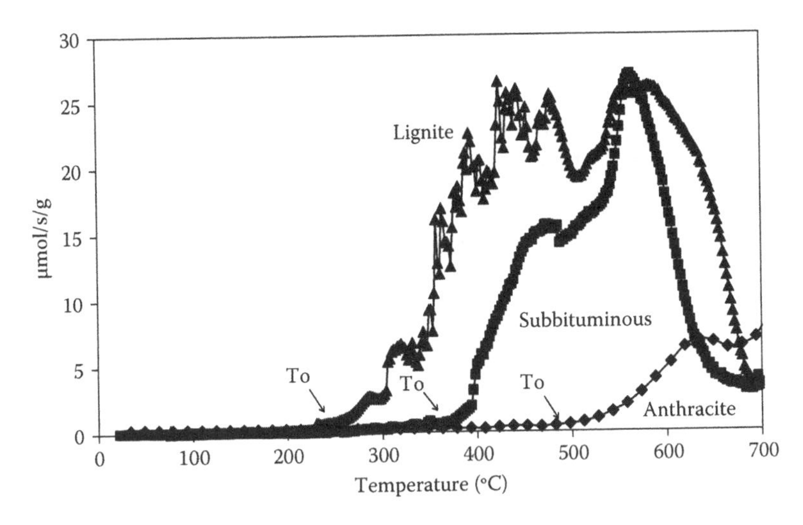

FIGURE 1.31 Evolution of hydrogen from anthracite (A), subbituminous coal (SB), and lignite (L) during chemical activation with NaOH (redrawn from Lillo-Ródenas, M.A., Juan-Juan, J., Cazorla-Amorós, D., and Linares-Solano, A. *Carbon* 42(7): 1371–1375, 2004. With permission).

TABLE 1.10
Temperatures of onset of chemical activation for different precursor–hydroxide mixtures

Precursor	Temperature (NaOH reaction) (°C)	Temperature (KOH reaction) (°C)
L	250	225
SB	375	325
A	475	375
LC	450	250
A-1000	575	550

reaction (1), the percentage of reacted carbon has been calculated for the different mixtures, and these data are compiled in Table 1.11. We observe that the degree of reaction in NaOH decreases as coal rank increases. The same trend is observed for KOH activation.

The importance of prior heat treatment of a carbon precursor can be observed in Figure 1.32 and in Tables 1.10 and 1.11 (comparing L and LC and A and A-1000). We see (Figure 1.32) that hydrogen evolution from the lignite (L) is much

TABLE 1.11
Hydrogen evolved in the activation experiments followed by mass spectrometry carried out over the mixture precursor–NaOH and percentage of reacted carbon

Precursor	H_2 Evolution (μ mol/g)	Percentage of Reacted Carbon
L	22759	9.0
SB	13563	6.0
A	3860	2.0
LC	11665	5.2
A-1000	680	0.3

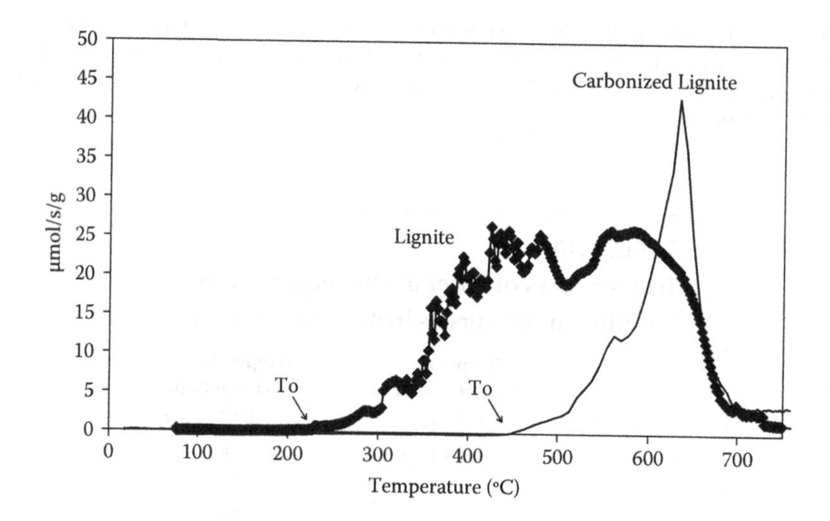

FIGURE 1.32 Evolution of hydrogen from lignite (L) and carbonized lignite (LC) during chemical activation with NaOH (redrawn from Lillo-Ródenas, M.A., Juan-Juan, J., Cazorla-Amorós, D., and Linares-Solano, A. *Carbon* 42(7): 1371–1375, 2004. With permission).

higher than from carbonized lignite (LC) and that heat treatment prior to activation increases the temperature of the onset of H_2 evolution (To).

All these results confirm that both the onset temperature and the amount of hydrogen evolved follow the same trend as the reactivity of the carbon precursor in oxygen-transfer carbon-gas reactions, as shown in Figure 1.33. This figure presents a thermogravimetric (TG) analysis of the different precursors done in air (60 mL/min), at 5°C/min. The TG curves follow the expected order for the reactivity of coals; the most reactive is the lignite, whereas the anthracite is the

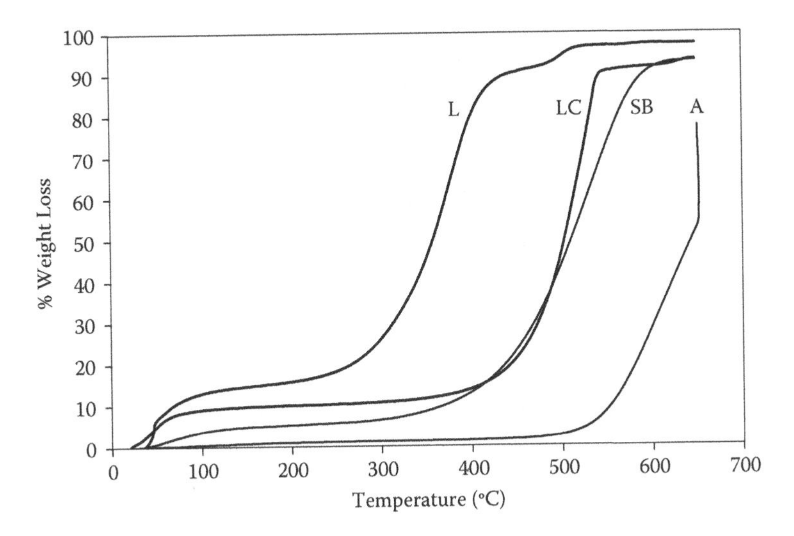

FIGURE 1.33 TG curves for air oxidation of anthracite (A), subbituminous coal (SB), lignite (L), and carbonized lignite (LC) (reprinted from Lillo-Ródenas, M.A., Juan-Juan, J., Cazorla-Amorós, D., and Linares-Solano, A. *Carbon* 42(7): 1371–1375, 2004. With permission from Elsevier).

least reactive [130]. The reactivity of the lignite (L) is much higher than that of the carbonized lignite (LC).

D. Influence of Precursor Structural Order

Some of the results discussed previously show that KOH and NaOH present similar activation behavior. However, there are some differences, as discussed in Section II.B. The relative effectiveness of the hydroxides depends on the nature of the precursor, especially on its structural order [100,101]; NaOH is apparently better for carbons without structural order, whereas KOH seems better for those having some structural order. These intriguing results could be related to the different behaviors of Na and K as intercalates in the graphene layers of the carbonaceous precursors (or the carbon products); such influence of metal insertion has been recently reported [98,101].

This section focuses on the importance of structural order of the carbonaceous precursor. With this purpose, different MWCNTs have been prepared at different temperatures and thus have different structural order [101].

Figure 1.34 presents the XRD results for MWCNTs prepared at 450, 500, and 600°C. The sharpening of the (002) band for quasi-graphitic carbon is interpreted as an increase in crystallinity/order despite its perhaps surprising broadness. Table 1.12 compiles the activation results for these three MWCNTs. We can observe that an increase in structural order makes the activation process difficult. This is in agreement with the observation that hydroxide activation is more

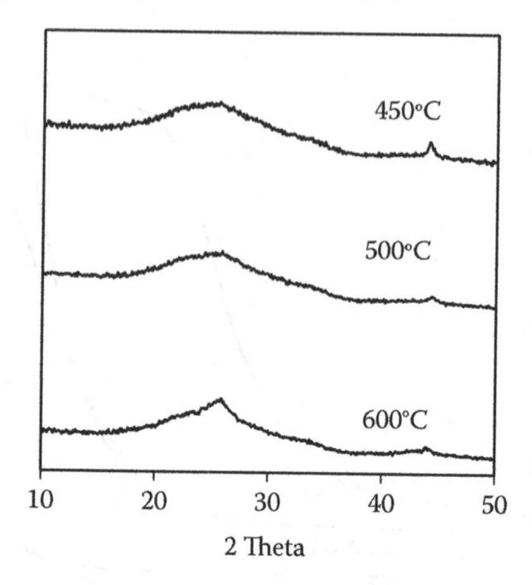

FIGURE 1.34 XRD of three MWCNTs prepared at 450, 500, and 600°C.

TABLE 1.12
Apparent BET surface area of MWCNTs with different crystallinity orders chemically activated with NaOH and KOH

	BET of Samples Activated with KOH (m^2/g)	BET of Samples Activated with NaOH (m^2/g)
MWCNT 450°C	1670	562
MWCNT 500°C	1220	386
MWCNT 600°C	868	377

difficult for carbonized (heat-treated) materials than for untreated materials, and it also helps explain why anthracite is more difficult to activate than lignite. In addition, the table allows us to further compare KOH and NaOH activation. In all the cases we observe that KOH produces higher adsorption capacities than NaOH. This seems to indicate that, although higher structural order is associated with poorer activation, the presence of structural order in MWCNT makes a difference, presumably favoring K intercalation more than Na intercalation. In this sense, recently [101] an insertion/intercalation study of metallic potassium and sodium in MWCNTs has been carried out using experiments with pure metallic K or Na as well as by analyzing the intercalation process during hydroxide activation. The results showed that carbon crystallinity makes activation difficult but favors more K intercalation between the graphene layers during the KOH activation process for ordered MWCNTs; such is not the case when NaOH is used.

In summary, chemical activation by hydroxides involves a main chemical reaction in which carbon atoms are removed from the carbon matrix and are transformed into an inorganic compound (i.e., carbonate), and a subsidiary process in which the metal atoms formed through reduction of hydroxide are thought to be inserted/intercalated between the graphene layers of the residual carbon. This process can play a certain role in the overall activation process, as discussed in the following text.

The reaction of carbon atoms with MOH (M: metal) to form the corresponding carbonate could contribute to porosity development through different routes: (1) by the removal of the most disorganized and thus most reactive parts of the sample; (2) the much larger molecular volume of carbonate (in relation to carbon atoms in the carbon matrix) would cause porosity changes and play an important role in porosity development; (3) additionally, insertion of metal atoms may have some effect through alteration of distribution of graphene sheets during the activation process, which does not necessarily imply a destruction of structural order, but could change the accessibility of the material. An example of such assessment can be found in a TEM study of KOH- and NaOH-activated carbons prepared from an anthracite [106]. Figure 1.35 shows the micrographs of the original anthracite and of an AC in which 34% of the carbon atoms were reacted: clearly, the chemically activated sample still retains, in part, the structural order of the precursor, in spite of the considerable amount of carbon reacted. The manner in which metal insertion/intercalation during the activation process alters the distribution of graphene sheets is not clear yet.

We should emphasize that in some cases (especially for graphite) these two processes (up to 40 to 50% reaction and accompanied by metal intercalation) do not necessarily result in porosity development (see Figures 1.4a and 1.4b). Additionally, if we recall the comparison between physical and chemical activation (see Section III), we can readily conclude that the BO of the carbon atoms (physical activation) or carbon atoms reacted (chemical activation) alone cannot explain the development of porosity, because much higher BO is necessary by gasification with CO_2 to get similar porosity than by the liquid-phase reaction with NaOH or KOH. Consequently, much more work is needed before a full understanding of the chemical activation process can be achieved.

V. APPLICATIONS OF KOH/NAOH-ACTIVATED CARBONS

In the previous sections, a detailed analysis of the chemical activation with hydroxides has been presented with special emphasis on the variables of the process and the final porosity of the ACs. Sample characterization by adsorption (combination of N_2 at 77 K and CO_2 at 273 K) has shown that hydroxide activation enables the preparation of ACs having both a highly developed MPV and a narrow MPSD. This section shows the importance of being able to control both these parameters to improve the performance of AC in existing and emerging applications.

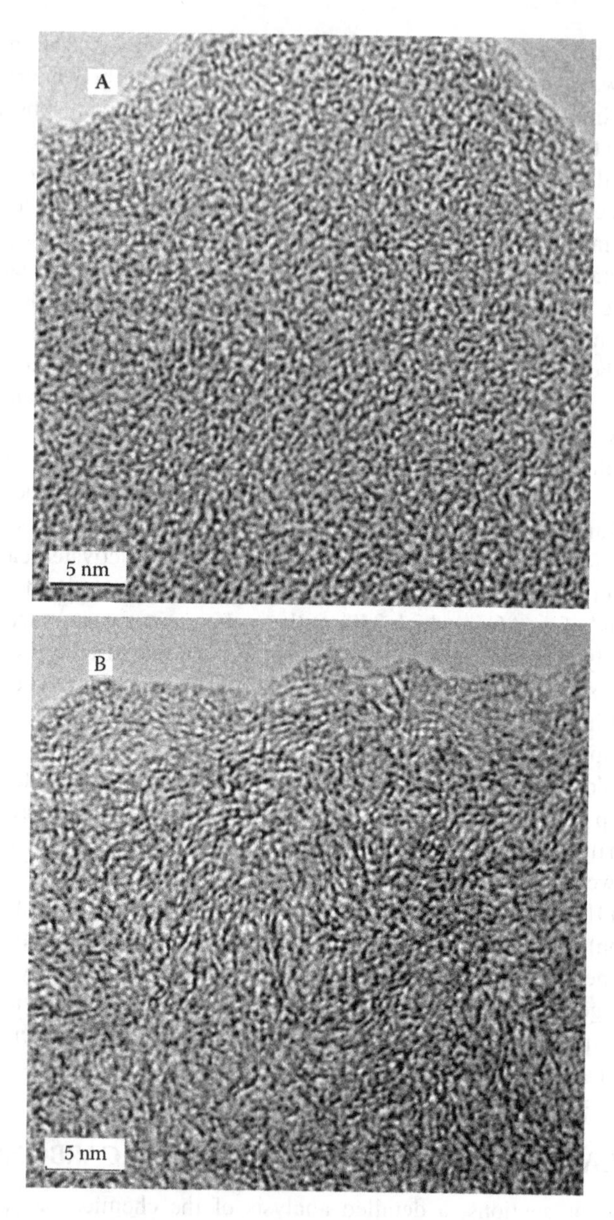

FIGURE 1.35 TEM photograph of (a) pristine anthracite and (b) anthracite-derived AC with a yield of 66% (BET area = 1594 m^2/g).

A. Removal of VOC at Low Concentrations

The removal of volatile organic compounds (VOCs) with ACs is of interest because these compounds can be very harmful for both human health and the environment, even at very low concentrations [6, 107–109, 131–138].

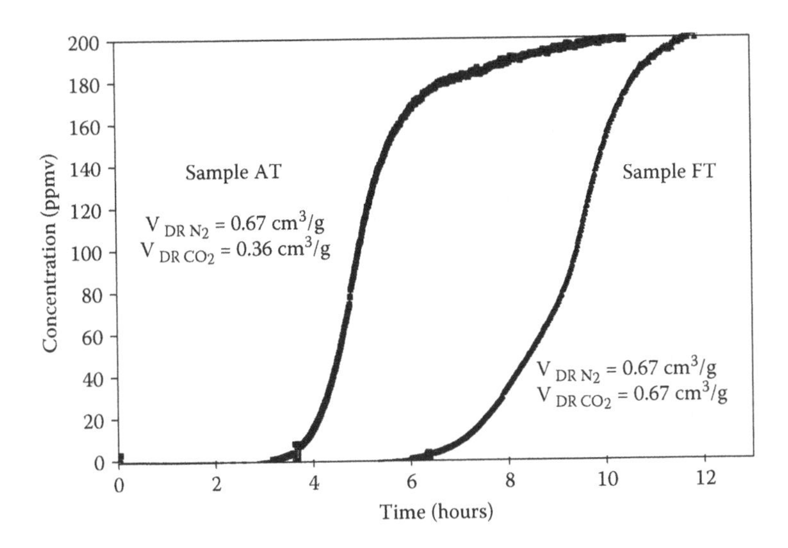

FIGURE 1.36 Benzene breakthrough curves for samples AT and FT, at room temperature and 200 ppmv benzene.

Thus, for example, adsorption of benzene and toluene at low concentrations (200 ppmv) has been carried out at room temperature in a fixed-bed reactor coupled to a mass spectrometer [107–109]. Two kinds of ACs have been studied: pristine ones and some thermally treated ones (including T in their nomenclature). The purpose of the thermal treatment was to remove most of the oxygen surface groups, to be able to analyze only the effect of porosity on VOC adsorption.

Figure 1.36 shows the breakthrough curves for two thermally treated samples (AT and FT). Both samples have the same N_2 micropore volume. Interestingly, their narrow micropore volumes (assessed by CO_2 adsorption at 273 K) can explain why sample FT has higher VOC uptake than sample AT. To confirm this finding, a large series of ACs has been used to correlate their VOC retention capacities with their textural properties [107–109,139].

Table 1.13 compiles the benzene and toluene adsorption capacities for pristine AC obtained by numerical integration of breakthrough curves similar to those of Figure 1.36. It includes the porous characterization of commercial samples developed for removal of hydrocarbons as well as of ACs prepared by different activation methods and activating agents.

Figure 1.37 shows that there exists a very good correlation between benzene adsorption capacity and the volume of narrow micropores for the pristine AC and for the heat-treated ACs [107,109]. This relationship, also valid for toluene, can hardly be found with other porosity parameters (e.g., surface area). We observe that this correlation is even better for the thermally treated AC (because the influence of the oxygen surface groups has been avoided) and that removal of surface oxygen groups increases the benzene adsorption capacity. It is important to underline that the adsorption capacities for benzene (as high as 34 g benzene/100 g AC)

TABLE 1.13
Adsorption capacities for some activated carbons tested for benzene and toluene uptake at low concentration (i.e., 200 ppmv)

Sample	Preparation Method	BET (m^2/g)	$V_{DR\ N_2}$ (cm^3/g)	$V_{DR\ CO_2}$ (cm^3/g)	Benzene Adsorption (g/100 g AC)	Toluene Adsorption (g/100 g AC)
A	Commercial H_3PO_4	1757	0.67	0.36	15	31
B	Commercial H_3PO_4	1297	0.54	0.30	12	25
C	Physically activated steam	883	0.35	0.26	12	19
D	Chemically activated NaOH	656	0.26	0.25	11	17
E	Chemically activated NaOH	932	0.39	0.40	15	25
F	Chemically activated NaOH	1594	0.67	0.67	22	—
G	Chemically activated NaOH	1872	0.80	0.73	23	38
H	Chemically activated KOH	2746	0.97	0.77	27	50

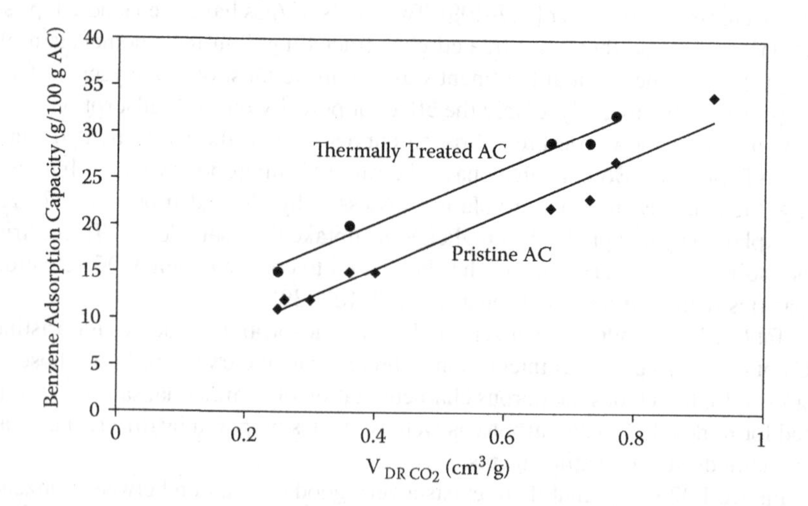

FIGURE 1.37 Relationship between volume of narrow micropores calculated by CO_2 adsorption and benzene adsorption capacity at 200 ppmv for several pristine and thermally treated activated carbons.

and toluene (as high as 64 g toluene/100 g AC) at 200 ppmv achieved by the AC prepared by hydroxide activation are higher than those of commercial samples and higher than those previously reported in the literature [109]. These results

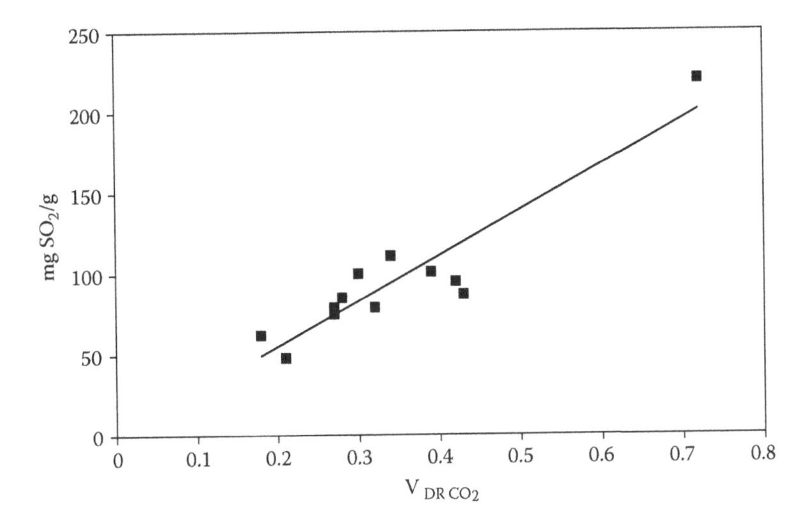

FIGURE 1.38 Relationship between volume of narrow micropores and adsorption of SO_2 for different carbon samples.

imply that, to improve benzene and toluene removal at low concentrations, we need to develop the narrow micropore volume as much as possible. In agreement with our conclusions from Section III, chemical activation with hydroxides seems to be the most suitable process.

B. SO_2 Removal

Most industrialized countries have problems with their SO_2 emissions. Thus, for several decades the interest in research for controlling SO_2 pollution has continued, use of ACs being one possible alternative for controlling these emissions [6, 60,140–142].

The effect of the porous structure of ACs on SO_2 retention has been the subject of intense research. These studies give information about the retention mechanism (both adsorption and oxidation) and about the effect that porosity has on the retention of SO_2. From our studies [140,141] it can be stated that a high volume of narrow micropores is also desirable to achieve high SO_2 adsorption capacity.

To analyze the SO_2 adsorption capacity until saturation [140,141], our experimental system consists of IR and UV gas analyzers (Rosemount Binos 1001 SO_2/O_2, Binos CO/CO_2). A 1.2-cm-diameter differential reactor has been used, with 0.2 g of sample. The gas flow was 620 mL/min, with a concentration of 2000 ppm SO_2, 5% O_2 in N_2.

Figure 1.38 presents the correlation obtained between the SO_2 adsorption capacity and the narrow micropore volume for different ACs (two carbon molecular sieves, a char, several chemically ACs, and several ACFs prepared by either steam or CO_2 gasification). The observed correlation strengthens the importance of narrow microporosity in this removal process.

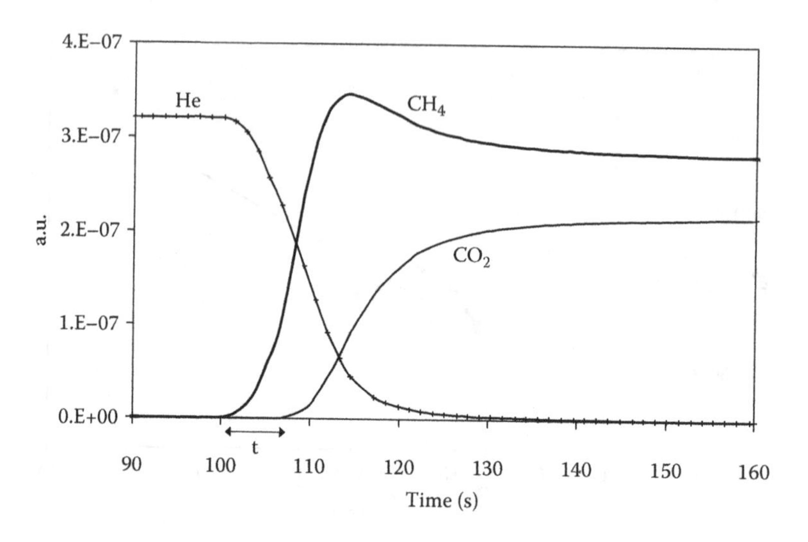

FIGURE 1.39 Breakthrough curves showing a delay time between CH_4 and CO_2.

C. CO_2 AND CH_4 SEPARATION

Separation of gases is a very important process in several industries (e.g., chemical, petrochemical, and related industries). Although cryogenics and absorption remain the most widely used processes, the last two decades have seen a tremendous growth in research activities and commercial applications of adsorption-based gas separation. Separation by adsorption is based on the selective accumulation of one or more components of a gas mixture on the surface of a microporous solid. The separation is achieved by one of three mechanisms: steric, kinetic, or equilibrium. Most processes operate by virtue of equilibrium (or competitive) adsorption of gases from binary or multicomponent mixtures [143].

Carbonaceous adsorbents are useful in separation processes because of their good kinetic properties and high adsorption capacities. Separation of CO_2 from CH_4 is required in several applications, such as landfill gas utilization, with approximately 50% each of CH_4/CO_2 [6], and tertiary oil recovery where the effluent gas contains approximately 80% CO_2 and 20% CH_4 as well as other light hydrocarbons [144].

The performance of different KOH-activated carbons for CO_2/CH_4 separation at room temperature and atmospheric pressure was accomplished in a modified BTR-Jr flow reactor (Autoclave Engineers) connected to a mass spectrometer. Figure 1.39 contains, as an example, the CH_4 and CO_2 breakthrough curves for a typical sample. Once He (that passed through the sample) was switched to the CH_4–CO_2 mixture, the CO_2 signal displayed a significant delay with respect to the CH_4 signal, indicating that the AC has a higher affinity for CO_2 than for CH_4. This detection delay was quantified and plotted for a series of KOH-activated carbons against their micropore volume ($V_{DR\,N_2}$) or their narrow micropore volume ($V_{DR\,CO_2}$).

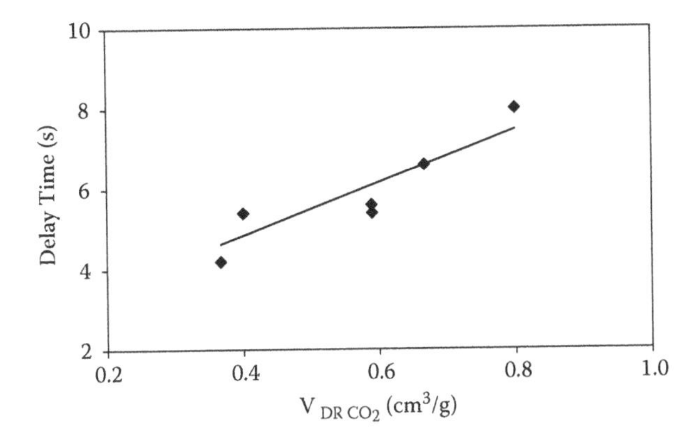

FIGURE 1.40 Relationship between delay time and $V_{DR\ CO_2}$ obtained for chemically activated carbons.

Figure 1.40 shows the trend with the narrow micropore volume characterized by CO_2 adsorption. The micropore volume deduced from N_2 adsorption at 77 K does not allow such correlation. According to this correlation, to improve the delay time and hence the separation effectiveness, the narrow microporosity development has to be maximized. As shown earlier, hydroxide activation is indeed a suitable process to obtain this type of AC.

D. CH_4 AND H_2 STORAGE

At a given pressure, a strong adsorption potential inside the micropores acting on gas molecules significantly increases the density of the adsorbed molecules in relation to the equilibrium normal gas-phase density. This phenomenon [16] can be exploited for enhancement of gas storage capacity through adsorption. This has been the main reason for the strong interest in using AC as a medium to reduce the pressure required to store weakly adsorbed compressed gases such as methane and hydrogen. The search for ACs able to store large amounts of natural gas at a reasonable pressure (3.5 to 4 MPa), as a substitute for natural gas compressed at much higher pressure (e.g., 21 MPa), has been very intense in the last few years [6,110–113,145–147].

In all previous studies [110–113,145–147] it has been concluded that, in general, the higher the surface area (or micropore volume), the higher the methane adsorption capacity. In our research group, a systematic study of the performance of KOH-activated carbons in methane storage has been carried out. We concluded that, in addition to surface area and packing density, MPSD is also important [110–113]. Thus, Figure 1.41 shows the methane uptake versus the apparent BET surface area corresponding to a series of KOH-activated carbons. The objective of this correlation was to extend the BET surface area range beyond 2000 m²/g. The linear relationship is seen to reach a maximum for ACs with a very high surface area. If only the apparent BET surface area or the micropore volume were responsible

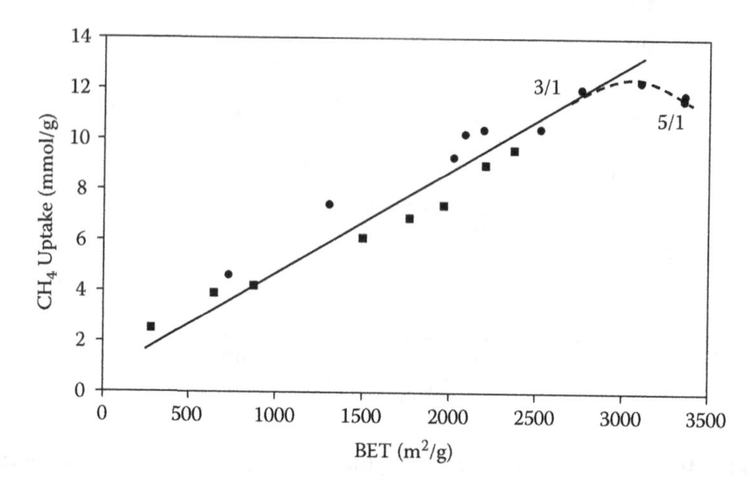

FIGURE 1.41 Methane adsorption capacity (up to 4 MPa) versus apparent BET surface area corresponding to a series of KOH-activated carbons.

TABLE 1.14
Porous texture characterization results and gravimetric methane adsorption capacity (up to 4 MPa) corresponding to two KOH ACs

Sample	BET (m^2/g)	$V_{DR\,N_2}$ (cm^3/g)	$V_{DR\,CO_2}$ (cm^3/g)	$V_{DR\,N_2}$ - $V_{DR\,CO_2}$ (cm^3/g)	CH_4 Uptake (mmol/g)
3/1	2758	1.35	0.72	0.63	12
5/1	3350	1.48	0.67	0.81	11.6

for the methane uptake, sample 5/1 (see Figure 1.41) should have a higher methane capacity than sample 3/1; however, the opposite behavior is observed. These results can be explained by analyzing the porous texture in more detail. Thus, Table 1.14 contains the porous texture characterization results and the gravimetric methane adsorption capacities (up to 4 MPa) for the same samples. The reason for 3/1 performing better than 5/1 lies in their very different MPSDs (see Figure 1.11), as evidenced also by values of $V_{DR\,N_2} - V_{DR\,CO_2}$ (Table 1.14). Sample 3/1 has a narrower MPSD, which results in a higher methane uptake, consistent with the enhanced adsorption potential argument [16].

From a practical standpoint, to increase the methane adsorption capacity we need to develop not only micropore volume but also control carefully the micropore size distribution. Thus, samples with high surface areas and very narrow micropore size distributions are required for this application. For this reason, in our studies of methane storage KOH instead of NaOH was used as the activating agent for the preparation of ACs because of the narrower MPSDs that can be obtained by KOH, as shown in Sections II and III.

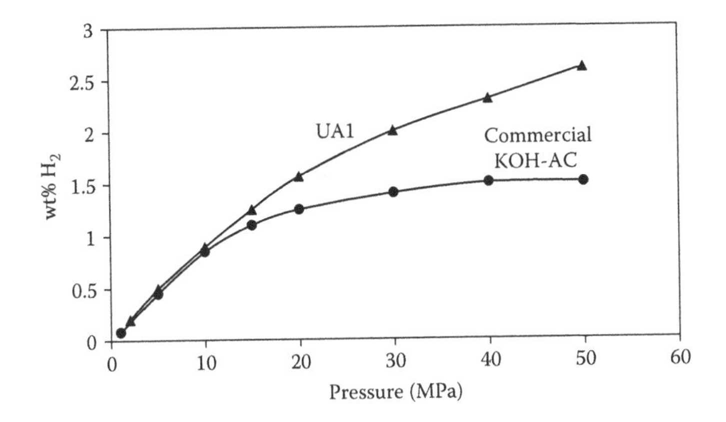

FIGURE 1.42 H_2 adsorption isotherms at 298 K corresponding to a commercially available KOH AC and to a KOH AC prepared in our laboratory (UA1).

TABLE 1.15
Porous texture characterization results corresponding to a commercially available KOH AC and to a KOH AC prepared in our laboratory

Sample	BET (m^2/g)	$V_{DR\ CO_2}$ (cm^3/g)
Commercial	3100	0.64
UA1	3012	0.71

In relation to high-pressure storage of H_2 on AC at room temperature, it must be remarked that this requirement is much more difficult to satisfy than methane storage, because of the much lower amount of H_2 uptake (either per unit mass of AC or per unit volume). However, to optimize these smaller H_2 uptakes, the characteristics of the adsorbent should be very similar to those analyzed for AC developed for methane storage.

Hydrogen storage at room temperature and up to 70 MPa has been studied for a series of ACs from different precursors and prepared by different activation processes [148,149]. An example of the behavior of two ACs prepared by KOH activation (in our laboratory [UA1] and commercially) can be seen in Figure 1.42. These two samples exhibit different behavior, not only in the amount of H_2 adsorbed but also in the shape of the isotherm. Table 1.15 compiles the characteristics of these two samples having similar BET surface areas, but different narrow MPVs and MPSDs. Sample UA1 has a higher narrow MPV (narrower MPSD) and also a higher H_2 uptake, reaching a value of about 3 wt%. This result is in agreement with theoretical and experimental values [148,150], showing that a maximum in the density of the adsorbed H_2 for a series of AC occurs for a pore size in the vicinity of 0.6 nm.

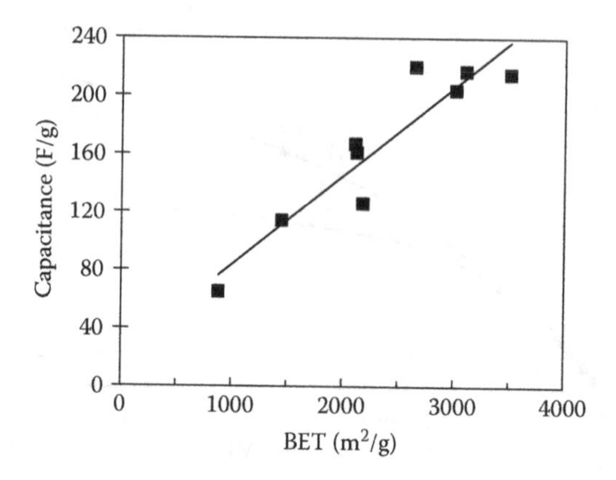

FIGURE 1.43 Capacitance versus apparent BET surface area for a series of KOH ACs prepared in our laboratory and a commercially available AC (Maxsorb-A).

E. Electrical Double-Layer Capacitors

There is a new generation of electrochemical components for energy storage known as *supercapacitors* or *electric double-layer capacitors (EDLCs)* [18]. These supercapacitors, which bridge the gap between conventional electrolytic capacitors and batteries on account of their unique combination of high energy density and high power density, are still the object of intense study. The main interest is to improve the quality and performance of the AC used as an electrode. An important research objective is to develop ACs with higher adsorption capacity in order to increase the energy stored [95,96,151–158].

In general, it is believed that there is a direct relationship between the apparent BET surface area and the double-layer capacitance (DLC) of porous carbons [156–158]. Theoretically, such a correlation should be expected. In practice, the situation is more complicated.

Considering that high-surface-area ACs with different pore size distributions and different surface chemistries can be prepared by KOH activation, the performance as EDLCs in nonaqueous electrolyte of a series of KOH-activated carbons has been measured. The details of sample preparation conditions and EDLC measurements are provided elsewhere [114]. Figure 1.43 summarizes the relationship between BET surface area and capacitance (potential range 2 V-4 V versus Li/Li^+) in 1.0 $LiClO_4$/PC. The capacitance increases with surface area, reaching a maximum of 220 F/g for the sample with more than 2500 m^2/g. Although a general trend between capacitance and BET surface area does exist, it is not a perfectly linear relationship, indicative that EDLC depends, to a lesser extent, on other characteristics of the porous carbon material. Nevertheless, considering that most of the samples are microporous materials, these results confirm (contrary to what is generally accepted) that micropores (pores < 2 nm) contribute significantly to DLC. A more detailed analysis of the contribution of the different properties of AC to EDLC has been offered elsewhere [114].

F. Space Cryocoolers

The European Space Agency's (ESA) Darwin Mission is a future space interferometer meant to search for terrestrial planets in orbit around other stars. The interferometric imaging of astrophysical objects will be accomplished via four free-flying telescopes and a central hub. To guarantee proper mechanical stability, any vibration of the optical system (with its integrated 4.5 K cryocoolers) cannot be tolerated. To reach such a low temperature, a two-stage vibration-free sorption He/H_2 cooler has to be designed with a suitable AC [159]. In this section, an example corresponding to the development of such an adsorbent is presented.

A sorption cooler has two parts: (1) a *cold stage*; and (2) a *sorption compressor.* The cold stage consists of a counterflow heat exchanger and a Joule–Thomson expansion valve. This cold stage works as a typical refrigerator. The high-pressure gas needed comes from the sorption compressor. A sorption compressor can be described as a thermodynamic engine that transfers thermal energy to the compressed gas in a system without moving parts. Its operation is based on the principle that large amounts of gas can be adsorbed on certain solids such as highly porous ACs. The amount of gas adsorbed is a function of temperature and pressure. If a pressure container is filled with an adsorbent and gas is adsorbed at low temperature and pressure, then high pressure can be produced inside the closed vessel by an increase in the adsorbent temperature. Subsequently, a controlled gas flow out of the vessel can be maintained at high pressure by a further increase in temperature until most of the gas is desorbed.

ACs are obviously very interesting candidates for this application. They have to satisfy three essential requirements: (1) large adsorption capacity per mass of adsorbent, (2) minimum void volume, and (3) very good mechanical properties.

In order to optimize AC properties for this application, studies with samples prepared from different raw materials and using different activation processes have been carried out. In order to predict their adsorption performance, helium adsorption isotherms at different temperatures and pressures have been obtained. From these results, KOH-activated monoliths were selected, because they combined special properties: high adsorption capacity, minimum void volume, high packing density, good thermal stability, and good mechanical properties. Figure 1.44 shows these isotherms corresponding to an AC prepared by anthracite activation. According to the conditions needed in the compressor stage, we are interested in obtaining a material with a maximum helium adsorption capacity at 2 bar and 50 K (adsorption stage) and a minimum adsorption capacity at 13 bar and 120 K (desorption stage).

From this example of space cryocoolers, as well as previous examples of other applications, the importance of the development of narrow microporosity during the preparation of AC has been demonstrated. In this chapter, the way to tailor this type of porosity has been shown; this is by chemical alkaline hydroxide activation of carbon precursors with careful control and thorough understanding of the variables affecting the carbon activation process.

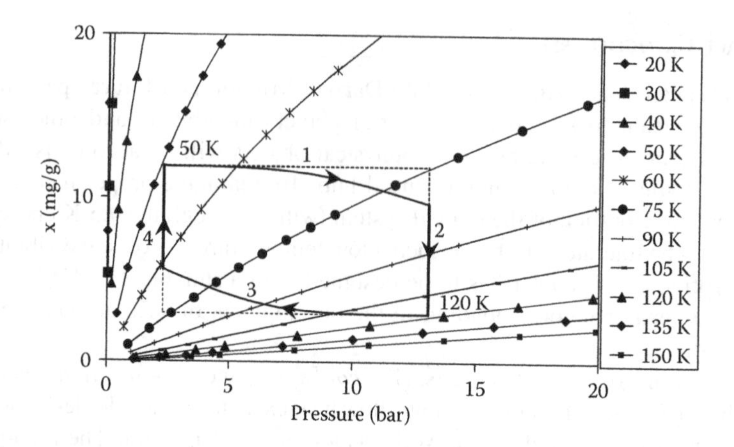

FIGURE 1.44 Helium adsorption isotherms at different temperatures corresponding to an activated carbon prepared by KOH activation of an anthracite. The numbers 1, 2, 3, and 4 correspond to each section of a complete cycle of a cell.

VI. CONCLUDING REMARKS

This chapter, developed from a plenary lecture presented at Carbon 2004 (Brown University) with the same title, summarizes and places into broader context the work that our research group has been doing since 1990 in the area of carbon activation with alkaline hydroxides.

Different aspects of hydroxide activation, such as the variables that most affect the activation process, the singular characteristics of the porosity developed, the main reactions occurring during activation, and some application performances of these ACs, have been analyzed.

The specific conclusions that should be emphasized from each section are summarized here for the reader's convenience:

In relation to the variables that control the hydroxide activation process:

1. Hydroxide activation is a complex process mostly influenced by the following variables: the nature of the precursor and of the hydroxide, the method of contacting hydroxide and precursor (e.g., impregnation or physical mixing), the hydroxide/carbon ratio, the heat treatment temperature and time, and the inert gas purging flow rate.
2. The effects of these variables need to be known to control and select the desired AC characteristics, in terms of mechanical properties, packing density, ash content, micropore volume, apparent surface area, and MPSD.
3. The adsorption capacity increases mainly with an increase in the hydroxide/carbon ratio, the heat treatment temperature, and the N_2 flow rate. All of these increase both the micropore volume and the MPSD, except for the N_2 flow rate, which produces an increase only in the micropore volume.

4. Activation results with KOH and NaOH are not necessarily the same. The relative effectiveness depends on the precursor used, especially on its structural order; NaOH appears to be better for carbons without structural order (e.g., subbituminous coal, lignite, and lignocellulosic materials), whereas KOH appears to be better for those having some structural order (e.g., anthracite, heat-treated coals, MWCNT).
5. Hydroxide activation can be carried out easily by mixing the hydroxide pellets with any type of carbon precursor (except graphite). This mixing method requires less time and work than impregnation and is especially useful for the less corrosive NaOH (giving better results than impregnation).
6. For comparable adsorption capacities, ACs prepared by KOH have a narrower pore size distribution than those prepared by NaOH.

In relation to the singular characteristics of the porosity of these ACs:

1. The most important conclusion about hydroxide activation is the type of microporosity that can be obtained. When ACs with similar adsorption capacities are compared, we always observe that those prepared with hydroxides present narrower MPSDs. In this sense, it is especially attractive to be able to prepare ACs with high apparent BET surface area (~2000 m^2/g) and having at the same time a very narrow MPSD (with $V_{DR\ N_2} \approx V_{DR\ CO_2}$).
2. Both the micropore volume and the MPSD can be controlled by carefully selecting the variables of the process as a function of the desired characteristics of the AC. Thus, if both high micropore volume and wide MPSD are required, we recommend using a high hydroxide/carbon ratio, high heat treatment temperature, high heating rate, a relatively disordered carbon precursor (e.g., lignite or subbituminous coal rather than anthracite), and NaOH rather than KOH, preferably by physical mixing. Contrarily, if concurrent large micropore volume and narrow MPSD are required, we recommend to keep, as much as possible, a low hydroxide/carbon ratio, low heat treatment temperature, low heating rate, high N_2 flow rate, and select an ordered precursor (e.g., anthracite rather than lignite) as well as KOH rather than NaOH (especially when impregnation is used).

In relation to reactions occurring during activation:

1. The chemical activation of carbons is a complex solid–liquid reaction. The reaction takes place when the solid (carbon) reacts, at about 500°C, with a liquid (hydroxide), producing mainly H_2, alkaline metal, and carbonate, if the heat treatment temperature is lower than 750°C. This reaction is a redox process in which carbon is oxidized to carbonate and hydroxide is reduced to alkali metal and hydrogen. This set of reaction products and the type of reactions taking place are not necessarily the same for all carbonaceous materials; thus, for example, lignocellulosic

materials, polyaromatic hydrocarbons, and pitches are expected to behave differently from more carbon-rich materials, such as coals and nanotubes. At higher heat treatment temperatures, CO_2 evolution from carbonate decomposition and CO production from carbon gasification by CO_2 become important.

2. Under most experimental conditions used, and according to the proposed reaction stoichiometry, the extent of the carbon–hydroxide reaction is controlled by the amount of hydroxide, which is the limiting reactant. Hydroxide/carbon ratios as high as 10/1 by weight (for NaOH) or 14/1 (for KOH) would be required to preclude the hydroxide being the limiting reactant.
3. As in any solid–liquid reaction, the reactivity of the carbon precursor is a key factor governing the activation process. Thus, in the case of coals, the lower the rank, the higher the reaction extent with the hydroxide. As the structural order of the carbon increases (i.e., anthracite versus lignite), the activation process becomes more difficult. In these cases, KOH is much more effective than NaOH, presumably because of its easier intercalation in the graphene layers of the carbon precursors.
4. More work is needed to understand the relationship between reacted carbon by hydroxides and porosity development.

In relation to applications, the main conclusion of this chapter is that the activation process involving hydroxides (both NaOH and KOH) is very suitable for preparing ACs with a wide range of characteristics. Thus, it allows us to select, according to their final uses, their MPVs and their MPSDs. Interestingly, activation of carbon materials by alkaline hydroxides allows us to obtain AC having both high MPV and narrow MPSD, which is required for some applications. In fact, this type of AC cannot be prepared by physical activation, and hardly by other chemical activating agents. As a result, in this chapter we have shown that hydroxide activation can enhance the performance of AC in classical applications (e.g., VOC abatement) as well as in new applications (e.g., methane or hydrogen storage or space cryocoolers).

ACKNOWLEDGMENTS

The authors thank MCYT (Projects PPQ 2002-01025 and PPQ 2003-03884) for financial support. The authors would like to thank all those colleagues with whom they collaborated for more than 15 years and who have made this chapter possible.

REFERENCES

1. Bansal, R.C., Donnet, J.B., and Stoeckli, F. *Active Carbon*. New York: Marcel Dekker, 1988.
2. Jankowska, H., Swiatkowski, A., and Choma, J. *Active Carbon*. New York: Ellis Horwood, 1991, pp. 29–38.

3. Marsh, H., Heintz, E.A., and Rodríguez-Reinoso, F. *Introduction to Carbon Technologies*. Alicante: Servicio de Publicaciones de la Universidad de Alicante, 1997.
4. Kyotani, T. In Yasuda, E., Inagaki, M., Kaneko, K., Endo, M., Oya, A., and Tanabe, Y., Eds. *Carbon Alloys: Novel Concepts to Develop Carbon Science and Technology*. Oxford: Elsevier Science, 2003, pp. 109–142.
5. Burchell, T.D. *Carbon Materials for Advanced Technologies*. Oxford: Elsevier Science, 1999.
6. Derbyshire, F., Jagtoyen, M., Andrews, R., Rao, A., Martin-Gullon, I., and Grulke, E.A. In Radovic, L.R., Ed. *Chemistry and Physics of Carbon 27*. New York: Marcel Dekker, 2001, pp. 1–66.
7. Radovic, L.R., Moreno-Castilla, C., and Rivera-Utrilla, J. In Radovic, L.R., Ed. *Chemistry and Physics of Carbon 27*. New York: Marcel Dekker, 2001, pp. 227–405.
8. Rodríguez-Reinoso, F. and Linares-Solano, A. In Thrower, P.A., Ed. *Chemistry and Physics of Carbon 21*. New York: Marcel Dekker, 1989, pp. 1–146.
9. Cazorla-Amorós, D., Alcañiz-Monge, J., and Linares-Solano, A. *Langmuir* 12(11): 2820–2824, 1996.
10. Stoeckli, F. and Ballerini, L. *Fuel* 70(4): 557–9, 1991.
11. Cazorla-Amorós, D., Alcañiz-Monge, J., de la Casa-Lillo, M.A., and Linares-Solano, A. *Langmuir* 14(16): 4589–4596, 1998.
12. Lozano-Castelló, D., Cazorla-Amorós, D., Linares-Solano, A., and Quinn, D.F. *J. Phys. Chem. B* 106(36): 9372–9379, 2002.
13. Lozano-Castelló, D., Cazorla-Amorós, D., and Linares-Solano, A. *Carbon* 42(7): 1233–1242, 2004.
14. Gregg, S.J. and Sing, K.S.W. *Adsorption, Surface Area and Porosity*. 2nd ed. London: Academic Press, 1982.
15. Ruthven, D.M. *Principles of Adsorption and Adsorption Processes*. New York: Wiley-Interscience, 1984.
16. Rouquerol, F., Rouquerol, J., and Sing, K. *Adsorption by Powders and Porous Solids*. London: Academic Press, 1999.
17. Yang, R.T. *Adsorbents: Fundamentals and Applications*. Wiley-Interscience, 2003.
18. Conway, B.E. *Electrochemical Supercapacitors: Scientific Fundamentals and Technological Applications*. New York: Kluwer Academic, 1999.
19. Rodríguez-Reinoso, F. and Molina-Sabio, M. *Carbon* 30(7): 1111–1118, 1992.
20. Ahmadpour, A. and Do, D.D. *Carbon* 34(4): 471–479, 1996.
21. Juan-Juan, J., Lozano-Castelló, D., Raymundo-Piñero, E., Lillo-Ródenas, M.A., Cazorla-Amorós, D., and Linares-Solano, A. In Cazorla, D. and J. Wirling, Eds. *Solutions in Dioxin and Mercury Removal*. Alicante: Publicaciones Universidad de Alicante, 2003, pp. 91–112.
22. Muñoz-Guillena, M.J., Illán-Gómez, M.J., Martin-Martinez, J.M., Linares-Solano, A., and Salinas-Martínez de Lecea, C. *Energy Fuels* 6(1): 9–15, 1992.
23. Illán-Gómez, M.J., García-García, A., Salinas-Martínez de Lecea, C., and Linares-Solano, A. *Energy Fuels* 10(5): 1108–1114, 1996.
24. Rodríguez-Mirasol, J., Cordero, T., and Rodríguez, J.J. *Energy Fuels* 7(1): 133–138, 1993.
25. Wu, F.C., Tseng, R.L., and Juang, R.S. *J. Colloid Interface Sci.* 283(1): 49–56, 2005.
26. Alcañiz-Monge, J., Cazorla-Amorós, D., Linares-Solano, A., Yoshida, S., and Oya, A. *Carbon* 32(7): 1277–1283, 1994.
27. Cazorla-Amorós, D., Ribes-Pérez, D., Román-Martínez, M.C., and Linares-Solano, A. *Carbon* 34(7): 869–878, 1996.

28. Teng, H. and Lin, H.C. *AIChE J.* 44(5): 1170–1177, 1998.
29. Benaddi, H., Legras, D., Rouzaud, J.N., and Béguin, F. *Carbon* 36(3): 306–309, 1998.
30. Molina-Sabio, M., Rodríguez-Reinoso, F., Caturla, F., and Sellés, M.J. *Carbon* 33(8): 1105–1113, 1995.
31. Díaz-Díez, M.A., Gómez-Serrano, V., González, C.F., Cuerda-Correa, E.M., and Macías-García, A. *Appl. Surf. Sci.* 238(1–4): 309–313, 2004.
32. Suárez-García, F., Martínez-Alonso, A., and Tascón, J.M.D. *Microporous Mesoporous Mater.* 75(1–2): 73–80, 2004.
33. Caturla, F., Molina-Sabio, M., and Rodríguez-Reinoso, F. *Carbon* 29(7): 999–1007, 1991.
34. Ibarra, J.V., Moliner, R., and Palacios, J.M. *Fuel* 70(6): 727–732, 1991.
35. Hayashi, J., Watkinson, A.P., Teo, K.C., Takemoto, S., and Muroyama, K. In Pajares, J.A. and Tascón, J.M.D., Eds. *Coal Science 1.* The Netherlands: Elsevier Science, 1995, pp. 1121–1124.
36. Carvalho, A.P., Gomes, M., Mestre, A.S., Pires, J., and de Carvalho, M.B. *Carbon* 42(3): 672–674, 2004.
37. Hayashi, J., Uchibayashi, M., Horikawa, T., Muroyama, K., and Gomes, V.G. *Carbon* 40(15): 2747–2752, 2002.
38. Ahmadpour, A. and Do, D.D. *Carbon* 35(12): 1723–1732, 1997.
39. Evans, M.J.B., Halliop, E., and MacDonald, J.A.F. *Carbon* 37(2): 269–274, 1999.
40. Teng, H. and Hsu, L.Y. *Ind. Eng. Chem. Res.* 38(8): 2947–2953, 1999.
41. Otowa, T., Tanibata, R., and Itoh, M. *Gas Sep. Purif.* 7(4): 241–245, 1993.
42. Lozano-Castelló, D., Lillo-Ródenas, M.A., Cazorla-Amorós, D., and Linares-Solano, A. *Carbon* 39(5): 741–749, 2001.
43. Lillo-Ródenas, M.A., Lozano-Castelló, D., Cazorla-Amorós, D., and Linares-Solano, A. *Carbon* 39(5): 751–759, 2001.
44. Perrin, A., Celzard, A., Albiniak, A., Kaczmarczyk, J., Marêché, J.F., and Furdin, G. *Carbon* 42(14): 2855–2866, 2004.
45. Baker, F.S. U.S. Patent 5,416,056, 1995.
46. Suárez-García, F., Martínez-Alonso, A., and Tascón, J.M.D. *J. Anal. Appl. Pyrolysis* 63(2): 283–301, 2002.
47. Teng, H., Yeh, T.S., and Hsu, L.Y. *Carbon* 36(9): 1387–1395, 1998.
48. Wennerberg, A.N. and O'Grady, T.M. U.S. Patent 4,082,694, 1978.
49. Lizzio, A.A. and Rostam-Abadi, M. *Fuel Process. Technol.* 34(2): 97–122, 1993.
50. Pandolfo, A.G., Amini-Amoli, M., and Killingley, J.S. *Carbon* 32(5): 1015–1019, 1994.
51. Lebgaa, D., Ehrburger, P., Papirer, J.B., Donnet, J.S., and Stoeckli, H.F. *Bull. Soc. Chim. France* 131: 763–773, 1994.
52. Verheyen, V., Rathbone, R., Jagtoyen, M., and Derbyshire, F. *Carbon* 33(6): 763–772, 1995.
53. Otowa, T., Nojima, Y., and Miyazaki, Y. *Carbon* 35(9): 1315–1319, 1997.
54. Agarwal, P. and Dollimore, D. *J. Therm. Anal.* 50(4): 525–531, 1997.
55. Qiao, W.M., Ling, L.C., Zha, Q.F., and Liu, L. *J. Mater. Sci.* 32(16): 4447–4453, 1997.
56. Sun, J., Brady, T.A., Rood, M.J., Lehmann, C.M., Rostam-Abadi, M., and Lizzio, A.A. *Energy Fuels* 11(2): 316–322, 1997.
57. Yoshizawa, N., Yamada, Y., and Shiraishi, M. *J. Mater. Sci.* 33(1): 199–206, 1998.
58. Ahmadpour, A., King, B.A., and Do, D.D. *Ind. Eng. Chem. Res.* 37(4): 1329–1334, 1998.
59. Lua, A.C. and Guo, J. *Energy Fuels* 12(6): 1089–1094, 1998.
60. Guo, J. and Lua, A.C. *Microporous Mesoporous Mater.* 32(1–2): 111–117, 1999.
61. Hu, Z.H. and Srinivasan, M.P. *Microporous Mesoporous Mater.* 27(1): 11–18, 1999.
62. Teng, H. and Wang, S.C. *Carbon* 38(6): 817–824, 2000.

63. Lee, S.H. and Choi, C.S. *Fuel Process. Technol.* 64(1–3): 141–153, 2000.
64. Hsu, L. and Teng, H. *Fuel Process. Technol.* 64(1–3): 155–166, 2000.
65. Salame, I.I. and Bandosz, T.J. *Ind. Eng. Chem. Res.* 39(2): 301–306, 2000.
66. Moreno-Castilla, C., Carrasco-Marín, F., López-Ramón, M.V., and Alvarez-Merino, M.A. *Carbon* 39(9): 1415–1420, 2001.
67. Tsai, W.T., Chang, C.Y., Wang, S.Y., Chang, C.F., Chien, S.F., and Sun, H.F. *Biores. Technol.* 78(2): 203–208, 2001.
68. Teng, H. and Weng, T.C. *Microporous Mesoporous Mater.* 50(1): 53–60, 2001.
69. Zou, Y. and Han, B.X. *Adsorption Sci. Technol.* 19(1): 59–72, 2001.
70. Lozano-Castelló, D., Cazorla-Amorós, D., and Linares-Solano, A. *Fuel Process. Technol.* 77–78: 325–330, 2002.
71. Lillo-Ródenas, M.A., Carratalá-Abril, J., Cazorla-Amorós, D., and Linares-Solano, A. *Fuel Process. Technol.* 77–78: 331–336, 2002.
72. Raymundo-Piñero, E., Cazorla-Amorós, D., Linares-Solano, A., Delpeux, S., Frackowiak, E., Szostak, K., and Béguin, F. *Carbon* 40(9): 1614–1617, 2002.
73. Yoshizawa, N., Maruyama, K., Yamada, Y., Ishikawa, E., Kobayashi, M., Toda, Y., and Shiraishi, M. *Fuel* 81(13): 1717–1722, 2002.
74. Huang, M.C., Chou, C.H., and Teng, H. *AIChE J.* 48(8): 1804–1810, 2002.
75. Park, S.J. and Jung, W.Y. *J. Colloid Interface Sci.* 250(1): 93–98, 2002.
76. Park, S.J. and Jung, W.Y. *J. Colloid Interface Sci.* 250(1): 196–200, 2002.
77. Frackowiak, E., Delpeux, S., Jurewicz, K., Szostak, K., Cazorla-Amorós, D., and Béguin, F. *Chem. Phys. Lett.* 361(1–2): 35–41, 2002.
78. Shimodaira, N. and Masui, A. *J. Appl. Phys.* 92(2): 902–909, 2002.
79. Girgis, B.S., Yunis, S.S., and Soliman, A.M. *Mater. Lett.* 57(1): 164–172, 2002.
80. Oh, G.H. and Park, C.R. *Fuel* 81(3): 327–336, 2002.
81. Carvalho, A.P., Cardoso, B., Pires, J., and de Carvalho, M.B. *Carbon* 41(14): 2873–2876, 2003.
82. Park, S.J. and Jung, W.Y. *J. Colloid Interface Sci.* 265(2): 245–250, 2003.
83. Shen, Z. and Xue, R. *Fuel Process. Technol.* 84(1–3): 95–103, 2003.
84. Diaz-Teran, J., Nevskaia, D.M., Fierro, J.L.G., Lopez-Peinado, A.J., and Jerez, A. *Microporous Mesoporous Mater.* 60(1–3): 173–181, 2003.
85. Guo, Y.P., Zhang, H., Tao, N.N., Liu, Y.H., Qi, J.R., Wang, X.C., and Xu, H.D. *Mater. Chem. Phys.* 82(1): 107–115, 2003.
86. Guo, Y.P., Qi, J.R., Jiang, Y.Q., Yang, S.F., Wang, Z.C., and Xu, H.D. *Mater. Chem. Phys.* 80(3): 704–709, 2003.
87. Guo, J. and Lua, A.C. *Chem. Eng. Res. Des.* 81(A5): 585–590, 2003.
88. Maciá-Agulló, J.A., Moore, B.C., Cazorla-Amorós, D., and Linares-Solano, A. *Carbon* 42(7): 1367–1370, 2004.
89. Gañan, J., González-García, C.M., González, J.F., Sabio, E., Macías-García, A., and Díaz-Díez, M.A. *Appl. Surf. Sci.* 238(1–4): 347–354, 2004.
90. Qi, J.R., Li, Z., Guo, Y.P., and Xu, H.D. *Mater. Chem. Phys.* 87(1): 96–101, 2004.
91. Wu, M.B., Zha, Q.F., Qiu, J.S., Guo, Y.S., Shang, H.Y., and Yuan, A.J., *Carbon* 42(1): 205–210, 2004.
92. Babel, K. and Jurewicz, K. *J. Phys. Chem. Solids* 65(2–3): 275–280, 2004.
93. Lua, A.C. and Yang, T. *J. Colloid Interface Sci.* 274(2): 594–601, 2004.
94. Yoon, S.H., Lim, S., Song, Y., Ota, Y., Qiao, W.M., Tanaka, A., and Mochida, I. *Carbon* 42(8–9): 1723–1729, 2004.
95. Mitani, S., Lee, S.I., Yoon, S.H., Korai, Y., and Mochida, I. *J. Power Sources* 133(4): 298–301, 2004.
96. Kim, Y.J., Horie, Y., Matsuzawa, Y., Ozaki, S., Endo, M., and Dresselhaus, M.S. *Carbon* 42(12–13): 2423–2432, 2004.

97. Stavropoulos, G.G. *Fuel Process. Technol.* 86(11): 1165–1173, 2005.
98. Maciá-Agulló, J.A., Moore, B.C., Cazorla-Amorós, D., and Linares-Solano, A. Chemical activation by KOH and NaOH of carbon materials with different cristalinity. *Proc. Carbon Conference 2003*, Oviedo (Spain), 2003.
99. Lillo-Ródenas, M.A., Cazorla-Amorós, D., and Linares-Solano, A. *Carbon* 41(2): 267–275, 2003.
100. Lillo-Ródenas, M.A., Juan-Juan, J., Cazorla-Amorós, D., and Linares-Solano, A. *Carbon* 42(7): 1371–1375, 2004.
101. Raymundo-Piñero, E., Azaïs, P., Cacciaguerra, T., Cazorla-Amorós, D., Linares-Solano, A., and Béguin, F. *Carbon* 43(4): 786–795, 2005.
102. Lozano-Castelló, D., Cazorla-Amorós, D., Linares-Solano, A., Hall, P.J., and Fernández, J.J. *Stud. Surf. Sci. Catal.* 128: 523–532, 2000.
103. Lozano-Castelló, D., Raymundo-Piñero, E., Cazorla-Amorós, D., Linares-Solano, A., Müller, M., and Riekel, C. *Carbon* 40(14): 2727–2735, 2002.
104. Lozano-Castelló, D., Raymundo-Piñero, E., Cazorla-Amorós, D., Linares-Solano, A., Müller, M., and Riekel, C. *Stud. Surf. Sci. Catal.* 144: 51–58, 2002.
105. Lozano-Castelló, D., Cazorla-Amorós, D., and Linares-Solano, A. *Chem. Eng. Technol.* 26(8): 852–857, 2003.
106. Lillo-Ródenas, M.A., Cazorla-Amorós, D., Linares-Solano, A., Béguin, F., Clinard, C., and Rouzaud, J.N. *Carbon* 42(7): 1305–1310, 2004.
107. Lillo-Ródenas, M.A., Carratalá-Abril, J., Cazorla-Amorós, D., and Linares-Solano, A. Optimisation of the properties of activated carbons for the adsorption of VOC's at low concentrations. *Proc. Carbon Conf. 2001*, Lexington (Kentucky), 2001.
108. Lillo-Ródenas, M.A., Carratalá-Abril, J., Cazorla-Amorós, D., and Linares-Solano, A. Usefulness of chemically activated anthracite for the abatement of VOC at low concentrations. *Proc. Coal Struct.* Gliwice (Poland), 2002.
109. Lillo-Ródenas, M.A., Cazorla-Amorós, D., and Linares-Solano, A. *Carbon* 43(8): 1758–1767, 2005.
110. Lozano-Castelló, D., Alcañiz-Monge, J., de la Casa-Lillo, M.A., Cazorla-Amorós, D., and Linares-Solano, A. *Fuel* 81(14): 1777–1803, 2002.
111. Lozano-Castelló, D., Cazorla-Amorós, D., Linares-Solano, A., and Quinn, D.F. *Carbon* 40(7): 989–1002, 2002.
112. Lozano-Castelló, D., Cazorla-Amorós, D., and Linares-Solano, A. *Energy Fuels* 16(5): 1321–1328, 2002.
113. Lozano-Castelló, D., Cazorla-Amorós, D., Linares-Solano, A., and Quinn, D.F. *Carbon* 40(15): 2817–2825, 2002.
114. Lozano-Castelló, D., Cazorla-Amorós, D., Linares-Solano, A., Shiraishi, S., Kurihara, H., and Oya, A. *Carbon* 41(9): 1765–1775, 2003.
115. Zhao, N., He, C., Jiang, Z., Li, J., and Li, Y. *Mater. Lett.* 61(3):681–5, 2007.
116. Lillo-Ródenas, M.A. Ph.D. thesis. University of Alicante, 2004.
117. Bailar, J.C., Emeléus, H.J., Nyholm, R.S., and Trotman-Dickenson, A.F. *Comprehensive Inorganic Chemistry*, Vol. I. 1st ed., Oxford: Pergamon Press, 1973.
118. Webb, P.A. and Orr, C. *Analytical Methods in Fine Particle Technology.* Micromeritics Instrument Corporation, 1997.
119. Martin-Gullon, I., Marco-Lozar, J.P., Cazorla-Amorós, D., and Linares-Solano, A. *Carbon* 42(7): 1339–1343, 2004.
120. Rodríguez-Reinoso, F., Linares-Solano, A., Torres-Puchol, G., and López-González, J.D. *Anales de Química* 78B: 166, 1982.
121. Linares-Solano, A., Martin-Gullón I, I, Salinas-Martinez de Lecea, C., and Serrano-Talavera, B. *Fuel*, 79(6): 635–643, 2000.
122. Béguin, F. and Setton, R. *Carbon* 10(5): 539–551, 1972.

123. Mochida, I., Nakamura, E., Maeda, K., and Takeshita, K. *Carbon* 13(6): 489–493, 1975.
124. Mochida, I., Nakamura, E., Maeda, K., and Takeshita, K. *Carbon* 14(2): 123–129, 1976.
125. Yamashita, Y. and Ouchi, K. *Carbon* 20(1): 41–45, 1982.
126. Yamashita, Y. and Ouchi, K. *Carbon* 20(1): 47–53, 1982.
127. Yamashita, Y. and Ouchi, K. *Carbon* 20(1): 55–58, 1982.
128. Chen, X. and McEnaney, B. Activation of meso-carbon microbeads. *Proceedings of Carbon Conference 2001*, Lexington (Kentucky), 2001.
129. Hu, Z. and Vansant, E.F. *Carbon* 33(9): 1293–1300, 1995.
130. Mahajan, O.P., Walker, P.L., Jr. Karr, C., Jr., Eds. Reactivity of heat treated coals. *Analytical Methods for Coal and Coal Products*, Vol. II. New York: Academic Press, 1978. pp. 465–94.
131. Foster, K.L., Fuerman, R.G., Economy, J., Larson, S.M., and Rood, M.J. *Chem. Mater.* 4(5): 1068–1073, 1992.
132. Ruhl, M.J. *Chem. Eng. Prog.* 89(7): 37–41, 1993.
133. Cal, M.P., Larson, S.M., and Rood, M.J. *Environ. Prog.* 13(1): 26–30, 1994.
134. Dimotakis, E.D., Cal, M.P., Economy, J., Rood, M.J., and Larson, S.M. *Environ. Sci. Technol.* 29(7): 1876–1880, 1995.
135. Yun, J.H. and Choi, D.K. *J. Chem. Eng. Data* 42(5): 894–896, 1997.
136. Cal, M.P., Rood, M.J., and Larson, S.M. *Gas Sep. Purif.* 10(2): 117–121, 1999.
137. Chiang, H.L., Huang, C.P., Chiang, P.C., and You, J.H. *Carbon* 37(12): 1919–1928, 1996.
138. Huang, Z.H., Kang, F.Y., Zheng, Y.P., Yang, J.B., and Liang, K.M. *Adsorption Sci. Technol.* 20(5): 495–500, 2002.
139. Lillo-Ródenas, M.A., Fletcher, A.J., Thomas, K.M., Cazorla-Amorós, D., and Linares-Solano, A. *Carbon* 44(8): 1455–1463, 2006.
140. Raymundo-Piñero, E., Cazorla-Amorós, D., and Linares-Solano, A. *Carbon* 39(2): 231–242, 2001.
141. Raymundo-Piñero, E., Cazorla-Amorós, D., Salinas-Martinez de Lecea, C., and Linares-Solano, A. *Carbon* 38(3): 335–344, 2000.
142. Davini, P., *Carbon* 39(9): 1387–1393, 2001.
143. Vansant, E.F. and Dewolfs, R. *Gas Sep. Technol.* Amsterdam: Elsevier, 1990.
144. Kapoor, A. and Yang, R.T. *Chem. Eng. Sci.* 44(8): 1723–1733, 1989.
145. Parkyns, N.D. and Quinn, D.F. In Patrick, J.W., Ed. *Porosity in Carbons.* Edward Arnold, 1995, pp. 293–325.
146. Menon, V.C. and Komarneni, S. *J. Porous Mater.* 5(1): 43–58, 1998.
147. Cook, T.L., Komodromos, C., Quinn, D.F., and Ragan. S. In Burchell, T.D., Ed. *Carbon Materials for Advanced Technologies.* New York: Pergamon, Elsevier Science, 1999, pp. 269–302.
148. Rzepka, M., Lamp, P., and de la Casa-Lillo, M.A. *J. Phys. Chem. B* 102(52): 10894–10898, 1998.
149. Jorda-Beneyto, M., Lozano-Castello, D., Suarez-Garcia, F., Cazorla-Amoros, D., and Linares-Solano, A. *Carbon* 45(2): 293–303, 2007.
150. de la Casa-Lillo, M.A., Lamari-Darkrim, F., Cazorla-Amorós, D., and Linares-Solano, A. *J. Phys. Chem. B* 106(42): 10930–10934, 2002.
151. Shi, H. *Electrochim. Acta* 41(10): 1633–1639, 1996.
152. Liu, X. and Osaka, T. *J. Electrochem. Soc.* 143(12): 3982–3986, 1996.
153. Liu, X. and Osaka, T. *J. Electrochem. Soc.* 144(9): 3066–3071, 1997.
154. Miller, J.M., Dunn, B., Tran, T.D., and Pekala, T.W. *J. Electrochem. Soc.* 144(12): 309–311, 1997.

155. Saliger, R., Fischer, U., Herta, C., and Fricke, J. *J. Non-Cryst. Solids* 225: 81–85, 1998
156. Tanahashi, I., Yoshida, A., and Nishino, A. *Denki Kagaku Oyobi Kogyo Butsuri Kagaku* 56(10): 892–897, 1988.
157. Morimoto, T., Hiratsuka, K., Sanada, Y., and Kurihara, K. *J. Power Sources* 60(2): 239–247, 1996.
158. Lin, C., Ritter, J.A., and Popov, B.N. *J. Electrochem. Soc.* 146(10): 3639–3643, 1999.
159. Burger, J., Holland, H., Ter Brake, M., Venhorst, G., Hondebrin, E., Meier, R.J., Rogalla, H., Coesel, R., Dierssen, W., Grim, R., Lozano-Castello, D., and Sirbi, A. *European Space Agency* (special publication) ESA SP (539): 379–384, 2003.

2 Template Approaches to Preparing Porous Carbon

Fabing Su, Zuocheng Zhou, Wanping Guo, Jiajia Liu, Xiao Ning Tian, and X. S. Zhao

CONTENTS

I. INTRODUCTION

Porous solids are of scientific and technological importance because of their ability to interact with atoms, ions, molecules, and supramolecules [1]. The presence of voids of controllable dimensions at the atomic, molecular, and nanometer scales in porous solids endows them with unique interfacial properties [2]. As a result, they are widely used as ion-exchangers, adsorbents, catalysts, and catalyst supports. Structurally, they are classified into microporous, mesoporous, and macroporous solids, with pore sizes of less than 2 nm, from 2 to 50 nm, and larger than 50 nm, respectively. Compositionally, they could be inorganic, polymeric, and composite [1–8].

Synthetic pathways to porous solids are manifold and often depend strongly on the targeted product. The early synthesis of porous solids was considered an art rather than science. A scientific approach to the synthesis of porous materials began in the first half of the 20th century [9], when the synthesis of materials such as activated carbons and silica gels was studied more systematically. Nowadays, one relies on scientific manipulation of preparation parameters and design principles, by which adjustment of the chemical and physical properties of the materials is possible. A notable example is the so-called template synthesis [9–11].

The concept of template synthesis of porous solids can be tracked back to the 1960s [12]. Over the past four decades or so, many template synthesis strategies have been demonstrated for producing porous solids, particularly ordered porous solids such as microporous zeolites [13], ordered mesoporous silicas and metal oxides [6,14–16], and ordered macroporous materials [17], including porous carbons [18–21]. Recently, Schüth classified the various templates into two categories [9], namely, endotemplates and exotemplates. The former category refers to a template that is an isolated entity incorporated into the growing solid whose removal leaves behind the desired pore system. The latter is a porous structure that provides a scaffold for another porous (or nonporous) solid.

Porous carbons (e.g., activated carbons) are an important family of porous solids that have a wide spectrum of applications because of their remarkable properties, such as high specific surface area, chemical inertness, abundant repertory of surface functional groups, good thermal stability, and low cost of manufacture. Their chemistry and physics have been reviewed [22–25]. The most common way to produce activated carbons is to carbonize a carbon-containing precursor, followed by activation or posttreatment [26]. Because of practical requirements of various applications, techniques for control over the pore size of activated carbons have been the subject of research for several decades [18]; for example, high burn-off activation, catalyst-assisted activation, and carbonization of polymer blends with thermally unstable components. For recent progress in the use of hydroxide activation, see Chapter 1 by Linares-Solano and coworkers in this volume [27]. However, none of these synthesis approaches is suitable for very precise control over pore structure, particle size, and morphology [19].

The discovery and development of various nanocarbon materials, such as fullerenes [28], nanotubes [29], nanocones [30], nanocages [31], membranes [32], microtrees [33], tubular graphite cones [34], nanoscrolls [35], and nanofibers [36], have significantly advanced the science and engineering of carbon nanostructures [37–39]. With the rapid emergence of new technologies such as energy storage and conversion, biomolecular engineering, drug delivery, and photonics, the time has come to develop novel porous carbons with more tailorable structural, morphological, and surface properties.

The exotemplate method has been successfully employed to prepare a variety of novel porous carbon structures with controllable properties [18–20,40–42]. To the best of our knowledge, Knox and coworkers [43] were the first to use exotemplates, including silica gel and porous glass, to prepare porous glassy carbon (PGC). Polymerization of a phenol–hexamine mixture in the pores of an exotemplate followed by carbonization at high temperatures under nitrogen atmosphere and subsequent removal of the template yielded the PGC, which was shown to display unique properties in liquid chromatographic separation [44]. This exotemplate method was soon adopted by other research groups to prepare various carbon structures, such as ultrathin graphite films and high-surface-area carbons [45–49], uniform carbon nanotubes [50–52], and composite materials [53–56].

Over the past few years, there have been rapid advances in using the template method to prepare ordered porous carbons, ranging from microporous to mesoporous carbons, and further, to macroporous carbons. Kyotani and coworkers [57–59] employed zeolite Y to prepare microporous carbons with a structural regularity similar to the template. The preparation of ordered mesoporous carbons (OMCs) composed of two-dimensional (2-D) regular arrays of uniform mesopores by employing ordered mesoporous silicas [60,61] as templates was demonstrated independently by two Korean research groups [62,63]. Subsequently, numerous OMCs have been described using different mesoporous silicas (e.g., MCM-48, SBA-15, SBA-1, HMS, and SBA-16) as templates. Ordered macroporous carbons were first demonstrated by Zakhidov and coworkers [64] using self-assembled colloidal crystals as templates. Furthermore, nanoparticle templates for making mesoporous and macroporous carbon materials were first described by Hyeon and coworkers [65] and Jaroniec and coworkers [66–68]. The exotemplate strategy has thus emerged as one of the most effective approaches to preparing microporous, mesoporous, and macroporous carbons with desired physical and chemical properties. It is opening new opportunities in making custom-designed novel porous carbon materials for a wide spectrum of applications. It is expected that soon the potential applications of templated porous carbons will be demonstrated in many more fields.

In this chapter, we provide an overview of the recent research and development in the preparation, characterization, and application of novel porous carbons using both the endotemplate and the exotemplate methods. A discussion of zeolite templates for microporous carbons is followed by that of ordered mesoporous silica templates for OMCs, nanoparticle templates for mesoporous carbons, sol–gel processed porous carbons, self-assembled colloidal crystal templates for ordered macroporous carbons, and colloidal sphere templates for hollow carbon spheres, as well as other templating approaches to preparing carbon nanostructures. Then,

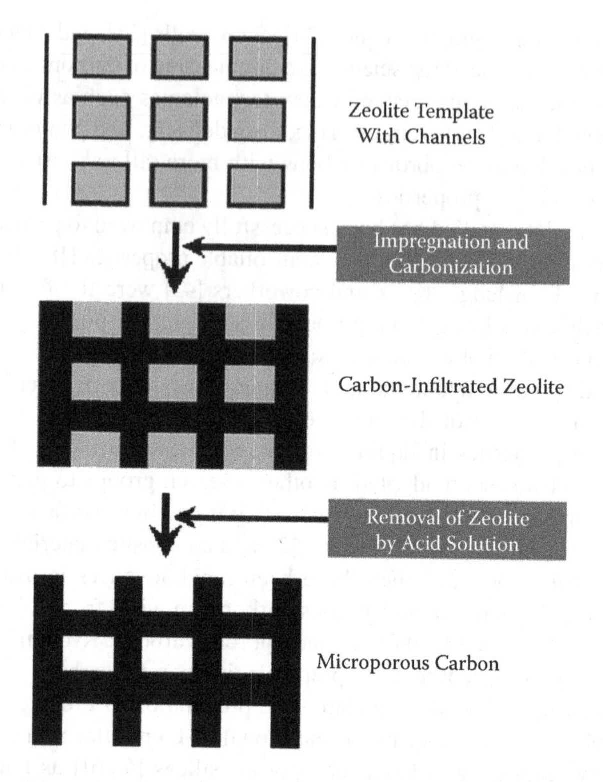

FIGURE 2.1 Schematic illustration of zeolite template approach to the preparation of microporous carbons.

we summarize some demonstrated potential applications. Finally, we offer an admittedly subjective view of the perspectives of template synthesis of carbon materials. Although a large number of reports in the literature are cited, we also use our own research data to illustrate the preparation and characterization of carbons with the assistance of various templates and their applications.

II. ZEOLITE-TEMPLATED MICROPOROUS CARBON

Zeolites are microporous crystalline aluminosilicates with a channel-like or cage-like pore structure with pore-opening sizes in the range of 0.3 to 1.5 nm [13]. The spatially periodic pore structure and well-defined nanospaces of zeolites offer opportunities to control the nanostructure and morphology of microporous carbon materials at the nanometer level. As schematically illustrated in Figure 2.1, zeolite pores can be filled with a carbon precursor such as furfuryl alcohol (FA). After a proper treatment, followed by removal of the zeolite framework, a carbon nanostructure with pores replicated from the zeolite framework is obtained. Over the past decade, many zeolite templates (e.g., zeolite Y, zeolite β, and ZSM-5) and carbon precursors (e.g., FA, phenol-formaldehyde, and sucrose) have been employed to prepare microporous carbons with a high specific surface area, a large pore volume, and various morphologies [57–59,69–84]. Using an appropriate preparative strategy, the pore-structural regularity of the zeolite can be transferred to the resultant carbon.

A. High-Surface-Area Carbon

Kyotani and coworkers [81] systematically demonstrated the preparation of microporous carbons using zeolite templates, whereas early studies [82–84] described the pyrolysis of carbon precursors to make carbon materials in the presence of zeolites. Subsequently, Mallouk and coworkers [85] employed zeolites Y, L, and β as templates to prepare microporous carbons with a specific surface area as high as 1580 m^2/g. It was reported that the zeolite template has a direct relationship with the structural and topological properties of the resultant carbon. Rodriguez-Mirasol et al. [71] described the preparation of microporous carbons with a wide distribution of pore sizes, well-developed mesoporosity, and high adsorption capacity. Zeolite Y was used as template, and a chemical vapor infiltration method was employed to deposit carbon in the template pores. It was found that the apparent surface area of the resultant carbons increased with increasing deposition temperatures. Meyers et al. [72] synthesized porous carbon materials with a surface area of about 1000 m^2/g using zeolites Y, β, and ZSM-5 as templates and acrylonitrile, FA, pyrene, and vinyl acetate as carbon precursors. The template-encapsulated carbon precursors were pyrolyzed at 600°C, and the resultant materials were observed to be composed of disordered carbon arrays.

Kyotani and coworkers made great progress in template synthesis of microporous carbons in terms of increasing surface area, pore volume, and enhancing structural regularity by combining chemical vapor deposition (CVD) with liquid-phase impregnation techniques. In an early attempt to prepare high-surface-area microporous carbon [69], these authors were able to make porous carbons with a BET (Brunauer Emmett Teller) surface area of over 2000 m^2/g and a total pore volume of 1.8 cm^3/g using the impregnation method. This was the first example of template synthesis of microporous carbons with a high surface area without involving an activation step. A two-step technique, namely, impregnation followed by CVD for enhancing carbon loading, was subsequently developed by the same research group [57–59] to prepare porous carbons having almost solely micropores, a BET area as high as 3600 m^2/g, and a micropore volume of 1.5 cm^3/g [70]. More recently, they prepared a microporous carbon with a BET area exceeding 4000 m^2/g [79] and a narrow pore size distribution (PSD) [78]. They also reported the preparation of porous carbons with a BET surface area of 5100 m^2/g and a micropore volume of 2.1 cm^3/g [86].

Without using the CVD method, we prepared microporous carbons with a BET area of 3680 m^2/g and a total pore volume of 2.02 cm^3/g [74]. Simply by impregnation of FA in the cages of commercial ammonium-form zeolite Y (NH_4Y), microporous carbons with different porosities, which are controlled by using different carbonization temperatures and carbon loadings, can be prepared.

It was observed that ammonium cations in the template participated in the reactions during the carbonization process, leading to the formation of nitrogen-containing functional groups in the resultant carbon [74].

B. Control over Pore-Structural Regularity

Previous studies [57–59,69–80] showed that without the assistance of a CVD technique, it is impossible to obtain a zeolite-templated carbon with high structural regularity, because of incomplete carbon infiltration and filling of the zeolite pores, implying that the porous carbon is not a true replica of the zeolite in spite of its morphological similarity to the zeolite template. To improve structural regularity of zeolite-templated carbons, Kyotani and coworkers [57–59] employed a two-step synthesis method, which involves thermal treatment of a zeolite–carbon composite, followed by subsequent carbon deposition using a CVD technique. The microporous carbon prepared with the zeolite Y template has a three-dimensional (3-D) nanoarray pore structure with a periodical order of 1.4 nm, as revealed by the appearance of an x-ray diffraction (XRD) peak at about 6° (see Figure 2.2a), which corresponds to the spacing of the (111) plane of the zeolite Y template. Recently, the pore-structural order of the microporous carbon was further improved, as can be seen from Figure 2.2b [79]. The key was reported to be the use of acetylene as the carbon precursor, whose molecular size is small in comparison with the diameter of the zeolite pores. Thus, carbon deposition occurred in the chemically controlled regime without serious pyrolytic decomposition on the external surface or blocking of pore mouths because of minor diffusion resistance [79].

It has been experimentally demonstrated that sufficient infiltration of carbon in the pores of a zeolite template is indispensable for preparing microporous carbon with a long-range structural order, as evidenced by the transmission electron microscopy (TEM) images shown in Figure 2.3. It must be pointed out that zeolites with a 3-D pore network, such as zeolites β and Y, must be used as templates to prepare ordered microporous carbons. The use of zeolite templates with a 1-D pore structure, such as zeolite L, leads to a structurally irregular carbon, as illustrated in Figure 2.4 [70]. Additionally, the crystal size of a zeolite template also affects the structural regularity of the resulting carbon, which is related to carbon precursor diffusion into the zeolite pores: small particle size favors the formation of ordered microporous carbon [87].

Barata-Rodrigues et al. [73] reported that the CVD technique does not help improve the structural regularity of a zeolite-templated carbon. We indeed found that, when wet impregnation is followed by a CVD treatment, the XRD peak at 6° 2θ can be observed, indicating the appearance of an ordered pore structure [75]. However, this XRD peak cannot be resolved for the carbon prepared without using the CVD technique [74]. An alternative route to the synthesis of ordered microporous carbons by using zeolite Y as template has also been described [88]. In addition, the use of other zeolite templates such as EMC-2 to improve structural regularity of replicated microporous carbons has been demonstrated

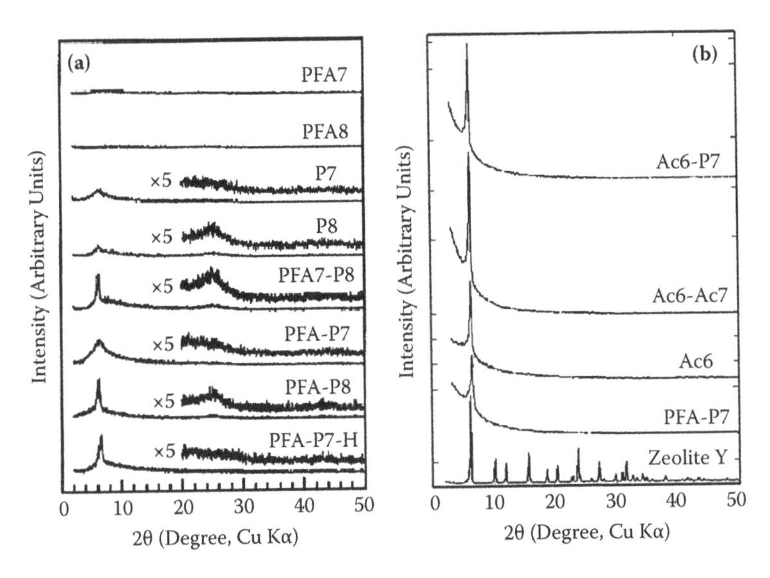

FIGURE 2.2 XRD patterns of zeolite Y-templated porous carbons prepared under different conditions. (a) PFA7: polymerization and carbonization of FA at 700°C; PFA8: polymerization and carbonization of FA at 800°C; P7: CVD of propylene at 700°C; P8: CVD of propylene at 800°C; PFA7-P8: polymerization and carbonization of FA at 700°C followed by CVD of propylene at 800°C; PFA-P7: polymerization of FA followed by CVD of propylene at 700°C; PFA-P8: polymerization of FA and CVD of propylene at 800°C; PFA-P7-H: polymerization of FA and CVD of propylene at 700°C followed by heat treatment at 900°C [59]. (b) Ac6-P7: CVD of acetylene at 600°C for 4 h followed by CVD of propylene at 700°C for 1 h; Ac6-Ac7: CVD of acetylene at 600°C for 4 h followed by CVD of acetylene at 700°C or 1 h; Ac6: CVD of acetylene at 600°C for 4 h; PFA-P7: polymerization of FA and CVD of propylene at 700°C for 1 h. (From Hou, P.X., Yamazaki, T., Orikasa, H., and Kyotani, T. *Carbon* 43: 2621–2627, 2005. With permission.)

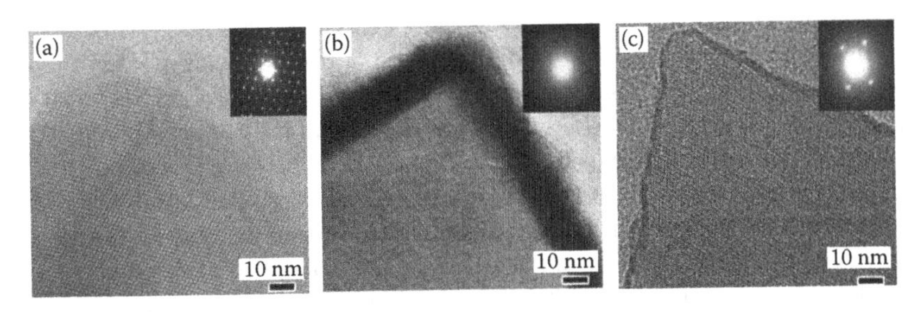

FIGURE 2.3 (a) HRTEM images of zeolite Y, (b) a zeolite Y-templated carbon prepared by impregnation and CVD, and (c) a zeolite Y-templated carbon prepared by impregnation and CVD followed by heat treatment. (From Ma, Z., Kyotani, T., Liu, Z., Terasaki, O., and Tomita, A. *Carbon* 40: 2367–2374, 2002. With permission.)

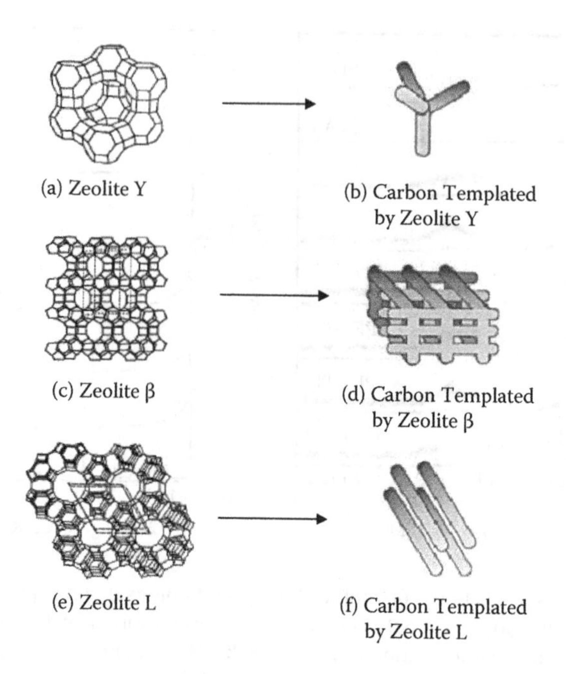

FIGURE 2.4 Schematic representation of the resultant carbon structures templated by using different zeolites. (From Johnson, S.A., Brigham, E.S., Ollivier, P.J., and Mallouk, T.E. *Chem Mater* 9: 2448–2458, 1997. With permission.)

[89]. The microporous carbon replica displayed three well-resolved XRD peaks corresponding to (100), (002), and (101) reflections of the EMC-2, proving the complete transfer of the hexagonal long-range pore ordering of zeolite to the carbon framework [89].

C. The Formation Mechanism

The porosity of a zeolite-templated carbon is determined not only by the zeolite framework structure, but also the subsequent processing conditions [59,69,73,85]. We investigated the evolution of pore structure at different carbonization temperatures [74]. Figure 2.5a shows the nitrogen adsorption/desorption isotherms of the carbons synthesized at different temperatures, with FA as the carbon precursor and NH_4Y zeolite as template. The isotherms are all of type I according to the IUPAC classification, indicating that they are microporous materials. Figure 2.5b presents the PSD curves of the carbons, computed using the density functional theory (DFT) method. It is seen that all carbons display two peaks at about 1.2 (labeled with a star) and 1.5 nm (labeled with +), respectively, indicating the presence of two types of micropores. It is believed that the first type was templated by the zeolite framework, whereas the second type was generated from the carbon matrix because of the release of small gaseous molecules such as CO_2, H_2, and H_2O during the carbonization process [85,90]. Using benzene as carbon precursor,

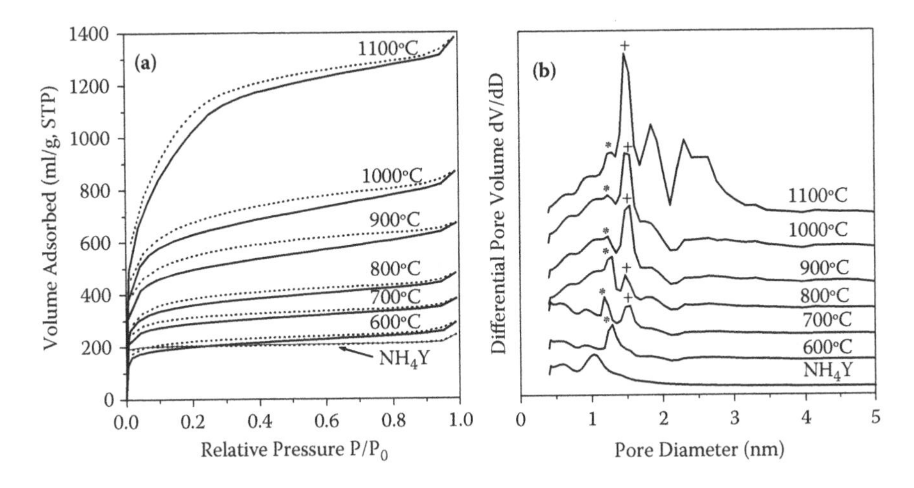

FIGURE 2.5 (a) Nitrogen adsorption–desorption isotherms (solid line: adsorption, dotted line: desorption), and (b) DFT–PSD curves of porous carbons synthesized at different carbonization temperatures, together with that of zeolite NH_4Y template. (From Su, F., Zhao, X.S., Lv, L., and Zhou, Z. *Carbon* 42: 2821–2831, 2004. With permission.)

the formation of the second type of micropores can be avoided [75]. Therefore, it can be concluded that the formation of the carbon pore structure is largely determined by the zeolite framework at low temperatures and by the carbon precursor at high temperatures. The peaks at about 1.8 and 2.2 nm (Figure 2.5b) for samples carbonized at 800°C or above were attributed to the collapse of pore walls at high temperature or to shrinkage of the zeolite framework [59].

D. Morphology Evolution

It has been shown that the morphology of a zeolite template is carried through to the resultant carbon [59–70,71,73,74,80,85]. Figure 2.6 presents the field-emission scanning electron microscope (FESEM) images of carbon samples prepared using the zeolite NH_4Y template [74]. It is clearly seen that the carbons preserved the morphology of the template particles even at a carbonization temperature of 1100°C, at which the structure of the zeolite will be largely destroyed. It should be noted that the particle size of the carbon is obviously smaller than that of the template, showing substantial carbon shrinkage. The images of a single carbon particle prepared at different carbonization temperatures (Figures 2.6b, 2.6d, and 2.6f) reveal that a dense carbon prepared at 600°C gradually evolved into a spongelike porous carbon prepared at 1100°C. A large number of irregular pores are seen in the sample prepared at 1100°C, which were probably formed on account of carbon fracturing and cracking at high temperatures.

Figure 2.7 shows the FESEM images of porous carbons prepared at 1000°C using different mass ratios of sucrose to zeolite NaY template [77]. As can be seen, proper control of this ratio (R) allowed us to obtain carbon structures with

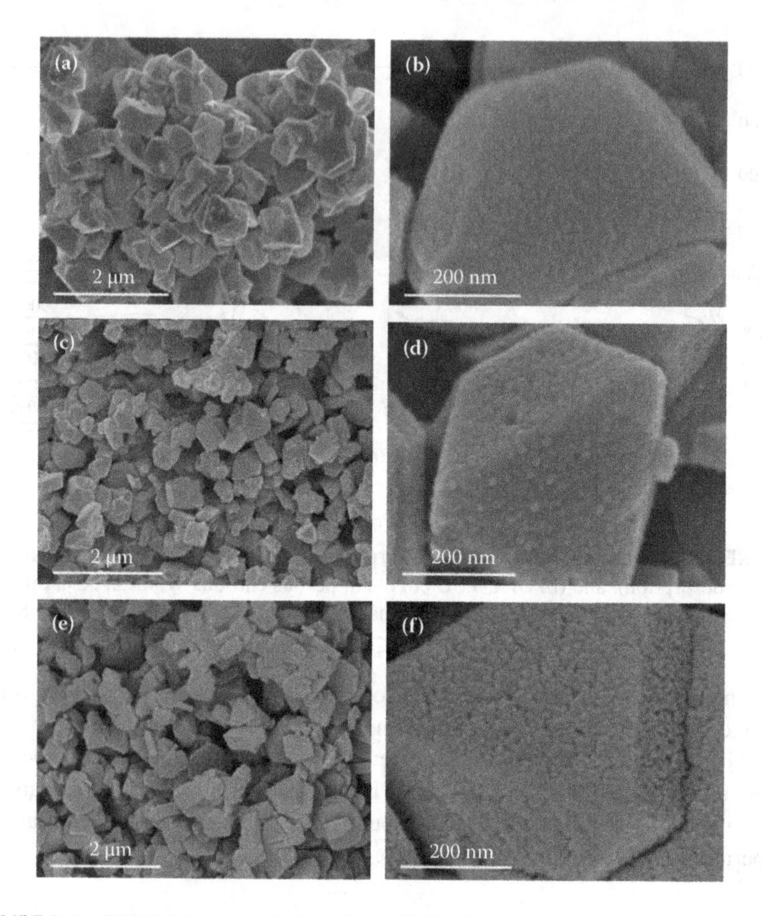

FIGURE 2.6 FESEM images of (a) zeolite NH_4Y, (b) a carbon prepared at a CVD temperature of 600°C, (c, d) a carbon prepared at a CVD temperature of 900°C, and (e, f) a carbon prepared at a CVD temperature of 1100°C. (From Su, F., Zhao, X.S., Lv, L., and Zhou, Z. *Carbon* 42: 2821–2831, 2004. With permission.)

various morphologies. A crumpled-wall nanostructured porous carbon with high surface area (1752 m^2/g) and mesopore volume (0.72 cm^3/g) was prepared with R = 1 (see Figure 2.7c). Carbons with a regular particle morphology were prepared when R = 0.5 (see Figure 2.7d). The morphological similarity of carbon to the template indicates a true exotemplating mechanism.

E. Control over Surface Chemistry

The surface chemistry of a carbon is known to play a crucial role in many applications, most notably in adsorption and catalysis [91–94]. It has been observed that the surface of templated porous carbons mainly has oxygen-containing functional groups, such as carboxyl, carbonyl, quinone, and hydroxyl [76,94,95]. Such groups, especially the carboxyls, can be eliminated by high-temperature

FIGURE 2.7 FESEM images of porous carbons prepared at 1000°C with different mass ratios of sucrose to zeolite Y (R): (a) R = 1:0, (b) R = 1:0.5, (c) R = 1:1, (d) R = 1:2. (From Su, F., Lv, L., and Zhao, X.S. *Stud Surf Sci Catal* 156: 557–565, 2005. With permission.)

treatment under inert atmosphere [95], thus rendering the surface more hydrophobic. In addition, other heteroatom-containing groups can be incorporated into template-synthesized carbons. Using the NH_4Y template, carbons with nitrogen-containing groups have been prepared [74]. Ammonia from decomposition of NH_4^+ possibly participated in the reaction during carbonization, forming nitrogen functional groups. Nitrogen-doped microporous carbon with a highly ordered structure has also been prepared by using acetonitrile [96,97]. Pyrrole has been used as a carbon precursor to prepare nitrogen-incorporated microporous carbons with nitrogen content as high as 9 wt% [98].

III. MESOPOROUS-SILICA-TEMPLATED MESOPOROUS CARBON

Ordered mesoporous materials synthesized using surfactant templates [60,61,99] have been widely used as hard templates to prepare various mesoporous carbons. Two Korean research groups were the first to independently demonstrate the use of ordered mesostructured silicas as hard templates [62,63]. Because they have larger pore sizes than zeolites, ordered mesoporous silicas with a 3-D pore system (e.g., MCM-48, SBA-15, and SBA-16) offer more flexibilities for control over the structural, morphological, and surface properties of templated mesoporous carbon.

A. MCM-48 AS A TEMPLATE

MCM-48 possesses a body-centered 3-D pore structure with $Ia\bar{3}d$ space group symmetry [100]. Depending on the synthesis and processing conditions, it can

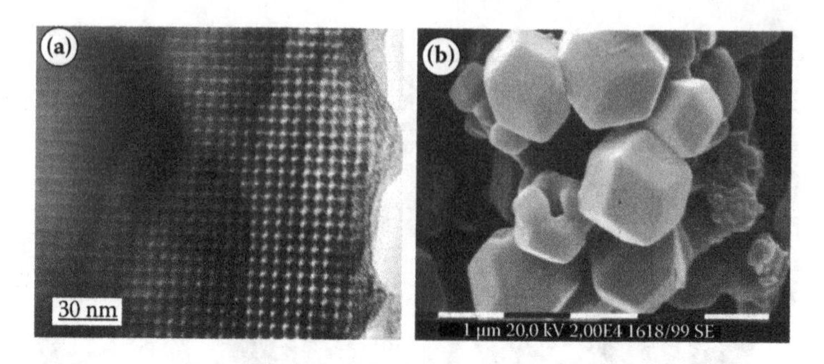

FIGURE 2.8 (a) TEM and (b) SEM images of carbon CMK-1. (From Ryoo, R., Joo, S.H., and Jun, S. *J Phys Chem* B 103: 7743–7746, 1999. With permission.)

exhibit mesopore sizes ranging from 3.7 to 4.5 nm. This material was first employed as a template to prepare OMC materials [62,63]. However, it has been shown that the thickness of its amorphous silica pore walls is not uniform [101].

Ryoo et al. [62] impregnated the pores of MCM-48 with a sucrose solution containing sulfuric acid, followed by carbonization at 900°C and removal of the silica framework to produce an OMC sample designed as CMK-1. Other carbon precursors, including glucose, xylose, FA, and phenolic resin, can also be used. The surface area of CMK-1 was reported to be in the range of 1300–1800 m^2/g [102]. The carbon contained not only micropores 0.5 to 0.8 nm in diameter but also about 3.0-nm mesopores [62]. The micropore and mesopore volumes were reported to be 0.3 and 1.1 cm^3/g, respectively. The micropores are formed due to pyrolysis of the carbon precursor prior to removal of silica pore walls [103], whereas mesopores are formed after template removal. The mesopore size is very much dependent on the unit cell of MCM-48 and, thus, it can be tailored [103]. Sample CMK-1 was reported to have good mechanical and thermal stability under inert atmosphere [104]. Figure 2.8a shows a TEM image. Uniform mesopores are clearly seen. The scanning electron microscope (SEM) image in Figure 2.8b reveals that the carbon particles retained the crystal morphology of the template [62]. The low-angle XRD pattern of the carbon–silica composite was found to be different from that of the template because of lattice contraction and intensity loss (see Figure 2.9). Lattice contraction usually occurs when MCM-48 is heated to high temperatures, whereas intensity loss is attributed to pore filling with carbon. According to Ryoo and coworkers, the pore structure can undergo a systematic transformation to a new ordered structure (cubic $I4_132$) upon removal of the silica framework [62]. Kruk et al. [103] speculated that the transformation is related to the disconnected nature of the two interwoven parts of the CMK-1 framework and may involve their mutual displacement to create some contacts between them. On the basis of continuous DFT analysis, Solovyov et al. [105] proposed that, in the course of the formation of CMK-1, the inverse carbon framework formed within the pores of the template was displaced with respect to one another without significant distortion after dissolution of the silica walls. The pore structure of CMK-1 can be described as an ordered interwoven assembly of

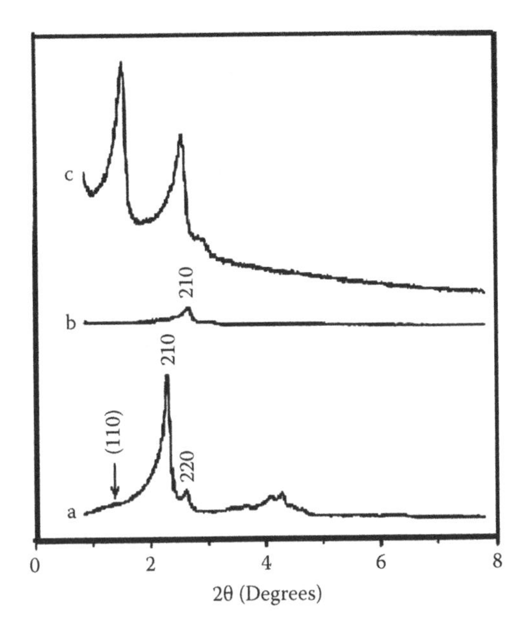

FIGURE 2.9 Low-angle XRD patterns of (a) mesoporous MCM-48, (b) MCM-48–carbon composite, and (c) CMK-1. (From Ryoo, R., Joo, S.H., and Jun, S. *J Phys Chem* B 103: 7743–7746, 1999. With permission.)

two enantiomeric subframeworks reproducing the shape of the MCM-48 mesopores. This is illustrated in Figure 2.10 [105].

Another OMC having a cubic $Ia\bar{3}d$ pore structure of the MCM-48 template was named CMK-4 [106]. The space group of the carbon does not change after complete template removal. The carbon arrays formed in MCM-48 channels are partly connected by the pores [104]. The 3-D pore structure of the carbon replica can be reconstructed using large-pore mesoporous silica with cubic $Ia\bar{3}d$ symmetry (see Figure 2.11) as a template [107]. In this way, the carbon pore structure continuously forms in the pore space of the silica, as shown in Figure 2.12.

Surface-modified MCM-48 was also employed as a template to prepare OMCs. A carbon designated as SUN-1 [63] was prepared by using MCM-48 with surface-implanted aluminum species, which served as acidic sites for catalytic polymerization of phenol and formaldehyde. Mesocellular carbon foams with uniform ultralarge mesocells, interconnected through uniform windows, were synthesized by using a mesocellular silica foam template modified with aluminum [108]. Yoon et al. [109] reported the synthesis of highly ordered mesoporous carbons (OMCs) using a surface-silylated MCM-48 template and divinylbenzene (DVB) as the carbon precursor; they found that surface modification greatly improved the pore-structural order and mechanical stability of the resultant carbon. In a subsequent study, the same authors showed that the surfactant occluded in the pores of as-synthesized mesoporous MCM-48 can serve as a carbon precursor to be directly converted into OMC [110]. Similarly, Kim et al. [111] described

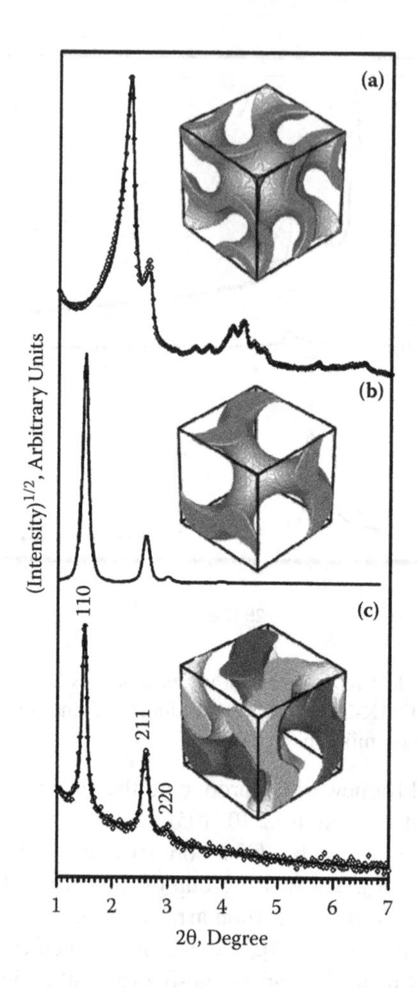

FIGURE 2.10 XRD powder patterns and unit cells of (a) MCM-48, (b) single enantiomeric carbon subframework, and (c) CMK-1. (From Solovyov, L.A., Zaikovskii, V.I., Shmakov, A.N., Belousov, O.V., and Ryoo, R. *J Phys Chem B* 106: 12198–12202, 2002. With permission.)

an in situ carbonization method for the fabrication of carbon nanotubes. A carbon film was formed on the ordered pore surfaces by carbonizing as-synthesized mesoporous silica with surfactant P123 in the pores. Removal of the silica framework yielded the carbon nanotubes. Additionally, MCM-48 has been used as a template to prepare other interesting carbon structures [112–114].

B. SBA-15 as a Template

The drawback of using MCM-48 as a template is the difficulty of tailoring its pore size in a broader range [19,103]. Therefore, recent research attention has been focused on using the SBA-15 template, which possesses relatively large pores

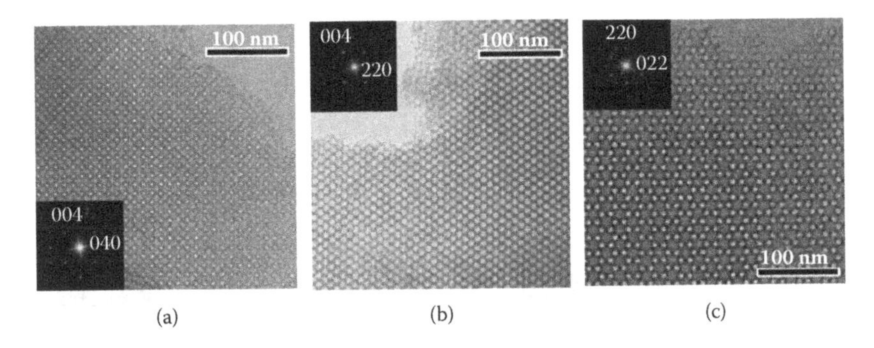

FIGURE 2.11 HRTEM images of a large-pore mesoporous silica taken with incident beam parallel to (a) [100], (b) [110], and (c) [111] directions. The insets are Fourier diffractograms. (From Sakamoto, Y., Kim, T.-W., Ryoo, R., and Terasaki, O. *Angew Chem Int Ed*. 43: 5231–5234, 2004. With permission.)

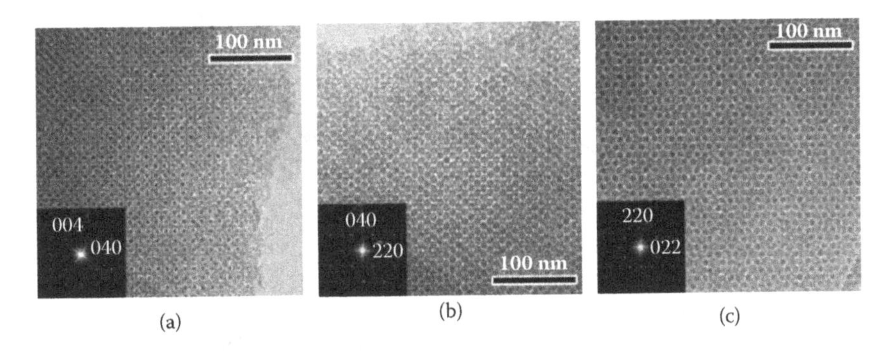

FIGURE 2.12 HRTEM images of a templated carbon taken with incident beam parallel to (a) [100], (b) [110], and (c) [111] directions. The insets are Fourier diffractograms. (From Sakamoto, Y., Kim, T.-W., Ryoo, R., and Terasaki, O. *Angew Chem Int Ed*. 43: 5231–5234, 2004. With permission.)

and a simple pore structure [61,115]. SBA-15 can be synthesized in large quantities in the presence of amphiphilic poly(alkylene oxide)-type triblock copolymers [116,117]. The diameter of its pores can be controlled over a wide range [118]. Figure 2.13 shows SEM images of SBA-15. Because of its unique porous structure, SBA-15 has been extensively employed as a template to prepare OMCs that consist of carbon nanorods generated from template mesopores connected by even smaller carbon rods produced from small micropores, as schematically illustrated in Figure 2.14 [119]. Thus, an ordered 3-D mesoporous carbon structure whose pore size is determined by the pore wall thickness of the silica template can be obtained.

Jun et al. [120] described the fabrication of an OMC, CMK-3, by utilizing the SBA-15 silica template and sucrose as the carbon precursor. Figure 2.15a shows the TEM images of CMK-3 viewed along and perpendicular to the pore axis.

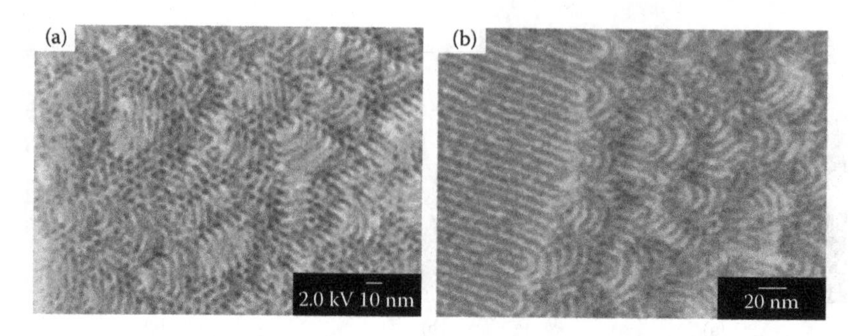

FIGURE 2.13 SEM images of SBA-15 mesoporous silica: (a) cross-sectional view and (b) view along the pore channel direction. (From Che, S., Lund, K., Tatsumi, T., Iijima, S., Joo, S.H., Ryoo, R., and Terasaki, O. *Angew Chem Int Ed*. 42: 2182–2185, 2003. With permission.)

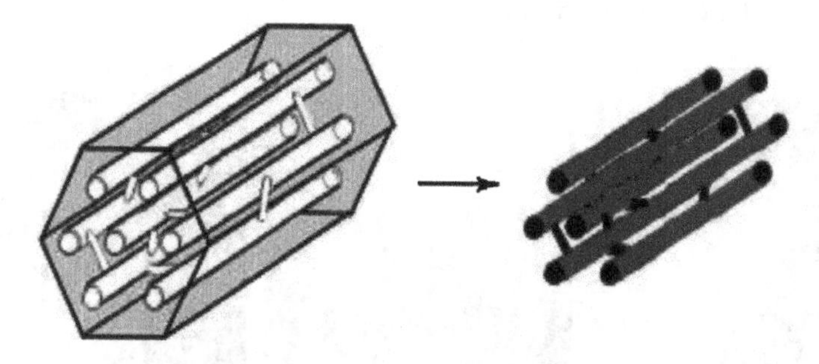

FIGURE 2.14 Schematic illustration of formation of OMC with interconnected pores using SBA-15 as a template. (From Che, S., Lund, K., Tatsumi, T., Iijima, S., Joo, S.H., Ryoo, R., and Terasaki, O. *Angew Chem Int Ed*. 42: 2182–2185, 2003. With permission.)

Unlike CMK-1 [62], here the ordered pore structure (see Figure 2.15b) is a true replica of SBA-15, without involving structural transformation that occurs when MCM-48 is used as a template [62,121]. The overall porosity of CMK-3 is a combination of uniform pores in spaces between the ordered carbon nanorods and micropores within these nanorods. The carbon is mostly mesoporous, with a quite narrow PSD centered at about 4.5 nm, as calculated using the Barrett–Joyner–Halenda (BJH) method [120]. The carbon nanorods in CMK-3 are about 7 nm in diameter, and the centers of two adjacent rods are about 10 nm apart, whereas the surface-to-surface distance is about 3 nm, based on XRD and TEM results [120]. The carbon nanorods are interconnected by carbon spacers [122], as illustrated in Figure 2.14. Because the pore size of SBA-15 can be tuned simply by changing the synthesis conditions, the mesopore size of templated carbon can be tuned in the 3.0- to 5.2-nm range [123].

The role of the spacers replicated from the micropores of SBA-15 is to integrate the carbon nanorods, thus constructing a pore structure of the OMC. A study showed [19] that exposure of SBA-15 to high temperatures resulted in the depletion of its micropores connecting the mesopores, thus eliminating the

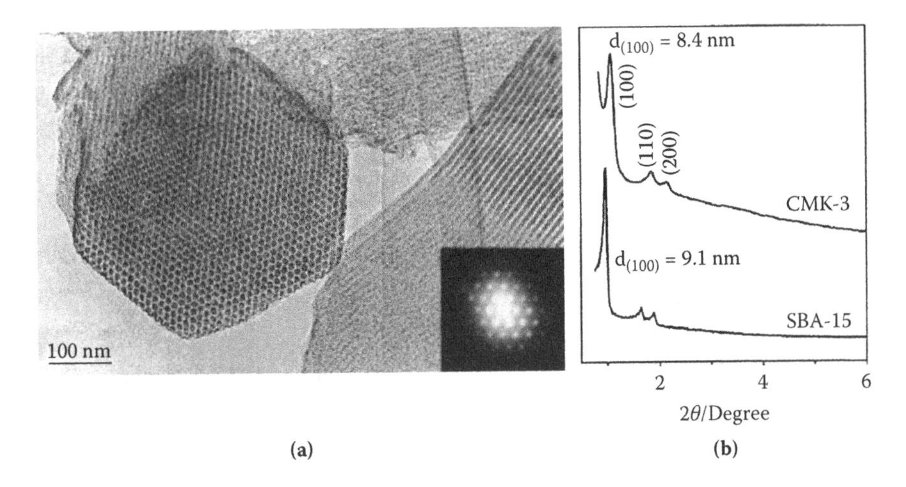

FIGURE 2.15 (a) TEM image of CMK-3 carbon and (b) low-angle XRD patterns of CMK-3 and SBA-15 template. (From Jun, S., Joo, S.H., Ryoo, R., Kruk, M., Jaroniec, M., Liu, Z., Ohsuna, T., and Terasaki, O. *J Am Chem Soc* 122: 10712–10713, 2000. With permission.)

carbon spacers [19] and resulting in structural collapses of carbon upon template removal [124]. As the SBA-15 pore wall is thicker than that of MCM-48, the pore size of CMK-3 is larger than that of CMK-1. Importantly, the thickness of the pore walls of SBA-15 can be manipulated by varying its synthesis conditions and, thus, SBA-15 is considered to be an ideal template for OMCs with controlled pore diameters [125,126].

Carbon "nanopipes" can be prepared if the pore walls of the SBA-15 template are coated with a layer of carbon film instead of completely filling the pores, as illustrated in Figure 2.16 [127]. Such carbon tubes were named CMK-5 [128,129]. The nanopipes retain the hexagonally ordered pore arrangement because of the presence of carbon spacers interconnecting the nanopipes (see Figure 2.14). The surface area of CMK-5 ranges from 1500 to 2200 m^2/g, depending on the template wall thickness [19,127]. Figure 2.17 shows the high-resolution SEM images of a CMK-5 sample with an inner diameter of 4.2 nm [119]. The images clearly show a morphology similar to that of the SBA-15 template (see also Figure 2.13).

OMCs with a bimodal pore system were prepared also with the SBA-15 template by varying the concentration of carbon precursor [130–132]. Both the external and internal diameters of carbon tubes of CMK-5 can be tuned by adjusting the concentration of an FA solution. Direct carbonization of an as-synthesized SBA-15 also produced a porous carbon with a bimodal pore system [133]. Solovyov et al. [134] performed a complete study on the properties and formation mechanisms of OMCs in the presence of the SBA-15 template.

The particle morphology of a hard template is inherited by the resultant carbon, as demonstrated in Figure 2.18. The morphologies of SBA-15 and the OMCs are very similar: both are made up of bundles of bamboo-shaped primary particles of about 400-nm diameter and 10-μm length [135]. Rodlike SBA-15 silica

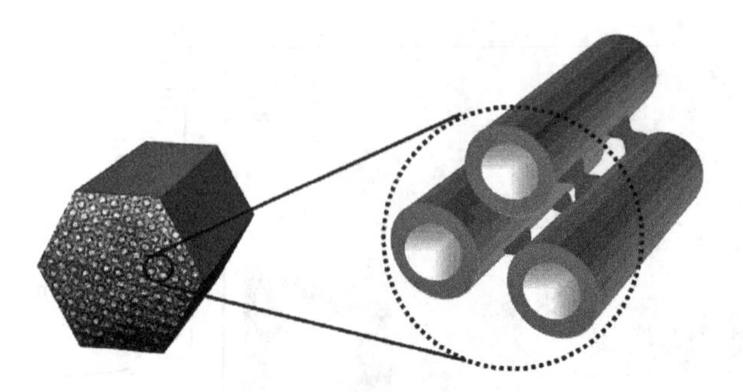

FIGURE 2.16 Schematic illustration of the nanostructure of CMK-5 carbon. (From Kruk, M., Jaroniec, M., Kim, T.W., and Ryoo, R. *Chem Mater* 15: 2815–2823, 2003. With permission.)

was employed to prepare mesoporous carbon rods with well-ordered two-dimensional hexagonal pore arrays [136]. Hollow carbon spheres with mesopores were fabricated using spherical SBA-15 as a template [137]. Carbons with particle sizes between 10 nm and 10 μm were prepared by using SBA-15 templates of different particle sizes [138].

Fuertes and Nevskaia [139] developed a vapor deposition polymerization (VDP) method to prepare OMCs. Carbon precursor FA was infiltrated into the pores via vapor-phase adsorption at room temperature. When ordered SBA-15 silica was used as a template, the resultant carbon possessed a unimodal pore structure similar to that of CMK-3. However, when a disordered mesoporous silica was used as a template, mesoporous carbon with a well-defined bimodal pore system (mesopores centered at 3 and 12 nm) was obtained, as can be seen from Figure 2.19. A mechanism responsible for the formation of such carbons was subsequently proposed [140] based on the degree of carbon infiltration, which can be controlled with the VDP method. Kruk et al. [141] described a polymerization method for carbon infiltration, which was believed to ensure uniform filling [142] and avoid the formation of nontemplated carbon.

The fabrication of OMCs with a highly graphitic nature (or a turbostratic nature) has been a recent research objective. Despite the excellent pore-structural ordering and well-defined PSD, the OMCs already discussed are essentially amorphous and nongraphitizable. In some applications, such as fuel cells and lithium ion batteries, graphitic porous carbons are desirable. Kim et al. [143] reported a synthesis strategy for fabricating graphitic carbons by in situ conversion of aromatic compounds to mesophase pitch and, further, to graphitic carbons. Their TEM results showed that the carbon framework consists of discoid graphene sheets. Various aromatic compounds, including acenaphthene, acenaphthylene, indene, indane, and naphthalenes, were considered possible carbon sources that could achieve the objective, as shown by the XRD data in Figure 2.20. Sample CMK-3G (Figure 2.20e), prepared using acenaphthene, is characterized by peaks at $2\theta = 26$, 45, 53, and 78, which presumably correspond to (002), (10), and (004) diffractions of potentially graphitic carbon. The (002) peak intensity and width

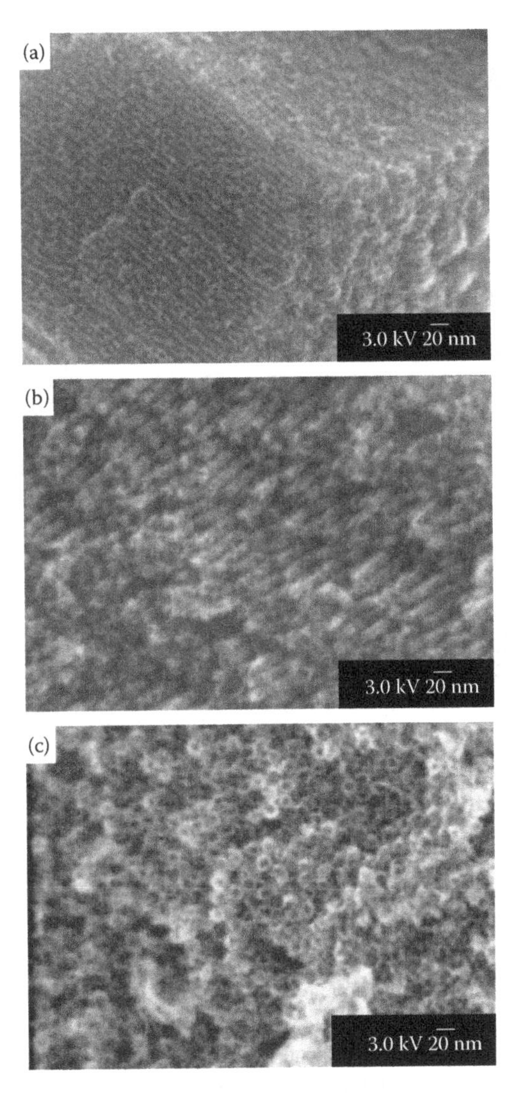

FIGURE 2.17 SEM images of CMK-5 carbon taken along different directions and using different accelerating voltages. (From Che, S., Lund, K., Tatsumi, T., Iijima, S., Joo, S.H., Ryoo, R., and Terasaki, O. *Angew Chem Int Ed.* 42: 2182–2185, 2003. With permission.)

are comparable to those of multiwalled carbon nanotubes (see Figure 2.20d) and are much more intense than those of CMK-3, which has an amorphous carbon framework (see Figure 2.20a), or those of an activated carbon (see Figure 2.20b) or carbon black (see Figure 2.20c). In addition, it was observed that CMK-3G exhibited remarkably improved mechanical strength and thermal stability according to XRD and TGA analyses [143]. Other liquid-phase graphitizable carbon precursors such as petroleum pitch [144], mesophase pitch [145], polyvinyl chloride (PVC) [146], naphthalene, anthracene, pyrene [147], and polypyrrole [148,149],

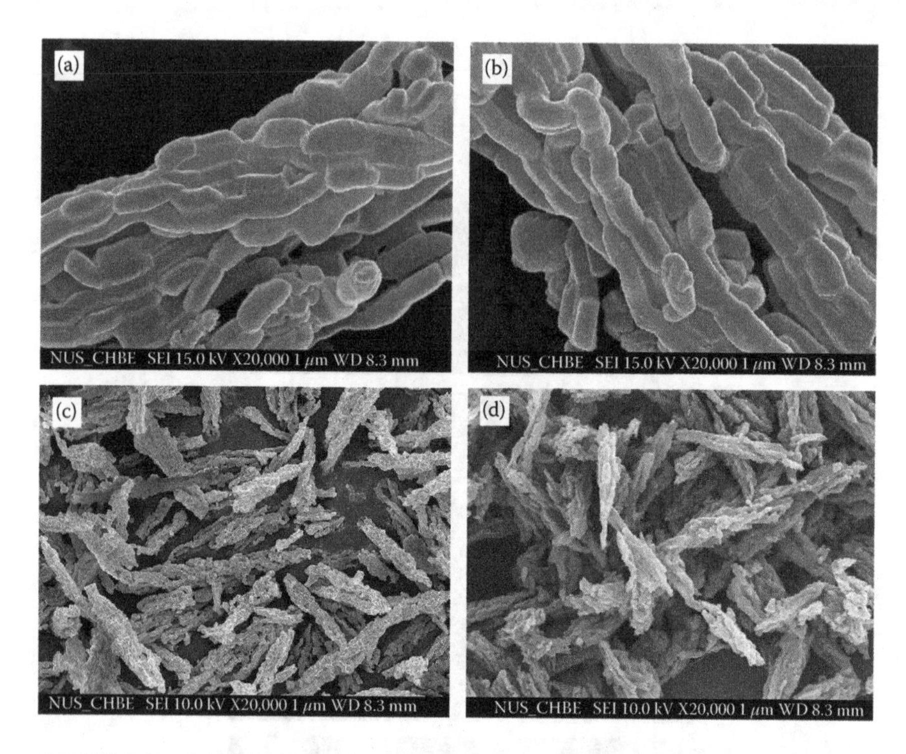

FIGURE 2.18 FESEM images of SBA-15 silica (a, c) and a mesoporous carbon (b, d) prepared using the SBA-15 template. (From Su, F., Zeng, J., Bao, X.Y., Yu, Y., Lee, J.Y., and Zhao, X.S. *Chem Mater* 17: 3960–3967, 2005. With permission.)

have also been employed to prepare OMCs with a nearly graphitic structure. Catalytic graphitization was also used to improve the graphitization degree of OMCs [150,151]. Although the use of such graphitizing carbon precursors allows one to prepare graphitic carbon, this method is time consuming because repeated infiltration and polymerization cycles are required to obtain an ordered carbon [152]. In addition, polymerization and pyrolysis during high-temperature carbonization often lead to emission of a large amount of small molecules (e.g., H_2O and CO), which can potentially deteriorate the pore structure of the template [153], thus having a negative effect on the structure of the resultant carbon [147,154]. By contrast, the CVD method, a well-established technique for preparing carbon molecular sieves [155], carbon nanofibers [156], and carbon nanotubes [157], has a number of advantages over the liquid-phase impregnation method, such as a high degree of pore filling [152] and easy control over the deposition of pyrolytic carbon, enabling the formation of dense pore walls, and potentially avoiding the formation of undesired microporosity [158,159].

The use of mesoporous pure-silica templates without the presence of any catalyst other than the surface silanol groups is desirable. With the pure-silica SBA-15 template, Mokaya and coworkers demonstrated the preparation of nitrogen-doped OMC materials [137,160,161]. They concluded that the CVD temperature is an

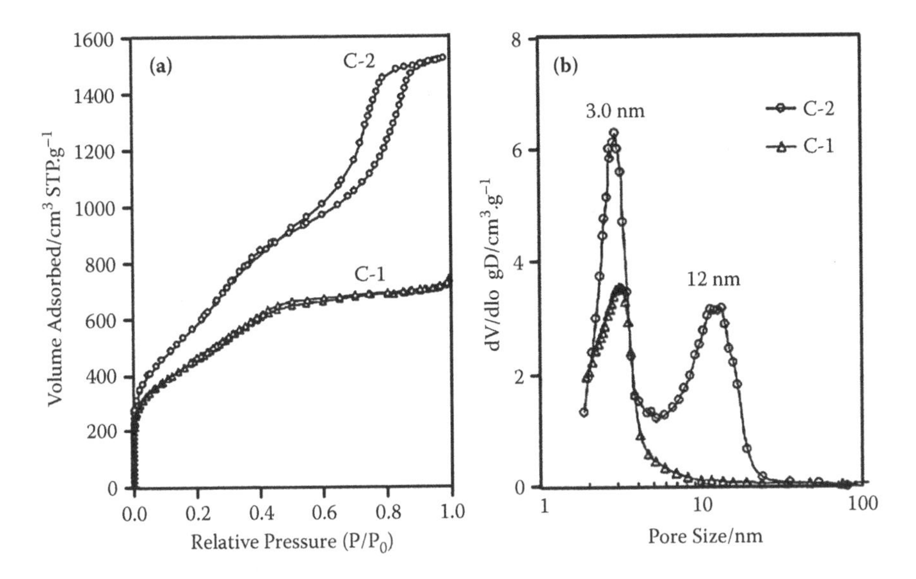

FIGURE 2.19 (a) N_2 isotherms and (b) PSD curves of an OMC sample (C-1) templated by ordered SBA-15 silica and disordered OMC sample (C-2) templated by a disordered mesoporous silica. (From Fuertes, A.B. and Nevskaia, D.M. *J Mater Chem* 13: 1843–1846, 2003. With permission.)

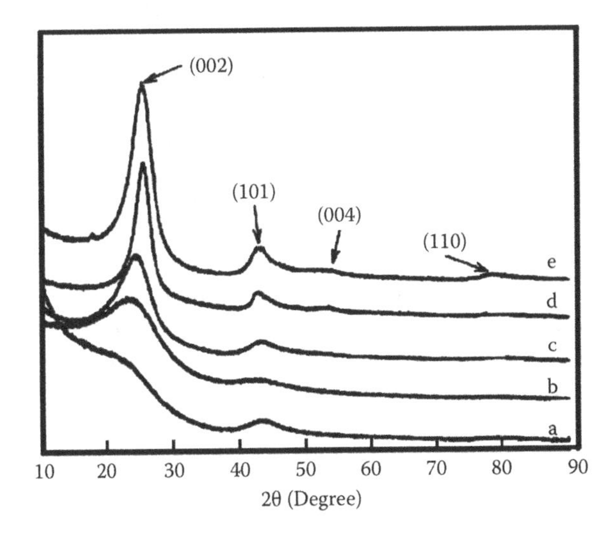

FIGURE 2.20 Wide-angle XRD patterns of different carbon materials: (a) CMK-3, (b) an activated carbon, (c) a carbon black, (d) multiwalled carbon nanotubes, and (e) CMK-3 prepared according to Reference 143. (From Kim, T.-W., Park, I.-S., and Ryoo, R. *Angew Chem Int Ed*. 42: 4375–4379, 2003. With permission.)

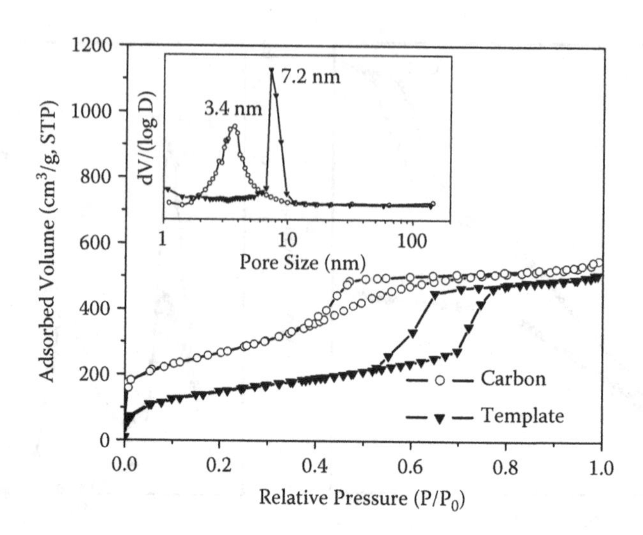

FIGURE 2.21 Nitrogen adsorption–desorption isotherms and BJH–PSD curves (inset) of pure-silica SBA-15 and an OMC sample (for clarity, the isotherm of carbon was vertically shifted by 50 cm^3/g). (From Su, F., Zeng, J., Bao, X.Y., Yu, Y., Lee, J.Y., and Zhao, X.S. *Chem Mater* 17: 3960–3967, 2005. With permission.)

important factor in controlling the degree of graphitization. Very recently, the same research group reported that highly graphitic N-doped mesostructured carbon materials can be prepared using various mesoporous pure-silica templates, including SBA-12, SBA-15, MCM-48, MCM-41, and HMS [162]. The use of a simple CVD nanocasting route to simultaneously control the morphology, pore size, and graphitization degree in structurally well-ordered mesoporous carbons was also described [163].

It should be noted that, according to the IUPAC recommendation [164], *graphitic carbon* is defined as all varieties of substances consisting of the element carbon in the allotropic form of graphite, irrespective of the presence of structural defects. The use of the term *graphitic carbon* is strictly justified only if a 3-D hexagonal crystalline long-range order can be detected in the material by diffraction methods (e.g., separation of two-dimensional [10] peak into [100] and [101] peaks at around 42 and 44° 2θ, respectively), independent of the volume fraction and the homogeneity of distribution of such crystalline domains. Although many researchers claim that their materials are ordered mesoporous graphitic carbons, such carbons should be more appropriately denoted *turbostratic* [165] rather than *graphitic*.

We also employed pure-silica SBA-15 as a template to prepare OMCs [135]. Figure 2.21 shows the nitrogen adsorption-desorption isotherms and BJH–PSD curves (inset) of SBA-15 silica, and a templated OMC sample. The presence of hysteresis loops indicates that both samples are mesoporous. The pore sizes derived from the adsorption branches were 7.2 nm for the template and 3.4 nm for the carbon. The surface area (676 m^2/g) and pore volume (0.72 cm^3/g) of the

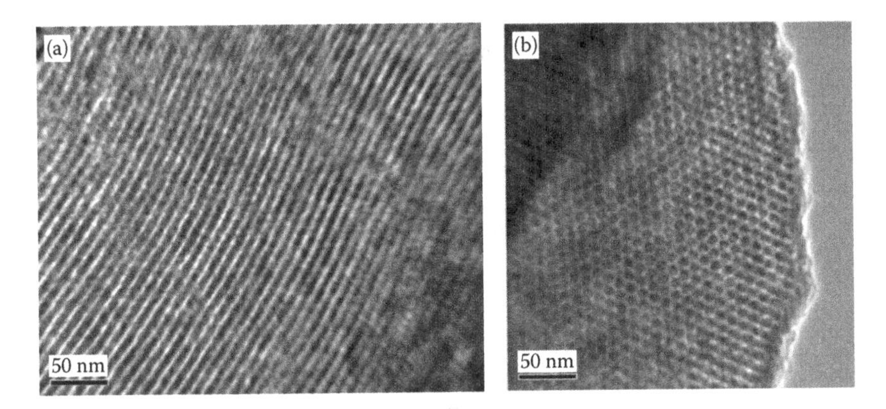

FIGURE 2.22 TEM images of an OMC prepared using CVD of benzene along (a) (001) and (b) (100) directions. (From Su, F., Zeng, J., Bao, X.Y., Yu, Y., Lee, J.Y., and Zhao, X.S. *Chem Mater* 17: 3960–3967, 2005. With permission.)

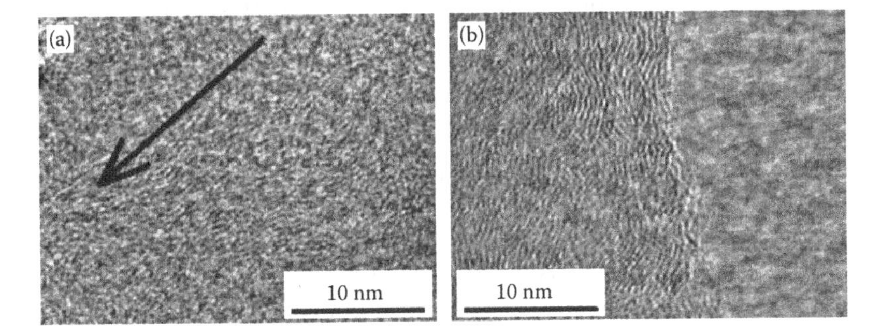

FIGURE 2.23 (a) FETEM lattice-fringe image of a templated carbon viewed along the (001) direction, and (b) of a carbon layer deposited on the external surface of the template. (From Su, F., Zeng, J., Bao, X.Y., Yu, Y., Lee, J.Y., and Zhao, X.S. *Chem Mater* 17: 3960–3967, 2005. With permission.)

OMC are lower than those of a CMK-3 sample (1520 m^2/g, 1.30 cm^3/g), which was prepared using sucrose as the carbon precursor [120], but slightly higher than those of graphitic mesoporous carbons prepared using mesophase pitch [145]. It is known that the mesopores of the amorphous carbon CMK-3 originated from the template framework, whereas a large number of micropores, which contributed significantly to the porosity, were created owing to the evolution of small gaseous molecules from the carbon precursor. With the CVD method, however, solid pyrolytic carbon fills the pores of the template, and the resultant carbon pore walls after template removal consist of stacked graphene layers, which are denser than amorphous carbon.

Figures 2.22a and 2.22b present the TEM images of OMC along the (001) and (100) directions, respectively, showing highly ordered hexagonal mesoporous arrays of the replicated carbon [135]. Figure 2.23 shows the lattice-fringe

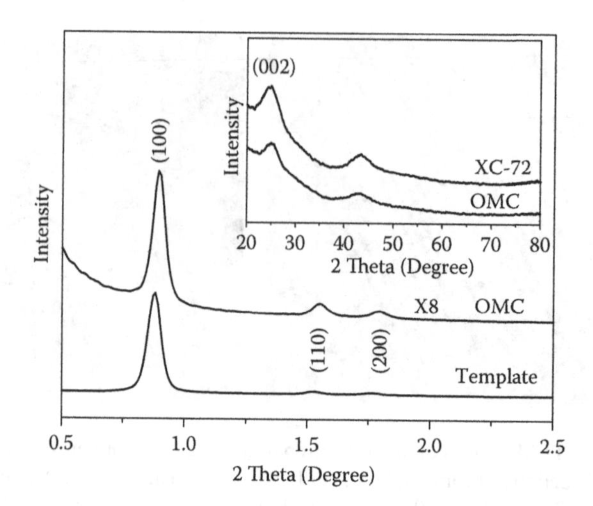

FIGURE 2.24 SAXS patterns of SBA-15 silica template and an OMC sample. (The inset shows the wide-angle XRD patterns of OMC and carbon black XC-72.) (From Su, F., Zeng, J., Bao, X.Y., Yu, Y., Lee, J.Y., and Zhao, X.S. *Chem Mater* 17: 3960–3967, 2005. With permission.)

images of the OMC taken along the long axis (001) direction (arrow direction in Figure 2.23a) and in the external surface region of the sample (Figure 2.23b). The direction of stacked graphene sheets, consisting of carbon nanorods parallel to the (001) direction, is different from what was observed by Kim et al. [143] and Yang et al. [145]. This direction is believed to have an influence on the carbon's electrochemical properties [145]. Figure 2.23b also shows the discrete stacked graphene sheets of the external carbon layer parallel to the external surface of the silica template: the turbostratic structure of the internal carbon seems to be different from that of the external carbon. Small-angle x-ray scattering (SAXS) patterns shown in Figure 2.24 reveal well-resolved (100), (110), and (200) peaks of OMC, demonstrating the presence of ordered 2-D hexagonal pore arrays [120]. The XRD patterns in Figure 2.24 (inset) show that graphene ordering of the OMC is comparable to that of the nongraphitic carbon black XC-72, but lower than that of the carbon derived from acenaphthene shown in Figure 2.20 [143].

On the basis of the preceding discussion, we proposed the following formation mechanism of OMCs [166]. At 900°C, the products of benzene pyrolysis are known to be polycyclic aromatic hydrocarbon species, such as biphenyl, m-, p-, and o-terphenyls, and anthracene [155]. Such species first adsorb on the pore surfaces of SBA-15. Subsequently, further aromatization and carbonization of the adsorbed species takes place to form the first layer of carbon film [167,168]. A layer-by-layer stacking then follows to create stacked graphene sheets. Thus, the favored direction of stacked graphene sheets is parallel to the long axis of the nanorods. Carbon deposition in the pores of the mesoporous silica template will stop when the pore size becomes smaller than the kinetic diameter of the benzene molecule (0.36 nm) [155,169]. As a result, further carbon deposition will occur on the external surface of the template, forming an external layer of dense carbon. Thus, precise control over deposition time can allow one to avoid the formation of the external carbon layer and to produce carbon nanopipes instead of nanorods. The CVD mechanism proposed here is different from that proposed by Mokaya and coworkers [137,160,161], who suggested that a carbon precursor be first in contact with the outer surface of SBA-15 before diffusing into the interior of the pore. Thus, the deposited carbon may block the pore channels.

It has been observed that increasing the pyrolysis temperature results in an increase in surface basicity [170] because of a decrease in the amount of acidic surface groups. Surface coating of CMK-3 and CMK-1 with poly(methyl methacrylate) (PMMA) was observed to enhance the adsorption of vitamin B_{12} because of hydrogen-bonding interactions of the amino groups in vitamin B_{12} with the carbonyl groups of the PMMA molecule [171]. A significant number of aryl groups substituted in the 4-position (Ar-R, R = chlorine, ester, and alkyl) were covalently attached to surfaces of the hexagonally structured mesoporous carbons (CMK-3, CMK-5) and cubically structured carbon (CMK-1) [172,173]. The primary hexagonal structure and its unit cell parameter were retained after chemical attachment of these surface functional groups. Well-ordered hexagonal mesoporous carbon nitride was also synthesized through a polymerization reaction between ethylenediamine and carbon tetrachloride [174]. Surface modification of OMC monolith by fluoroalkylsilane of $CF_3(CF_2)_7CH_2CH_2Si(OCH_3)_3$ (FAS-17) made this material superhydrophobic, allowing one to manipulate the surface properties of OMC materials [175].

Nonmagnetic α-Fe_2O_3 was incorporated into CMK-3 mesoporous carbon by postimpregnation [176]. Schüth and coworkers immobilized cobalt nanoparticles on CMK-3 carbon [177,178] according to the procedure shown in Figure 2.25. Lee et al. [179] reported the preparation of magnetic OMC (M-OMC) materials with iron oxide nanoparticles embedded in the carbon walls, as illustrated in Figure 2.26. Pyrrole monomer was incorporated into SBA-15 by vapor-phase infiltration. The resulting pyrrole–SBA-15 nanocomposite was dispersed in H_2O containing $FeCl_3$ to polymerize pyrrole. After recovering the poly(pyrrole)–SBA-15 nanocomposite by filtration, it was carbonized in a N_2 atmosphere at 700°C. The template was removed by boiling the composite in 1 M NaOH solution. Similarly, the preparation of magnetic mesocellular carbon foam using an in situ method was demonstrated [180]. The resulting material was found to possess large interconnected pores and well-dispersed magnetic nanoparticles. Nanocrystallites of NiO were encapsulated in the carbon wall of CMK-3 [181]. Metallic cobalt nanoparticles were encapsulated into the carbon wall of CMK-1 [182]. Ru nanoparticles sandwiched in the pore walls of OMCs were also prepared [183]. Additionally, with SBA-15 as a template, the fabrication of other nanostructured carbon materials, such as an ordered polymer network [184], carbon nanofibers [185], carbon monoliths [186], CNTs [111], and OMC particles covered by or bridged with CNTs [187,188], was demonstrated.

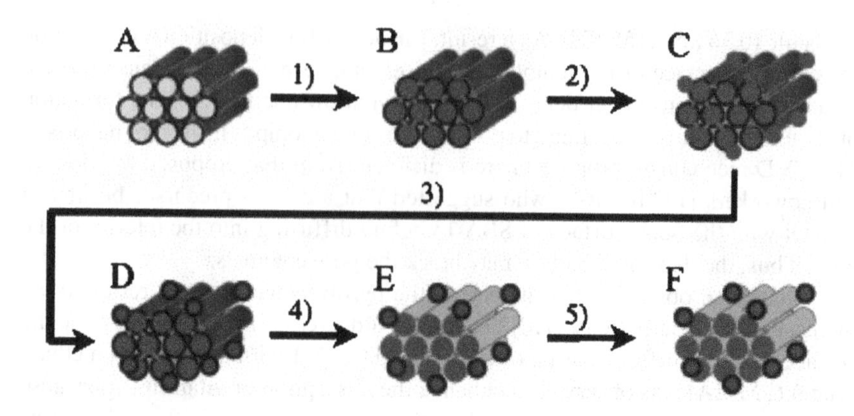

FIGURE 2.25 Illustration of the preparative steps for carbon CMK-3 with cobalt nanoparticles immobilized on the carbon surface: (A) ordered mesoporous silica SBA-15, (B) carbon–SBA-15 composite, (C) after deposition of cobalt nanoparticles on the composite surface, (D) protected cobalt nanoparticles on the composite, (E) magnetic OMC, and (F) after loading of Pd nanoparticles on the carbon surface. (1) Carbonization of carbon precursor in the pores of SBA-15, (2) deposition of cobalt nanoparticles on the surface, (3) coating of carbon on cobalt nanoparticles, (4) dissolution of SBA-15 silica to generate pores, and (5) loading of Pd nanoparticles on the pore surface. (From Lu, A.-H., Schmidt, W., Matoussevitch, N., Bönnemann, H., Spliethoff, B., Tesche, B., Bill, E., Kiefer, W., and Schüth, F. *Angew Chem Int Ed.* 43: 4303–4306, 2004. With permission.)

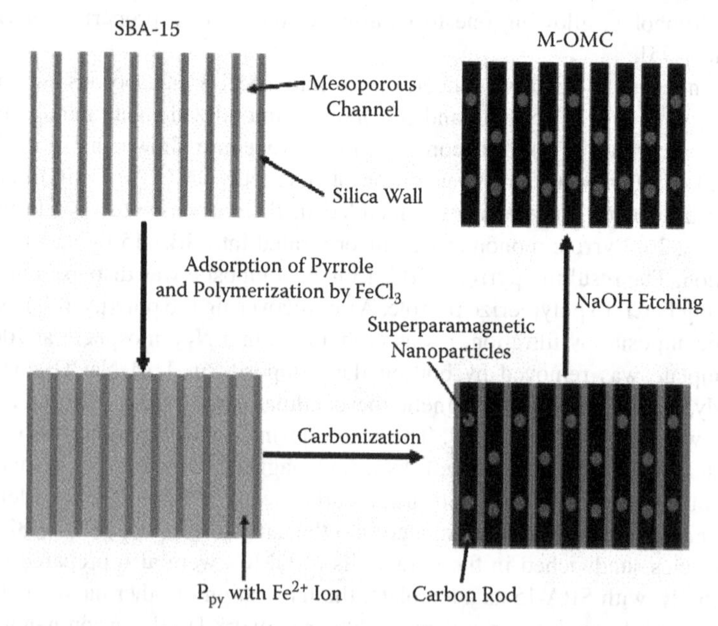

FIGURE 2.26 Schematic representation of the preparation of M-OMC. (From Lee, J., Jin, S., Hwang, Y., Park, J.-G., Park, H.M., and Hyeon, T. *Carbon* 43: 2536–2543, 2005. With permission.)

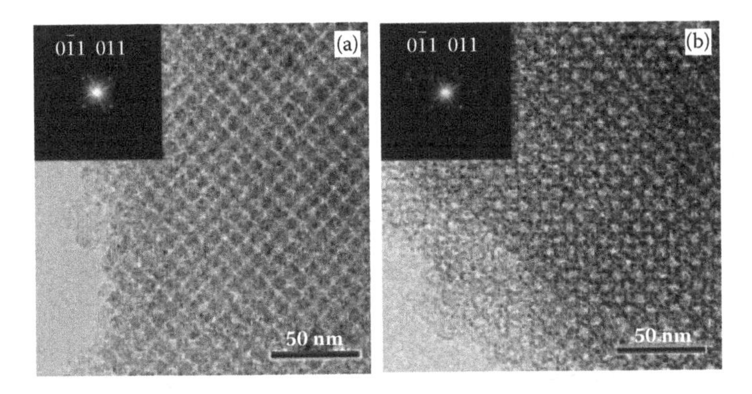

FIGURE 2.27 TEM images of (a) SBA-16-templated carbon and (b) surface-carbon-coated SBA-16. (From Kim, T.-W., Ryoo, R., Gierszal, K.P., Jaroniec, M., Solovyov, L.A., Sakamoto, Y., and Terasaki, O. *J Mater Chem* 15: 1560–1571, 2005. With permission.)

C. SBA-1 AND SBA-16 AS TEMPLATES

Templates such as SBA-1 are of interest because they have a $Pm\bar{3}n$ cubic structure with two different kinds of mesoporous cages [189]. Crystallographic analysis showed that SBA-1 is composed of cages (typically 3.3 × 4.1 nm) that are interconnected through two different kinds of uniform pores (mesopores and micropores). The pore diameter can be controlled within the range of 1.4 to 2.7 nm [190]. The SBA-16 is another type of 3-D mesoporous template with a well-ordered cubic $Im\bar{3}m$-type structure, a cavity size of 9.5 nm, and a pore entrance size of 2.3 nm [189]. The structure of OMCs prepared using this template has been studied [191]. It was shown that furfuryl alcohol and acenaphthene are better carbon precursors than sucrose for forming rigidly interconnected carbon bridges through narrow apertures of the cagelike siliceous SBA-16 mesostructure. The TEM images in Figure 2.27a clearly show an exact inverse carbon structure from the SBA-16 template. Figure 2.27b exhibits an array of interconnected hollow carbon particles that were formed after coating the SBA-16 mesopore walls with a thin carbon film, followed by silica dissolution. The use of FA as a carbon precursor allows one to control the degree of mesopore filling. Our recent studies [192,193] showed that the surfactant removal method plays an important role in the formation of carbon mesostructures: the use of solvent-extracted SBA-16 produced well-ordered mesostructured carbon, whereas calcined SBA-16 gave rise to poorly ordered carbon, as shown in Figure 2.28. This is related to pore entrance size and surface hydrophobicity of the mesoporous silica templates.

D. HMS AND MSU AS TEMPLATES

Hexagonal mesoporous silica (HMS) [194] is another template that has been used to make mesoporous carbons. It is known that HMS has a wormhole-like pore structure [195]. $AlCl_3$-grafted HMS was used to prepare carbon SUN-2 [196] with

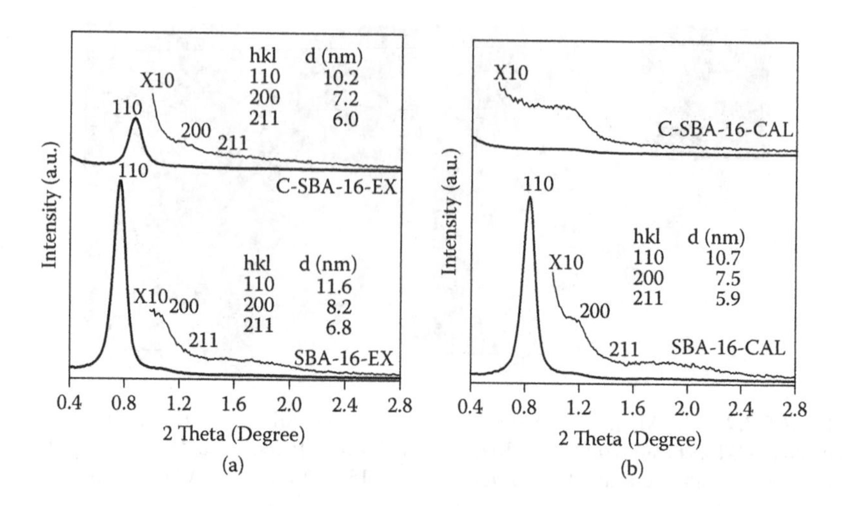

FIGURE 2.28 SAXS patterns of (a) solvent-extracted SBA-16 silica and templated porous carbon C-SBA-16-EX, and (b) calcined SBA-16 silica and templated porous carbon C-SBA-16-CAL. (From Guo, W., Su, F., and Zhao, X.S. *Carbon* 43: 2423–2426, 2005. With permission.)

a bimodal pore system, 0.6 and 2.0 nm in size. The XRD patterns suggested that SUN-2 has a similar pore structure as the template. As HMS materials can be synthesized to have different pore wall thickness, the resultant carbons can be tailored to possess pore sizes over a wide range, for example, 2.1 to 10.8 nm [197].

Another highly ordered mesoporous silica with large mesopores (7.6 to 11.9 nm), named MSU-H [198], was also used to template carbon (C-MSU-H), whose structure is identical to that of CMK-3 [199]. Because MSU-H can be synthesized under neutral conditions using inexpensive sodium silicate as the silica source, the preparation of C-MSU-H is considered to be cost-effective. A mesocellular carbon foam was synthesized using hydrothermally synthesized MSU-F silica as a template and FA as a carbon source [200]. This mesocellular carbon foam had large cellular pores (about 20 nm) and disordered uniform mesopores (about 4 nm). Using as a template a bimodal mesoporous silica consisting of 30- to 40-nm nanoparticles with 3.5-nm 3-D interconnected mesopores, a bimodal mesoporous carbon having 4-nm framework mesopores and about 30-nm textural pores was prepared [201]. Ordered carbons with varying graphitization degrees were prepared using naphthalene, anthracene, and pyrene as carbon sources and MSU-H silica as a template [147].

A new route to tailoring the pore size of mesoporous carbons with a spherical particle morphology was described by using MSU-1 silica as a template [202–204]. The pore size can be modulated continuously in the range of 2 to 10 nm by changing the preparation temperature of the mesoporous silica template. Figure 2.29 shows the SEM and TEM images of MSU-1 silica and mesoporous carbon templated by MSU-1. Indeed, the carbon was a replica of the template.

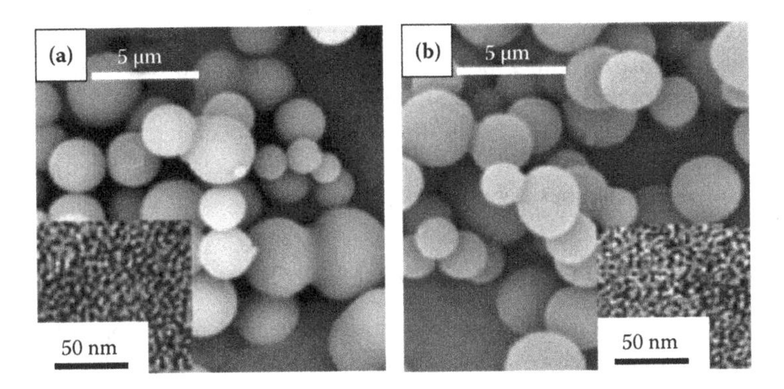

FIGURE 2.29 SEM and TEM (inset) images of (a) MSU-1 silica and (b) a carbon templated by MSU-1. (From Alvarez, S. and Fuertes, A.B. *Carbon* 42: 433–436, 2004. With permission.)

E. Other Mesoporous Silica Templates

Other mesoporous silica materials that have been employed as hard templates include silica xerogels [140,205] and mesoporous silica monoliths [206–209]. For the latter template type, carbon monoliths can be molded to a required shape and dimension without forming cracks and inducing shrinkage because the silica template has thick skeleton walls and large pores. Carbon monoliths exhibiting multimodal porosity were prepared by a one-step impregnation method with a silica monolith template containing bimodal porosity via volume- and surface-templating routes [208,209]. In addition to mesopores and macropores, the carbon monoliths also had microporosity. A study showed that preparation conditions had a great influence on the porosity of the resulting carbon [210]. Monolithic carbon aerogels with good pore interconnectivity were also prepared using magnesium acetate as a catalyst by polycondensation of resorcinol with formaldehyde [211].

The synthesis of spherical OMC particles using mesoporous silica spheres as templates was demonstrated by Lu and coworkers [212–213]. Large-pore mesoporous silica KIT-6 (cubic $Ia\bar{3}d$ symmetry) was used as a template [214, 215]. It was found that, although the structural properties of the resulting OMCs are mainly determined by the template used, their adsorption properties depend on the type of carbon precursor. A large-pore, cage-type mesoporous carbon was prepared using a 3-D, large cage-type face-centered cubic $Fm\bar{3}m$ mesoporous silica KIT-5 as a template [216]. Combining both nanoreplication (ordered mesoporous silicas as templates) and nanoimprint techniques (silica nanoparticles as templates, which can be mixed with a carbon precursor and introduced in the pores of mesoporous silica) and using dual silica templates yielded a simple way to synthesize OMCs with bimodal PSD (1.5 and 3.5 nm) [217]. Interestingly, poly(ethylene oxide)-polyacrylonitrile (PEO-*b*-PAN) diblock copolymer occluded in mesoporous silica was used as a carbon precursor to prepare mesoporous carbons [218]. Oriented mesoporous carbon nanofilaments were prepared by replication of silica-based mesoporous nanofilaments synthesized in the pore channels of anodic alumina membranes [219,220]. Tuning the pore size of such mesoporous carbon via a confined activation process was demonstrated using phosphoric acid [221], which is often used as an activation agent in the production of commercial activated carbons. Hollow carbon tubes with mesoporous walls and rectangular channels were fabricated with silica tubes as templates [222].

IV. NANOPARTICLE-TEMPLATED MESOPOROUS CARBON

Although mesoporous carbons can be prepared using ordered mesoporous silicas as exotemplates, the high cost of synthesizing mesoporous silicas due to the use of surfactants makes this approach practically infeasible. In addition, the preparation often involves multiple steps. Therefore, a simpler method is desirable. If one uses nanoparticles as endotemplates, a porous carbon with a pore size exactly matching the size of the nanoparticles could be synthesized directly. This idea was nicely demonstrated by Antonietti et al. [223] in preparing macroporous silica materials with latex particles as templates.

The use of nanosized silica particles as templates for fabrication of mesoporous carbons was pioneered by Hyeon's group [65]. Polymerization of resorcinol and formaldehyde (RF) in the presence of a silica sol solution (Ludox HS-40, average particle size 12 nm) yielded a RF gel–silica nanocomposite. Carbonization of the composite, followed by HF etching to remove the silica colloidal particles, yielded a nanoporous carbon, designated SMC1, which had pores in the range of 10 to 100 nm, a pore volume > 4 cm^3/g, and a surface area of about 1000 m^2/g. It is obvious that the resultant carbon was not a true replication of the individual silica nanoparticles, because of the inconsistency between the carbon pore size and silica particle size. According to the authors, this was due to aggregation of silica particles during the casting process. To address this issue, surfactant-stabilized silica particles were used as templates [224] to produce nanoporous carbons, designated SMC2, which exhibited a very narrow PSD centered at 12 nm, matching perfectly the size of silica particles. Subsequent studies [225,226] showed that the higher the silica sol content and pH, the higher the surface area and pore volume of the resulting carbons. Other silica particles such as a spherical silica sol (Ludox SM-30, 30 wt% silica, primary particle size 8 nm) and an elongated silica sol (SNOWTEX-UP, 20 wt% silica, particles 5 to 20 nm in width and 40 to 300 nm in length) were also employed as templates [227]. Mesoporous carbons with highly uniform and tunable mesopores were fabricated by one-step vapor-deposition polymerization using colloidal silica nanoparticles (Ludox SM-30) as templates and polyacrylonitrile as the carbon precursor [228].

Jaroniec's group reported the synthesis of mesoporous carbons by using Lichrosper Si 100 silica spherical particles as templates and a synthetic mesophase pitch or acrylonitrile as the carbon precursors [229]. Such carbons possess mesoporosity with negligible microporosity. Recently, a colloidal imprinting (CI) method for producing mesoporous carbons was also described by Jaroniec and coworkers, as schematically illustrated in Figure 2.30 [66–68]. The key to this

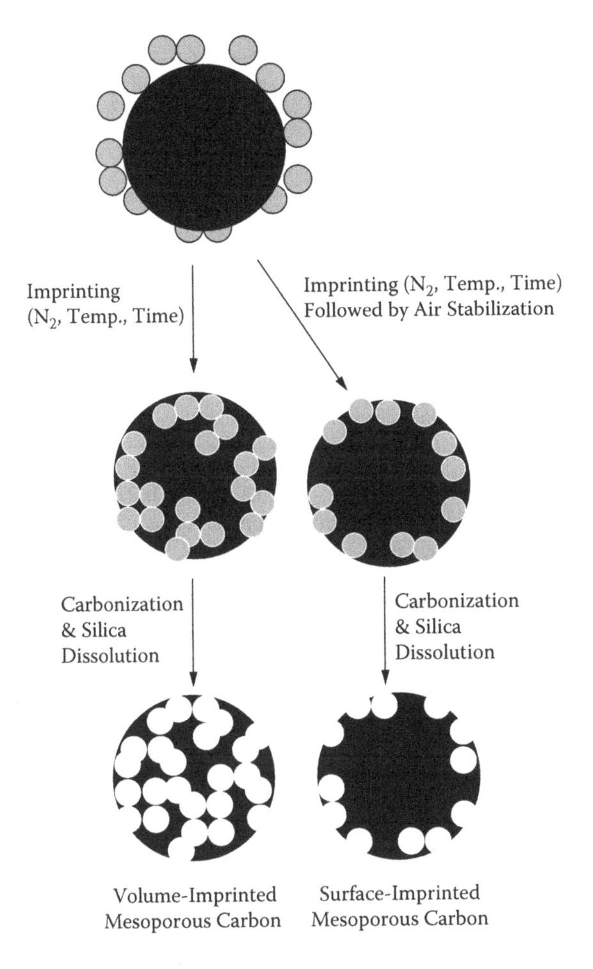

FIGURE 2.30 Schematic illustration of the synthesis of colloid-imprinted carbons. (From Li, Z. and Jaroniec, M. *Chem Mater* 15: 1327–1333, 2003. With permission.)

method is to incorporate spherical silica colloids into mesophase pitch particles. The CI method allows the formation of spherical pores. The colloid-imprinted carbons can be further graphitized at very high temperatures (say, 2400°C), and their porosity can be largely maintained [230]. The main additional advantage of using colloidal particles as templates lies in the fact that the raw materials (e.g., pitch and colloidal silica) are readily available and cost-effective.

The preparation of mesoporous carbon foams was described recently [231]. By mixing an RF sol with polystyrene (PS) microspheres, followed by pyrolysis, a carbon skeleton of mesoporous carbon foam was obtained.

Degussa TiO_2 (commercially known as P25) with particle sizes in the range of 20 to 30 nm can also be used as a template to prepare mesoporous carbon materials [232]. Figure 2.31 shows the adsorption–desorption isotherms and BJH–PSD curves of the carbons prepared by using P25 templates. The sample prepared in

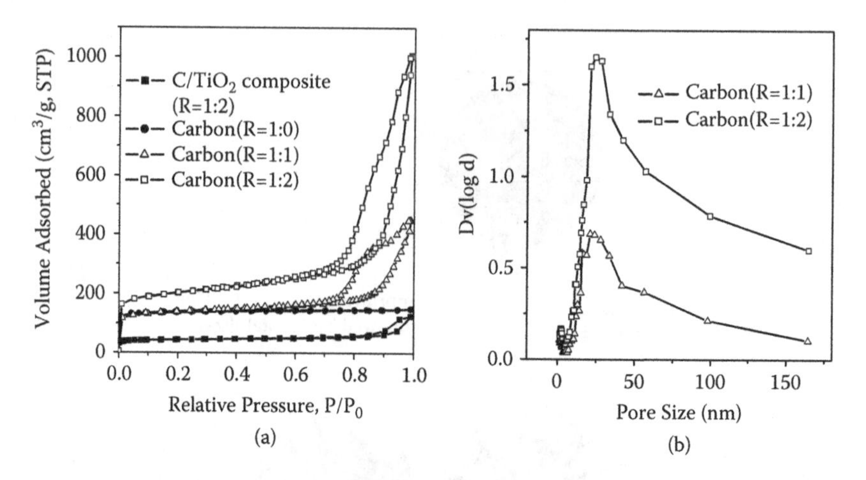

FIGURE 2.31 (a) Nitrogen adsorption isotherms and (b) BJH–PSD curves of mesoporous carbons templated by TiO_2 nanoparticles prepared using different mass ratios of FA/TiO_2 (R). (From Su, F. and Zhao, X.S. Unpublished work, 2006b.)

the absence of P25 does not have any mesoporosity, whereas the two samples prepared in the presence of P25 display a type IV isotherm with mesopore sizes centered in the range of 20 to 30 nm, which is consistent with the particle size of P25. The P25-templated porous carbon (R = 1:2) possesses a BET area of about 750 m^2/g, a pore volume of 1.6 cm^3/g, and a mesopore volume of 1.3 cm^3/g. We also demonstrated the synthesis of mesoporous carbon using furfuryl alcohol as a carbon precursor and acetic acid as a catalyst for polymerization. Figure 2.32 shows the images of P25 TiO_2, a composite of carbon–TiO_2, and mesoporous carbons obtained by carbonization of furfuryl alcohol at 800°C in the presence of P25, with R = 1:1 mass ratios of furfuryl alcohol/TiO_2, together with TEM images of the mesoporous carbon. The observed mesopores are centered at about 25 nm, and their geometry is similar to that of TiO_2 particles, confirming that the mesopores were replicated from the template particles.

In situ copolymerization of organic and inorganic monomers in the presence of a catalyst generating silica–polymer gel composites was shown [233–235] to be a time-saving method for making porous carbons. It is known that the sol–gel process often involves reaction of silicon alkoxides with water in the presence of an acidic or basic catalyst [233]. The reaction products under the acidic catalyst are high-molecular-weight polysilicates (a sol), which form a 3-D network filled with solvent molecules (a gel), and a silica-gel-interconnected framework is formed by aggregation of nanoparticles of colloidal silica having approximately uniform size [25]. Therefore, if the solvent contains a large amount of liquid carbon precursor, it is possible to prepare porous carbon by carbonization and removal of the silica gel. Kyotani and coworkers [234] reported the preparation of mesoporous carbons by copolymerization of FA and tetraethyl orthosilicate (TEOS). It was found that the pore sizes and size distribution are strongly dependent on the sol–gel reaction conditions. Under optimized conditions, it is possible to fabricate carbons with

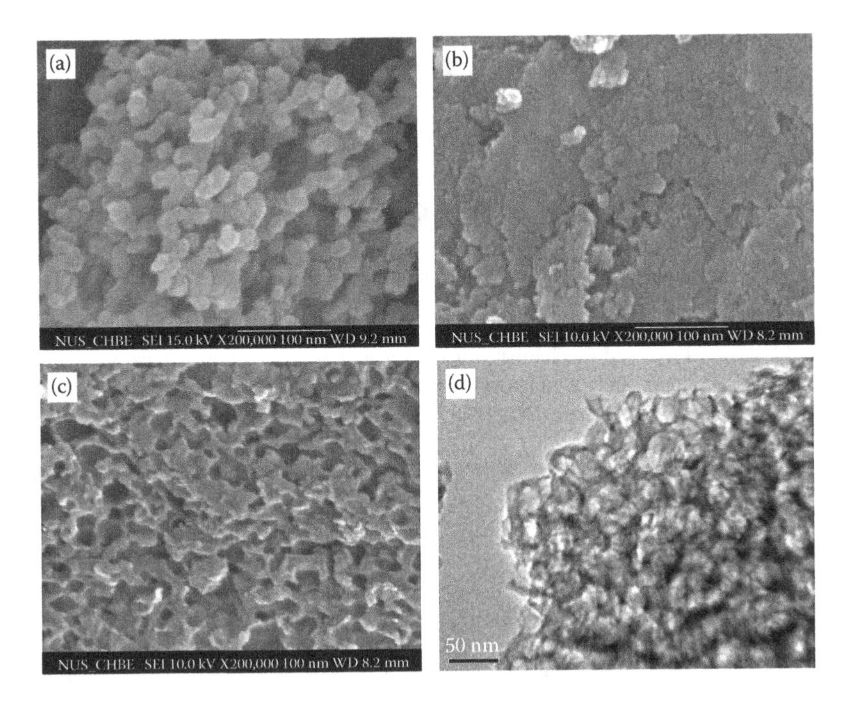

FIGURE 2.32 FESEM images of (a) TiO_2 nanoparticles, (b) carbon–TiO_2 composite, (c) a porous carbon prepared using TiO_2 nanoparticle template, and (d) TEM image of the carbon. (From Su, F. and Zhao, X.S. Unpublished work, 2006b.)

uniform mesopores of about 4 nm. Recently, Han et al. [235] demonstrated the preparation of mesoporous carbons using an in situ polymerization method. The key step was the simultaneous polymerization of sucrose and sodium silicate to prepare a homogeneous silica–carbon nanocomposite, in which HCl acted as a catalyst. The synthesis procedure was optimized under various reaction conditions. The effect of the silica template structure on the pore structure of the resultant mesoporous carbons was also investigated.

Continuous mesoporous carbon thin films were fabricated by direct carbonization of sucrose–silica nanocomposite films and subsequent removal of the silica [236]. The mesoporous carbon film with uniform and interconnected pores had a surface area of 2603 m^2/g and a pore volume of 1.39 cm^3/g. Subsequently, nanoporous carbons with bimodal PSD centered at about 2 and 27 nm in diameter were prepared by using both the TEOS-derived silica network and the colloidal silica particles as templates [237]. Figure 2.33 illustrates the preparation pathway. The pore sizes of the carbon are determined by the sizes of the added silica particles and the silica network. As the colloidal silica particles are commercially available with different diameters (e.g., 20 to 500 nm), this dual template synthesis process provides an efficient route to preparing nanoporous carbons with a controllable hierarchical pore structure.

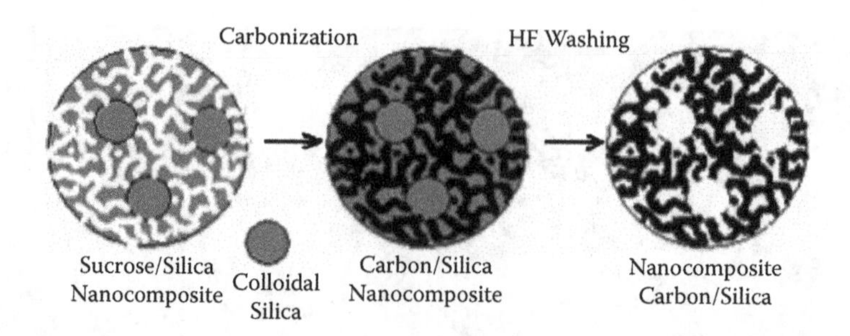

FIGURE 2.33 Proposed mechanism for the formation of nanopores in carbon using a colloidal silica nanoparticle template. (From Pang, J., Hu, Q., Wu, Z., Hampsey, J.E., He, J., and Lu, Y. *Micropor Mesopor Mater* 74: 73–78, 2004. With permission.)

FIGURE 2.34 TEM images of mesoporous carbons prepared by in situ carbonization of (a) P123 and (b) P103 surfactants. (From Kim, J., Lee, J., and Hyeon, T. *Carbon* 42: 2711–2719, 2004. With permission.)

Other reports of one-step synthesis of porous carbon materials can be found in the literature [238–241]. Of particular interest is the work of Hyeon and coworkers [238,241]. Sol–gel polymerization of silica in the presence of Pluronic P123 ($EO_{20}PO_{70}EO_{20}$) triblock copolymer and phenol generated a P123–phenol–silica nanocomposite. Carbonization followed by removal of the silica component produced the final mesoporous carbon materials. The carbon pore size was controlled by varying the ratio of phenol to copolymer. Mesocellular carbon foams were also synthesized by direct carbonization of as-synthesized silica–triblock copolymer nanocomposites [241], which were prepared by acid-catalyzed sol–gel hydrolysis of TEOS in the presence of a triblock copolymer acting as both a carbon precursor and a structure-directing agent. Figure 2.34 shows TEM images of the mesoporous carbons using triblock copolymers P123 and P103 as carbon precursors. This synthetic strategy is thought to be simpler and more cost-effective in comparison with the general hard template method.

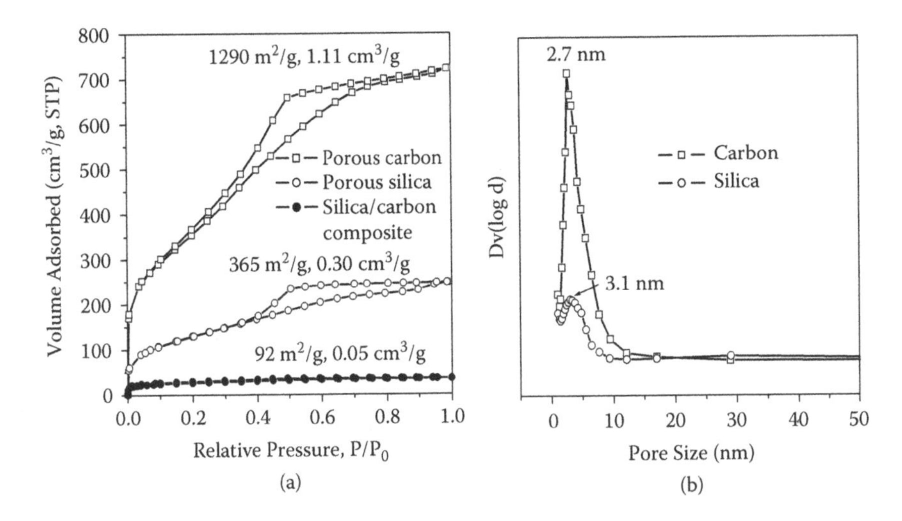

FIGURE 2.35 (a) Nitrogen adsorption isotherms and (b) BJH–PSDs of silica–carbon composite, mesoporous carbon, and mesoporous silica. (From Su, F. and Zhao, X.S. Unpublished work, 2006c.)

In a recent study [242], we employed the sol–gel method to prepare mesoporous carbons. Copolymerization of a mixture containing TEOS, sucrose (or furfuryl alcohol), HCl (or H_2SO_4), and H_2O at 60 to 80°C for 1 h resulted in a gel. Thermal treatment of the gel at 100°C for 10 h and 180°C for 5 h produced a pitchlike black solid. Carbonization at 800 to 1000°C yielded a silica–carbon composite. Removal of the silica component by acid washing produced the mesoporous carbon. If calcination in air instead of acid washing is used to treat the silica–carbon composite, mesoporous silica is obtained. Figure 2.35 shows the nitrogen adsorption–desorption isotherms and BJH–PSD curves (from adsorption branches) of a sample carbonized at 800°C, as well as those of mesoporous carbon and porous silica. The surface area, pore volume, and pore size values of the samples are also included in Figure 2.35. The porous carbon displays a similar isotherm to that of porous silica, but with a higher adsorption capacity than silica. Both samples seem to have a similar porous structure. Figure 2.36 shows TEM images of the mesoporous carbon and silica samples. The random pores of the carbon are smaller than those of the silica, in agreement with the adsorption results. Figure 2.37 shows the FESEM images of the composite (carbonization at 800°C), mesoporous carbon, and mesoporous silica. The mesoporous carbon and silica seem to be all composed of nanoparticles, with sizes ranging from 5 to 20 nm.

On the basis of these results, a templating mechanism was proposed and is schematically illustrated in Figure 2.38 [242]. Hydrolysis of TEOS catalyzed by acid produces a large number of globular primary silica particles, 1 to 2 nm in diameter. These colloidal particles will coagulate to form larger particles (secondary nanoparticles) [243]. Infiltration of a carbon precursor into the voids

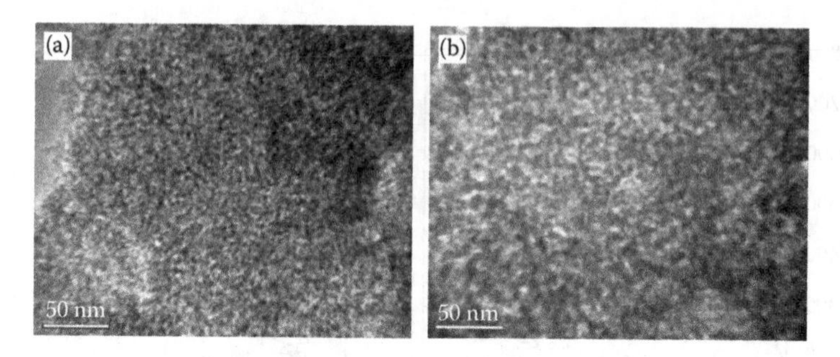

FIGURE 2.36 TEM images of (a) mesoporous carbon and (b) mesoporous silica. (From Su, F. and Zhao, X.S. Unpublished work, 2006c.)

between the primary and secondary particles, followed by carbonization, results in a silica–carbon nanocomposite. Removal of the silica component produces a porous carbon. On the other hand, calcination of the nanocomposite gives rise to porous silica. Thus, the porous structure of such silica-nanoparticle-templated porous carbons is controlled by the size of the primary and secondary nanoparticles [244–246].

V. COLLOIDAL-CRYSTAL-TEMPLATED MACROPOROUS CARBON

The colloidal-crystal-templating approach offers yet another new approach to preparing 3-D macroporous solid materials [6,17]. Spherical colloidal particles of submicrometer size can self-organize themselves into a colloidal crystal, the so-called opal [64], which can be utilized as an endotemplate to fabricate ordered macroporous carbons of two types: volume-templated carbon, which is an exact inverse replica of the opal template, and surface-templated carbon, which is formed by coating the colloidal spheres. Zakhidov et al. [64] were the first to use colloidal crystals as templates to prepare highly ordered 3-D macroporous carbon of both types. As schematically illustrated in Figure 2.39, for the volume-templating approach, a carbon precursor is infiltrated into interstitial spaces between colloidal spheres. Carbonization and removal of the opal template leave behind a 3-D periodic carbon structure (i.e., an inverse carbon). With this approach, macroporous carbon structures with a wide range of pore sizes have been produced.

Some parameters are more important than others for the final structure of the templated carbon. First, the template must have a high-quality crystal structure. Otherwise, the carbon framework will not display an ordered structure [247]. Second, the void infiltration degree of the opal template plays an important role in the morphology of the resulting carbon. To achieve perfect replication using the volume-templating method, the interstitial voids of the colloidal crystal must be completely filled with carbon. Normally, liquid-phase carbon precursors such as phenolic resin [64,248] and sucrose solution [249–251] are used for this purpose. Because of carbonization, the carbon matrix can shrink significantly. Thus, repeated impregnations are recommended to obtain a highly

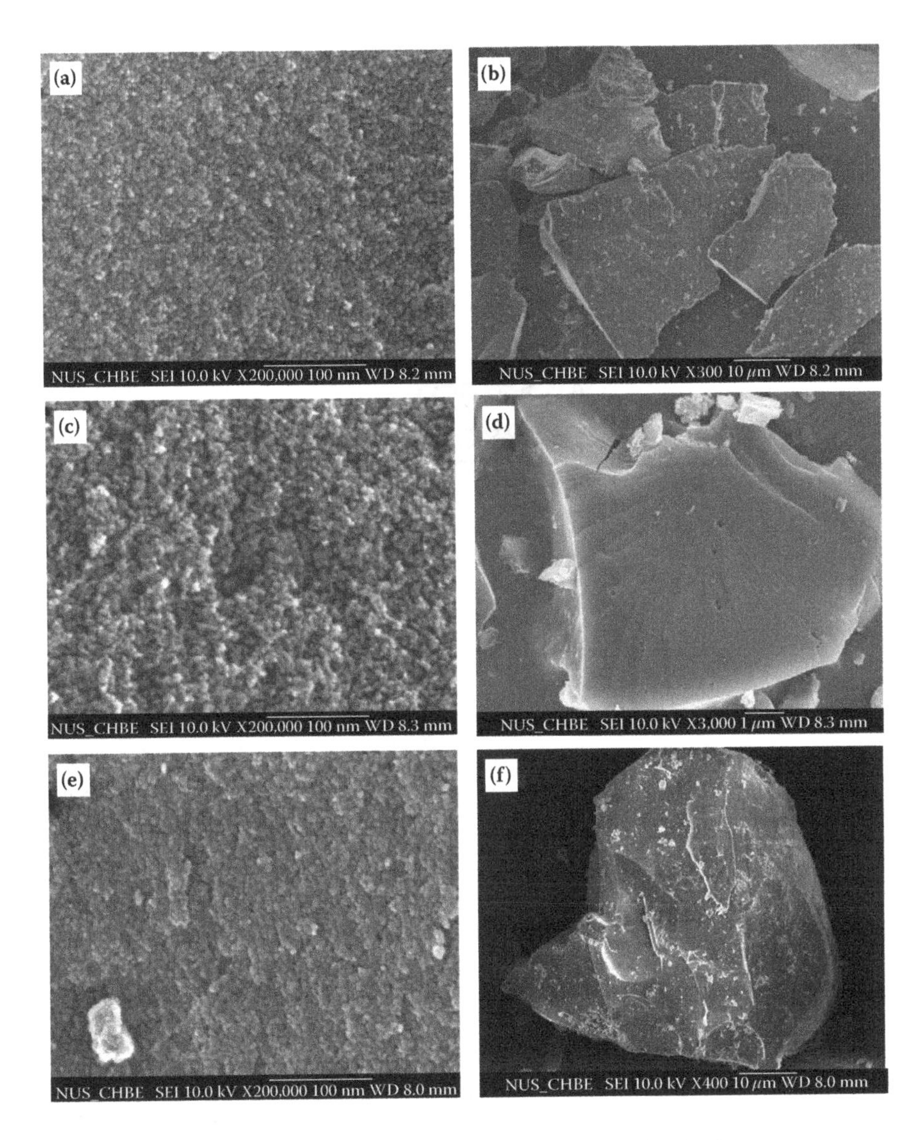

FIGURE 2.37 FESEM images of (a, b) carbon–silica composite, (c, d) mesoporous carbon, and (e, f) mesoporous silica. (From Su, F. and Zhao, X.S. Unpublished work, 2006c.)

ordered carbon structure. Recently, Perpall et al. [252] described a new carbon precursor, namely, bis-*ortho*-diynyl arene (BODA), whose use can substantially minimize carbon shrinkage during carbonization. Both sp^2- and sp^3-hybridized macroporous carbons can be prepared by using CVD and plasma techniques [64]. Third, carbon growth is determined by the surface properties of colloidal spheres and the chemical nature of the carbon precursor. Good adhesion of the precursor to the opal surface leads to a surface-templating mechanism [250]. Thus, surface modification of the spheres can greatly affect the position of carbon growth.

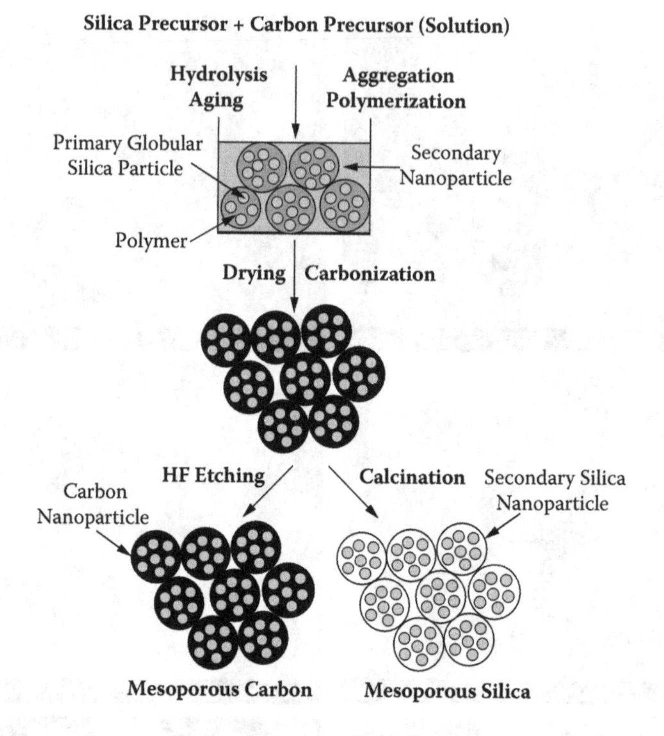

FIGURE 2.38 Schematic illustration of the formation of mesoporous carbon and silica using the sol–gel processing method. (From Su, F. and Zhao, X.S. Unpublished work, 2006c.)

For example, Yu et al. [251] used Al-implanted silica spheres as a template and thus controlled the initiation sites of the acid-catalyzed carbonization reaction of phenol and formaldehyde. An Al-grafted silica array resulted in surface coating (surface templating), as shown in Figure 2.40a. In this case, polymerization was initiated at the acidic sites on the Al-implanted silica surface. However, when an acid catalyst is mixed with a carbon precursor solution, polymerization is thought to occur everywhere to fill up the entire space between the particles. Figure 2.40b clearly shows the formation of an ordered macroporous carbon framework by the volume-templating mechanism with complete filling of all the voids in the silica spheres. Each of the spherical pores is interconnected through small channels. Ordered uniform porous carbon frameworks with pore sizes in the range 10 to 1000 nm were synthesized using removable colloidal silica spherical crystalline templates by carbonization of phenol and formaldehyde [248]. The porous carbons possessed different porous structures, which can be easily controlled by predetermining the sizes of the silica spheres.

Theoretically, colloidal-crystal-templated macroporous structures should have a relatively low surface area because of the dominant presence of large pores. However, macroporous carbons prepared by this mechanism can possess a high surface area because a considerable number of micropores may be present as

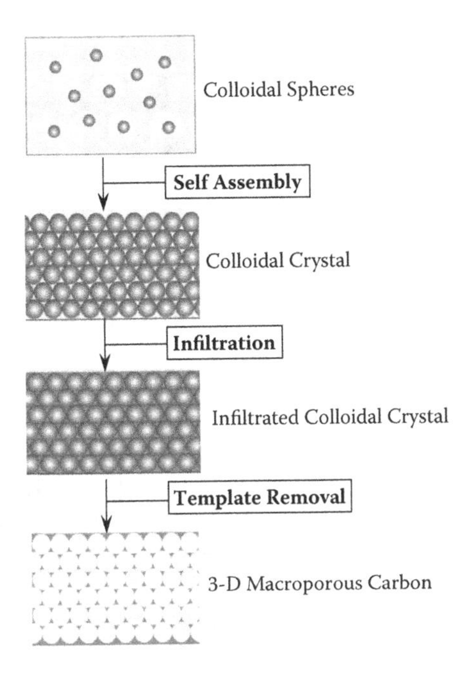

FIGURE 2.39 Schematic representation of the volume-templating approach using a colloidal-crystal template.

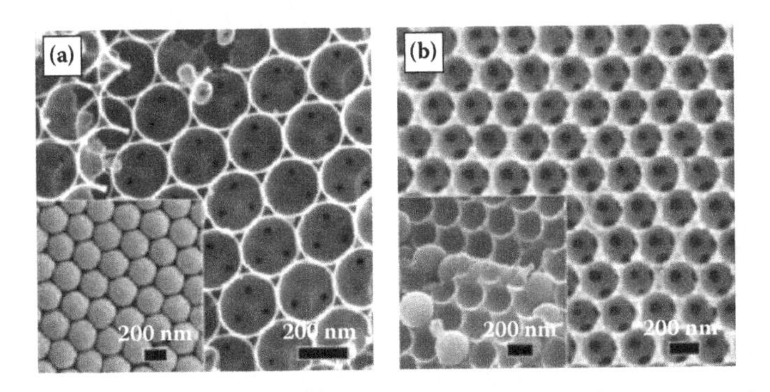

FIGURE 2.40 SEM images showing macroporous carbons fabricated using (a) the surface-templating and (b) volume-templating approaches with 250-nm silica spheres as templates. (The insets are the carbon–silica composites before removal of silica spheres.) (From Yu, J.S., Kang, S., Yoon, S.B., and Chai, G. *J Am Chem Soc* 124: 9382–9383, 2002. With permission.)

well in the ordered carbon framework. For example, a macroporous carbon with a pore size of 62 nm possesses a BET surface area of 750 m^2/g [253].

Recently, a dual template method using both polystyrene spheres and silica particles as templates was reported to result in the preparation of ordered, uniform, macroporous carbons with mesoporous walls [254]. Figure 2.41a shows the

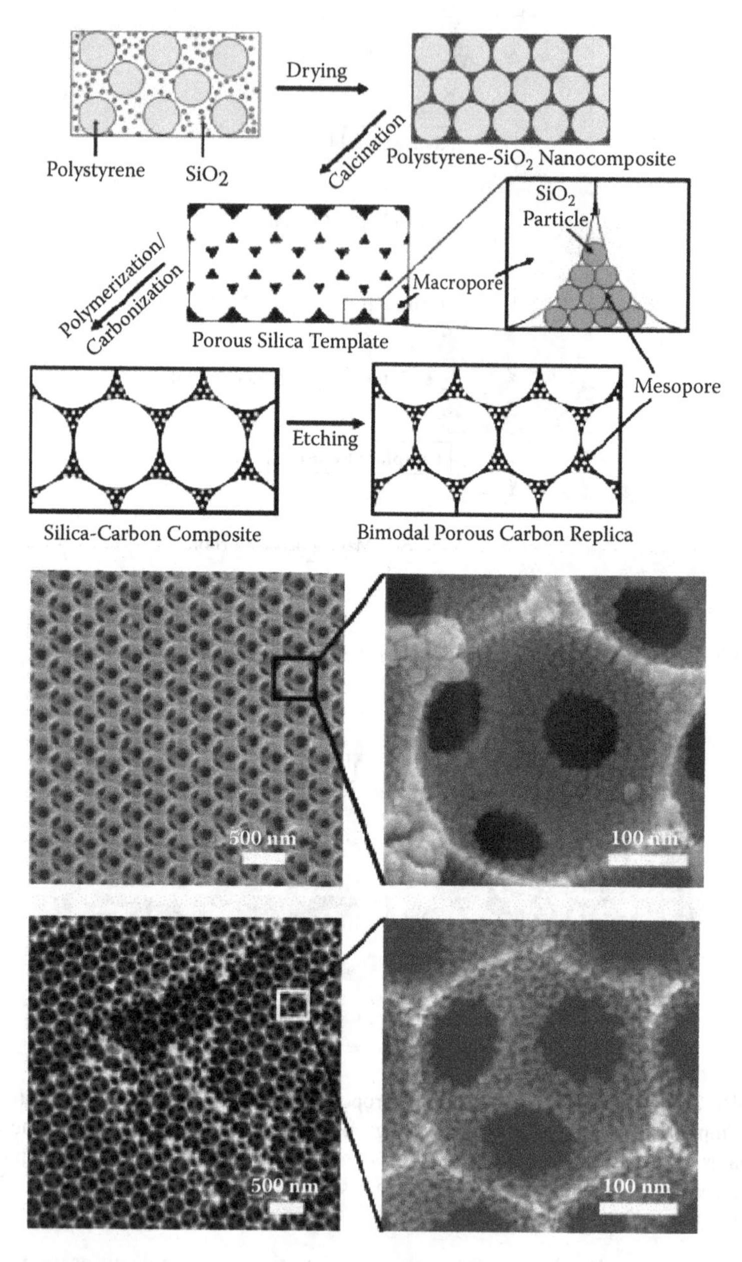

FIGURE 2.41 Schematic illustration of (a) the preparation steps of 3-D macromesoporous carbon using PS spheres and silica nanoparticles as templates (above) and (b) SEM images of silica template composed of silica nanoparticles and templated porous carbon (below). (From Chai, G.S., Shin, I.S., and Yu, J.-S. *Adv Mater* 16: 2057–2061, 2004. With permission.)

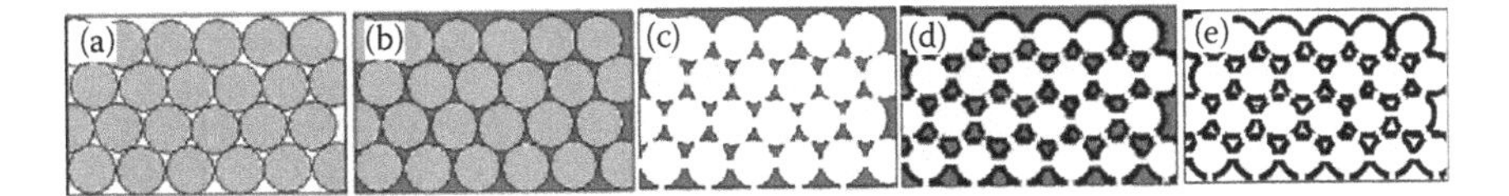

FIGURE 2.42 Schematic illustration of fabrication of 3-D macroporous carbon using the surface-templating approach: (a) PS colloidal crystal, (b) silica-infiltrated PS colloidal crystal, (c) an inverse silica opal after removal of PS spheres, (d) carbon-coated inverse silica opal, and (e) 3-D macroporous carbon after removal of silica.

synthesis procedure, in which divinylbenzene is used as the carbon precursor. Figure 2.41b displays SEM images of the template and porous carbon. An ordered 3-D interconnected macroporous carbon with mesopores can be prepared using resorcinol–formaldehyde resin as a carbon source and uniform silica particles as a template [255].

Using the surface-templating method, we succeeded in the preparation of ordered macroporous carbons with a 3-D interconnected pore structure using an inverse silica opal as a template [256]. The preparation procedure is depicted in Figure 2.42. First, monodisperse polystyrene (PS) spheres, synthesized according to Shim and coworkers [257], are fabricated into colloidal crystals using a flow-controlled vertical deposition method [258]. Second, the interstices between the PS spheres are infiltrated with a silica sol, which is prepared by mixing tetraethyl orthosilicate, ethanol, and 0.1 M HCl solution with a volume ratio of 1:3:0.1. After further hydrolysis and condensation of the silica sol, the PS spheres are removed by calcination in air at 500°C for 5 h, yielding a 3-D macroporous inverse silica opal as the template. Third, pyrolytic carbon deposition on the surface of the inverse silica opal is carried out via CVD of a hydrocarbon gas as a carbon precursor (benzene). Etching of the silica template by HF solution obtains 3-D interconnected nanostructured macroporous carbon.

Figure 2.43a is a cross-sectional image of the macroporous inverse silica opal template, in which the spherical voids have a diameter of about 220 nm, packed in a face-centered cubic (fcc) structure. The 3-D ordered structure of the carbon (GMC1000) is seen in Figure 2.43b: the long-range order of the carbon with an fcc arrangement of spherical voids is inherited from the macroporous silica template. Figure 2.43c is a higher-magnification image of sample GMC1000, demonstrating that the carbon was replicated via a surface-templating mechanism. Figure 2.43d shows the presence of imperfect graphene layers. The tightly packed graphene layers and the (002) diffraction peak in Figure 2.44 suggests the synthesized carbon has a turbostratic nature with a relatively high crystallinity. On the contrary, carbon C1000-2, synthesized at 1000°C using zeolite Y as the template and furfuryl alcohol as the carbon precursor [74], is much more disordered. The surface area of GMC1000 was about 45 m^2/g.

The 3-D macroporous carbons discussed earlier usually possess low crystallinity, partly because their carbonization temperature is relatively low (<1000°C). Very recently, Jaroniec and coworkers reported that a highly ordered, and indeed graphitized, nanoporous carbon with spherical pores can be prepared by heat

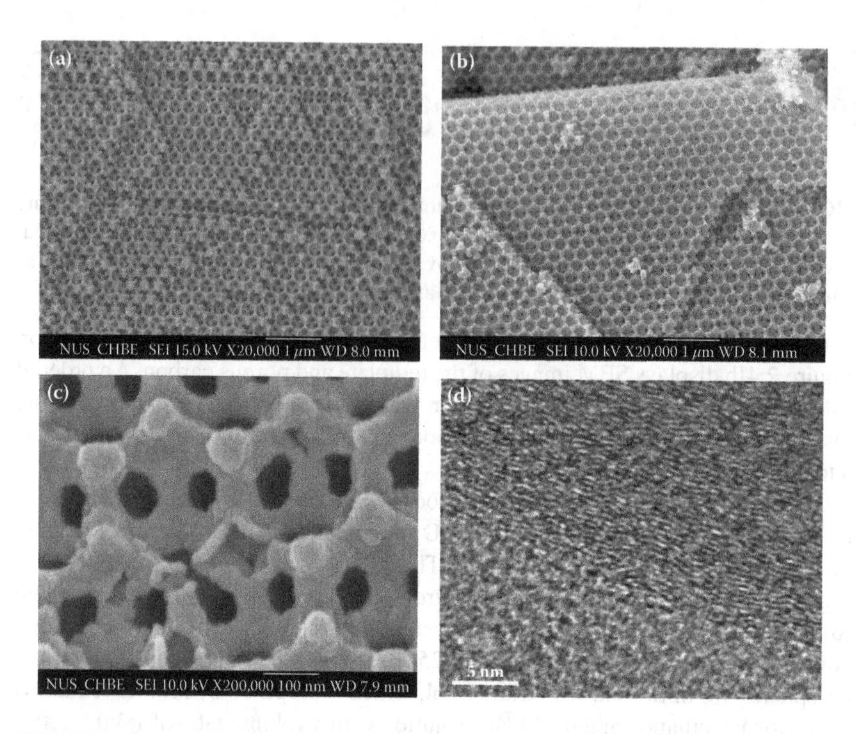

FIGURE 2.43 FESEM images of (a) inverse silica template, (b, c) 3-D macroporous carbon (GMC1000) of different magnifications, and (d) lattice-fringe TEM image of the carbon framework. (From Su, F., Zhao, X.S., Wang, Y., Zeng, J., Zhou, Z., and Lee, J.Y. *J Phys Chem B* 109: 20200–20206, 2005. With permission.)

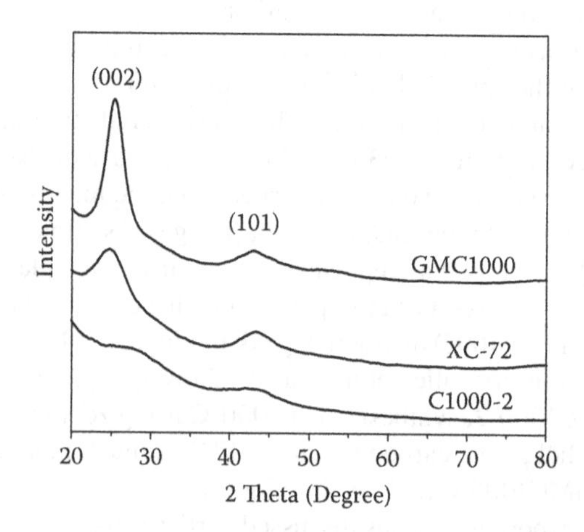

FIGURE 2.44 XRD patterns of a macroporous carbon GMC1000, carbon black XC-72, and an amorphous microporous carbon C1000-2. (From Su, F., Zhao, X.S., Wang, Y., Zeng, J., Zhou, Z., and Lee, J.Y. *J Phys Chem B* 109: 20200–20206, 2005. With permission.)

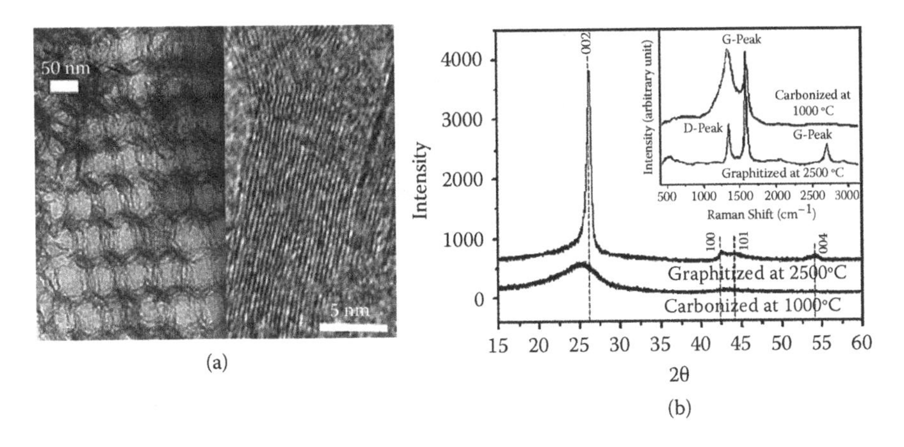

FIGURE 2.45 (a) TEM images, (b) XRD patterns, and Raman spectra (the inset) of an ordered nanoporous carbon graphitized at 2500°C under argon. (From Yoon, S.B., Chai, G.S., Kang, S.K., Yu, J.-S., Gierszal, K.P., and Jaroniec, M. *J Am Chem Soc* 127: 4188–4189, 2005. With permission.)

treatment of a mesophase pitch-based carbon at 2500°C [259]. Figure 2.45a shows TEM images of such a carbon at different resolutions. The left image reveals the presence of ordered and interconnected spherical pores created by dissolution of the colloidal crystal template. The right image reveals that the carbon pore walls are composed of a somewhat ordered stacking of graphene layers, giving rise to high crystallinity. The XRD pattern in Figure 2.45b gives an interlayer spacing of about 0.33 nm and there is some either of (10) splitting into (101) and (100) peaks. A method of assembling integrated multifunctional titania-coated 3-D ordered macroporous carbon was also demonstrated [260]. Ordered macroporous carbons containing small metal particles (Co, Ni, Cu, Pt) and a hierarchical pore structure were fabricated using the templating method [261,262].

VI. OTHER TEMPLATING APPROACHES TO PREPARING CARBON

The discovery of fullerenes in 1985 [28] and carbon nanotubes in 1991 [29] has spurred worldwide interest in carbon nanostructures for various applications. These nanostructures have been shown to be promising catalyst supports [261], nanodevices [263], energy storage materials [264], and drug delivery systems [265]. Hollow carbon nanospheres are a class of nanostructured carbons. The template method [266], media-reduction method [267], and shock-compression method [268] have all been used to produce such structures. The template method can be used to make hollow carbon spheres of diameters ranging from tens [269] to hundreds of nanometers [270]. For this purpose, silica spheres are normally utilized as templates, because they can withstand high-temperature processing such as carbonization. In addition, synthesis of silica spheres of various diameters has been well established [271]. Polymer spheres such as polypyrrole (PPy) and PS can also be used as templates [269,272]. Although these are organic in nature,

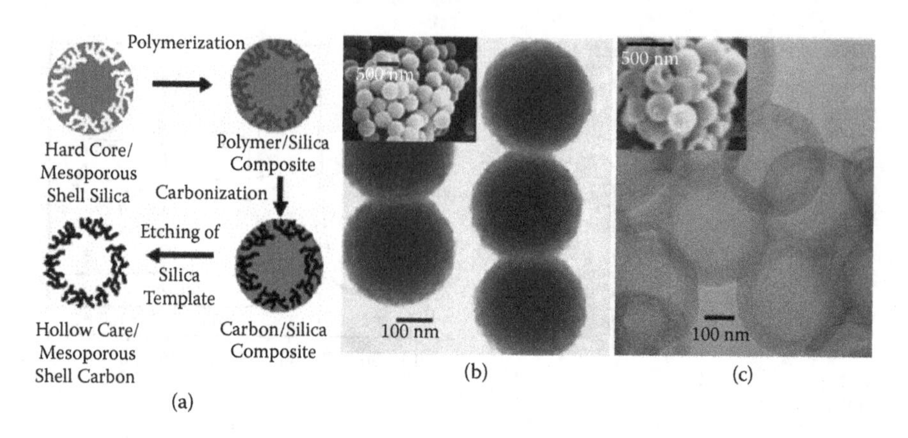

FIGURE 2.46 (a) Schematic illustration of the fabrication steps of hollow carbon capsules with a mesoporous shell. (From Yoon, S.B., Sohn, K., Kim, J.Y., Shin, C.H., Yu, J.S., and Hyeon, T. *Adv Mater* 14: 19–21, 2002. With permission.) (b) TEM and SEM (inset) images of SCMS silica template with a core diameter of 220 nm and a shell thickness of 60 nm. (From Kim, M., Yoon, S.B., Sohn, K., Kim, J.Y., Shin, C.H., Hyeon, T., and Yu, J.S. *Micropor Mesopor Mater* 63: 1–9, 2003. With permission.) (c) TEM and SEM (inset) images of carbon capsules with a core diameter of 220 nm and a shell thickness of 55 nm. (From Kim, M., Yoon, S.B., Sohn, K., Kim, J.Y., Shin, C.H., Hyeon, T., and Yu, J.S. *Micropor Mesopor Mater* 63: 1–9, 2003. With permission.)

they display very different thermal behavior from carbon and thus can be removed by solvent washing or calcination. Various methods for making core-shell carbon structures are available [266,269,270,273–276]. For example, silica spheres with solid-core/mesoporous-shell (SCMS) were used as templates to prepare hollow carbon capsules with mesoporous shell, as shown in Figure 2.46.

Using silica spheres as templates, we fabricated hollow carbon spheres using the CVD method with benzene as a carbon precursor [277]. Figure 2.47a shows that the silica sphere template has a diameter of about 650 nm. Figures 2.47b, 2.47c, and 2.47d display SEM images at different magnifications of hollow carbon spheres fabricated at 1000°C. Their surface area is <100 m^2/g, much smaller than that of hollow core/mesoporous shell (HCMS) carbon capsules [274], shown in Figure 2.46. Hollow carbon spheres with a single shell, deformed shell, double shell, and nitrogen-doped carbon shell were reported recently [278].

By manipulating the adhesion between the polymer template and the carbon precursor (to be infiltrated into the voids of the porous polymer), Jiang et al. [279] designed a "lost-wax approach" to generate a variety of highly monodisperse, inorganic, polymeric, and metallic solids and hollow spheres. Recently, we demonstrated the fabrication of various nanostructured carbon materials using the internal-surface-templating method [280,281]. Using sucrose or FA as precursors, a range of carbon nanostructures was demonstrated [280]—including hollow spheres with open windows (Figure 2.48a), hollow spheres without open windows (Figure 2.48b), walnutlike carbon hollow spheres with separated cells (Figure 2.48c), and carbon capsules (Figure 2.48d)—by controlling fabrication

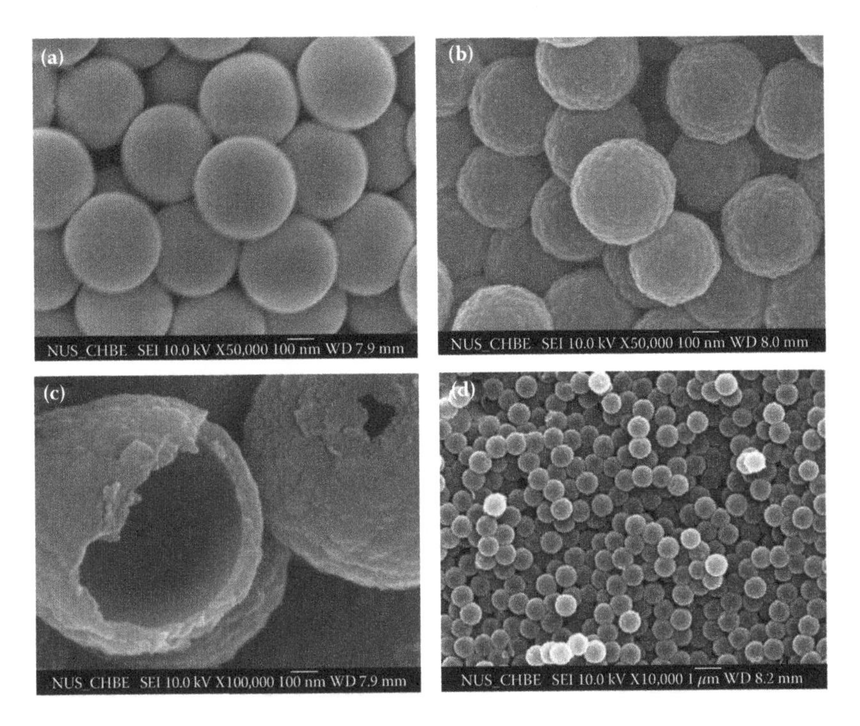

FIGURE 2.47 FESEM images of (a) a silica sphere template and (b, c, and d) hollow carbon spheres of different magnifications. (From Wang, Y., Su, F., Lee, J.Y., and Zhao, X.S. *Chem Mater* 18: 1347–1353, 2006. With permission.)

parameters such as the neck size of a 3-D macroporous silica template, the concentration of the carbon precursor, the infiltration time, and the sequence of the polymerization/carbonization/etching procedure.

It is noteworthy that most template approaches to the synthesis of porous carbons involved inorganic (or hard) templates and organic carbon precursors. More recently, supramolecules have also been employed as templates to synthesize ordered porous polymers, and subsequently, ordered porous carbons. This novel synthesis route avoids the use of hard templates and thus simplifies the HF etching step. Tanaka et al. [282] reported a direct synthesis approach to preparing OMCs with a channel structure from an organic–organic nanocomposite. The strategy is to use an organic–organic interaction between a thermosetting polymer and a thermally decomposable surfactant to form a periodic ordered nanocomposite. The thermosetting polymer is carbonized into a carbonaceous pore wall. RF and triethyl orthoacetate (EOA) were used as carbon coprecursors, and triblock copolymer Pluronic F127 was used as a surfactant. The resultant materials (referred to as COU-1) are shown in Figure 2.49. The choice of an appropriate set of thermosetting polymers and surfactants is the most significant factor in controlling the structure. Zhang et al. [283] reported a dilute aqueous route for direct synthesis of novel bicontinuous cubic mesostructured mesoporous polymer (FDU-14) and carbon (C-FDU-14) with $Ia\bar{3}d$ symmetry from an organic–organic self-assembly

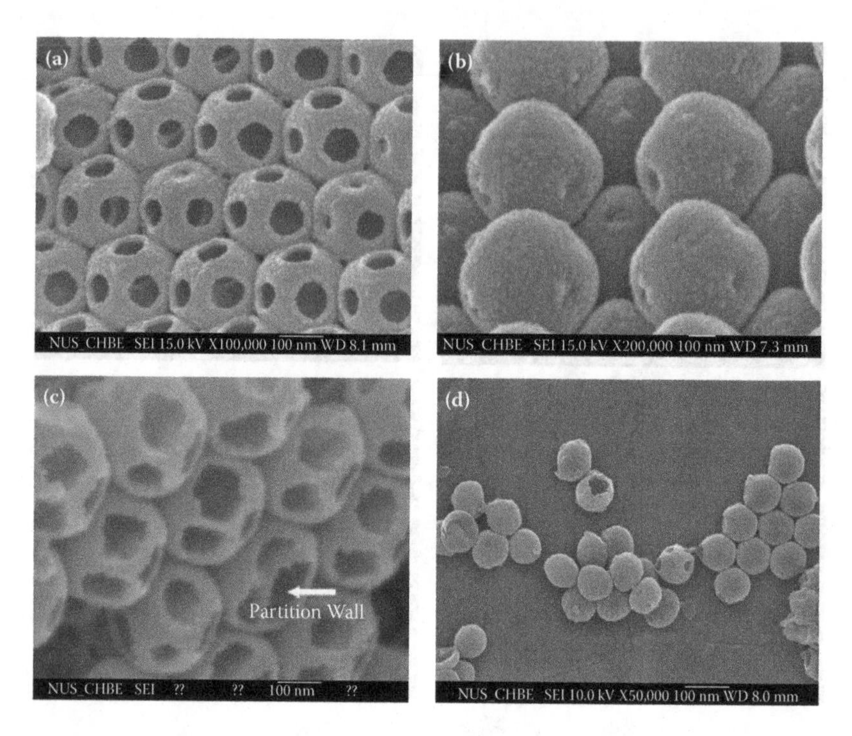

FIGURE 2.48 FESEM images of (a) hollow carbon spheres with open windows, (b) hollow carbon spheres without open windows, (c) walnutlike carbon hollow spheres, and (d) carbon capsules. (From Zhou, Z., Yan, Q., Su, F., and Zhao, X.S. *J Mater Chem* 15: 2569–2574, 2005. With permission.)

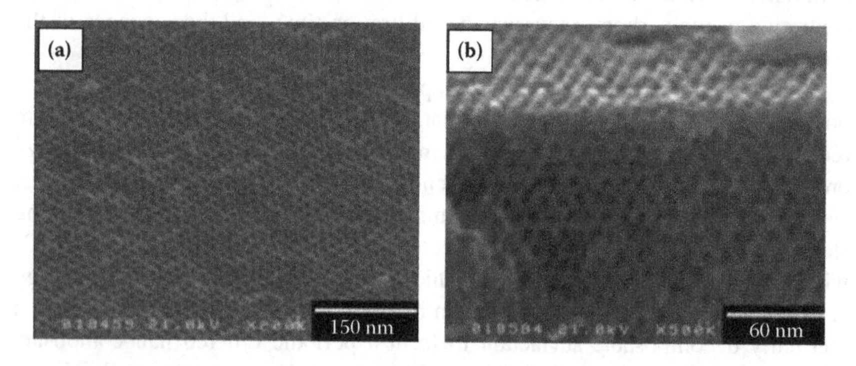

FIGURE 2.49 FESEM images of porous carbon carbonized at (a) 600°C and (b) 800°C. (From Tanaka, S., Nishiyama, N., Egashira, Y., and Ueyama, K. *Chem Commun* 2125–2127, 2005. With permission.)

of triblock copolymer P123 as a template and using soluble low-molecular-weight polymers of phenol and formaldehyde (resol: Mw = 500 to 5000) as precursors, followed by a thermopolymerization process. The choice of resol as a precursor is the key to the successful organization of organic–organic mesostructures, because

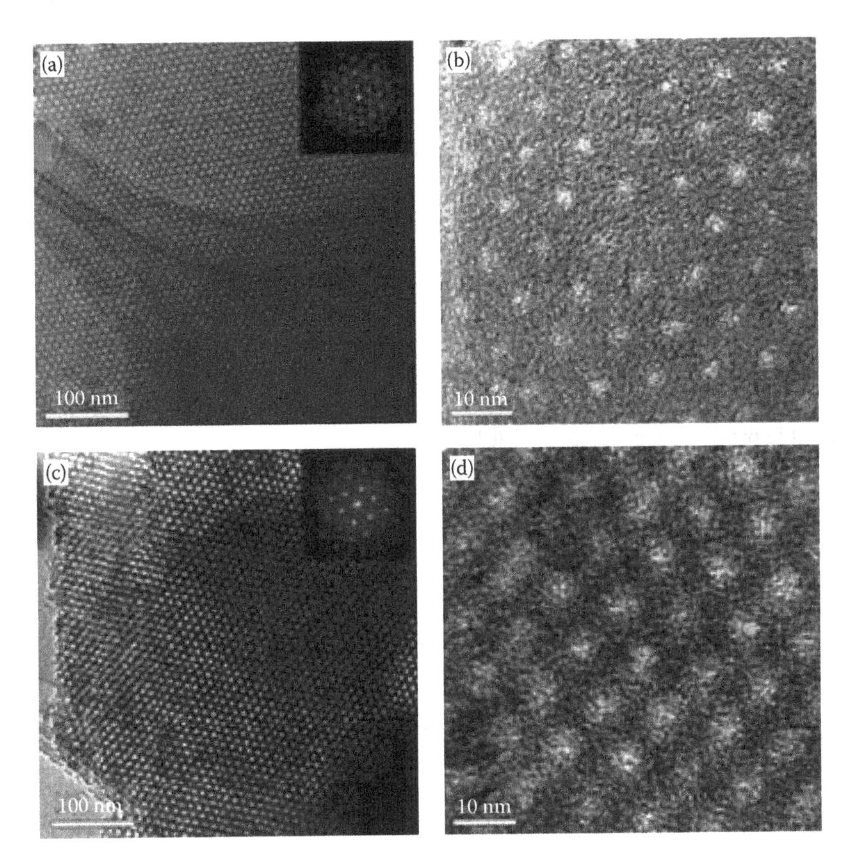

FIGURE 2.50 (a) TEM image of C-FDU-15 calcined at 1400°C, (b) HRTEM image of FDU-15 calcined at 1400°C, (c) TEM image of C-FDU-16 calcined at 1400°C, and (d) HRTEM image of C-FDU-16 calcined at 1400°C. (From Meng, Y., Gu, D., Zhang, F., Shi, Y., Yang, H., Li, Z., Yu, C., Tu, B., and Zhao, D. *Angew Chem Int Ed*. 44: 7053–7059, 2005. With permission.)

resol has a large number of hydroxyl groups that can interact strongly with triblock copolymers through H bonds and form 3-D covalently bonded connections. Lately, a 2-D hexagonal (*p6mm*) mesoporous carbon (C-FDU-15) and a well-ordered cubic ($Im\bar{3}m$) mesostructured carbon (C-FDU-16) were synthesized using different surfactants as templates [284]. It was demonstrated that the structures of the assembled frameworks can be controlled by varying the amphiphilic surfactant/resol ratio or choosing different surfactants. The 2-D hexagonal and 3-D cubic structures can be obtained by using F127 ($EO_{106}PO_{70}EO_{106}$) as a template, whereas a lamellar mesostructure can be prepared using P123 ($EO_{20}PO_{70}EO_{20}$) as a template. The open mesoporous carbon frameworks have high surface areas, narrow PSDs, large pore volumes, and ultrastability (>1400°C). TEM images (Figure 2.50) show that C-FDU-15 with well-ordered hexagonal channel arrays and C-FDU-16 with body-centered cubic symmetry have a large domain regularity even after calcination at 1400°C. The approach is quite similar to that using mesoporous silica

(MCM-41, SBA-15, and so on) and is considered to be more suitable for industrial production compared to the evaporation-induced self-assembly (EISA) method. Dai's group reported a simple methodology to develop well-oriented porous carbon films through carbonization of phenolic resin based on stepwise self-assembly of block copolymer (polystyrene-blockpoly(4-vinylpyridine)) [285]. The orientation of ordered carbon nanopores was successfully aligned normal to the substrates using a solvent annealing process. The block copolymers in these processes play two important roles: (1) directing the formation of the phenolic resin nanostructure and (2) serving as templates for nanopores. Nanostructured carbons in the form of monoliths, fibers, sheets, and films were readily synthesized via self-assembly of triblock copolymer (Pluronic F127) as a structure-directing agent and a mixture of phloroglucinol and formaldehyde as carbon precursors, under mild and widely variable processing conditions [286].

It is believed that this direct synthesis method (or soft template) will expand the possibility of synthesizing a variety of ordered porous carbon structures, facilitate their mass production, and lead to the fabrication of novel carbon and carbon–polymer porous materials.

VII. APPLICATIONS OF TEMPLATED POROUS CARBON

The tailor-designed physicochemical properties of porous carbons prepared using the template method make them promising candidates in many applications, including adsorption, catalysis, chromatography, energy storage and conversion, and photonics.

A. Adsorption

Porous carbon microbeads fabricated using silica gel beads as a template were shown to exhibit high adsorption capacity for volatile organic compounds, including benzene, toluene, and xylene [234]. The reason for this is that the beads can expand during adsorption and contract to the original state upon desorption [234]. Template-synthesized microporous carbons were also found to display molecular sieving properties [56]. Mesoporous carbons templated by silica nanoparticles were observed to have a higher adsorption capacity for dyes (e.g., a bulky Direct blue 78 dye) and humic acids than commercial activated carbons [225,226]. The adsorption properties of methyl mercaptan on OMC materials were studied using a dynamic adsorption method in a fixed bed [170]. Results showed, as expected, that adsorption was strongly influenced by the pore structure and surface chemistry of the carbon. In recent work [94], we investigated these effects in phenol adsorption and found that, after surface modification by thermal treatment under nitrogen, phenol adsorption capacity dramatically increased. It was also observed that N-doped microporous carbons displayed high adsorption affinity toward water vapor [96,98]. An adsorption study of fullerenes C60 and C70 on CMK-3 showed that mesoporous carbon can be an effective and reversible adsorbent for the separation of fullerenes [287].

Template-synthesized porous carbons with high pore volume and a well-developed pore network are potential adsorbents for gas storage. Thus, for example, microporous carbons prepared with zeolite Y as a template possess a methane storage capacity of 0.255 g/g at 298 K and 3.5 MPa [86]. High-pressure adsorption at 303 K on mesoporous carbon CMK-1 showed that the adsorbed layer of supercritical methane is highly compressed by the strong interaction potential from adjacent pore walls, and there may be an intermediate state between supercritical fluid and liquid [288]. It was also reported that as much as 0.412 g/g of methane can be stored in CMK-3 at 275 K and 7 MPa in the presence of water [289]. Such sorption capacity is about 30% higher than the highest sorption capacity attained by activated carbon under otherwise same conditions and equal to or higher than storage capacity under compression at 20 MPa. Enhanced methane storage is considered to be due to the formation of methane hydrate in the pores of CMK-3.

Hydrogen storage properties of several templated carbon materials have been studied as well. Results demonstrated that adsorption capacity at 77 K correlates well with micropore volume (or BET surface area) [290,291]. The contribution of ultramicropores (whose pore diameter is below 1 nm) was significantly dominant. Additionally, templated carbon has shown a potential as an electrode for electrochemical hydrogen storage: a hydrogen storage capacity of up to 1.95 wt% was recorded in 6 M KOH solution [292], which is much higher than that obtained with other nanostructured carbon materials and $LaNi_5$ alloy. Template-prepared carbon materials were also observed to possess excellent cycling capacity and high-rate capability because of highly developed porosity with ultramicropores (<0.7 nm) for hydrogen storage and interconnected mesopores for facilitating fast mass transfer [293].

B. Catalysis

As catalyst supports, template-synthesized porous carbons offer potential advantages over traditional porous carbons. Thus, Pt and Pd catalysts supported on template-synthesized mesoporous carbon were found to perform better in hydrogenation reactions of nitrobenzene, 2-ethylanthraquinone, and 4-isobutylacetophenone in comparison with commercial catalysts [294]. Pd supported on Cocontaining OMC exhibited a high catalytic activity in octane hydrogenation [178]. Another [295] showed that mesoporous carbon is a good support for heteropolyacids. Palladium-containing mesoporous carbons were prepared using a sol–gel copolymerization method [296]. In comparison with a commercial activated carbon, the template-synthesized carbon displayed a higher catalytic activity in Heck reactions of iodobenzene and styrene.

C. Electrochemical Energy Conversion

Carbon materials are well known as good working electrodes in electrochemistry [297,298]. The demand for advanced energy conversion and storage devices has intensified studies on nanostructured carbon materials [299]. The templating strategy for making materials with desired characteristics provides opportunities to substantially improve electrochemical energy conversion. Template-synthesized carbon materials have proved their unique performances in lithium-ion batteries, electrical double-layer capacitors (EDLCs), and fuel cells because of their surface chemistry, pore structure, and electrical conductivity characteristics.

a. Lithium-Ion Batteries

The selection of carbon as the negative electrode substantially promotes its commercialization in lithium-ion batteries [300]. Reversible intercalation of lithium into graphene layers of the carbon host lattice avoids the problem of lithium dendrite formation and results in a large improvement in terms of cyclability and safety. In principle, all carbon materials, including natural and synthetic graphites, graphitizable carbons, low-temperature and nongraphitizing carbons, as well as doped carbons, can be lithiated to a certain extent [301]. However, more demanding electrochemical properties such as reversible capacity (lithium reversibly incorporated into graphene layers) and irreversible capacity (faradic losses during first charge–discharge cycle), and the details of the current–voltage curves during charging and discharging, depend more critically on the chemical and physical properties of carbon. Therefore, much effort is being expended to find and synthesize high-performance carbon materials.

Lithium-ion batteries with OMC as the electrode material have been reported to display good stability and relatively high columbic efficiency in charge–discharge cycles: for example, high energy capacity, about 1100 mAh/g (Li_3C_6), and good cycling performance of CMK-3 carbon [302]. An almost constant resistance and lithium ion diffusion coefficient were found when the potential was lower than the critical potential, with a charge capacity of 616 mAh/g and a reversible capacity of 131 mAh/g at first cycle [303]. This carbon material also presents a higher reversible capacity than carbon nanotubes and can be charged/discharged at a high current rate. Such unusual electrochemical performance is thought to be due to the 3-D ordered mesoporous structure and large surface area. Use of ordered, nanostructured, tin-based oxide carbon composite (ONTC) as a negative electrode was also reported [304]. ONTC was fabricated by full deposition of tin-based oxide into nanochannels of mesoporous carbon CMK-3. These novel composites showed improved cycling performance relative to pure tin oxide material. Use of tin-decorated OMC (CMK-3) as a negative electrode material for rechargeable batteries was also investigated, and the electrode was observed to deliver a reversible capacity of 400 mAh/g (C + Sn) at 100% cycling efficiency [305], which is much higher than that of conventional graphite electrodes.

Lee et al. studied the electrochemical properties of colloidal-crystal-templated macroporous carbon materials in secondary batteries [306]. They found that the characteristics of 3-D ordered macroporous (3-DOM) carbon composed of interconnected pores with wall thickness of a few tens of nanometers enhances the kinetic performance. The authors suggested that improved performance originates from (1) a short solid-state diffusion length for lithium ions, (2) a large

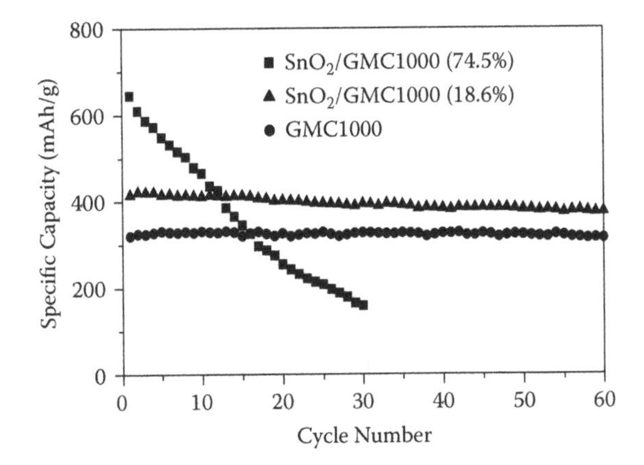

FIGURE 2.51 The cycling performances of macroporous carbon GMC1000 before and after coating of SnO_2. (The discharge potential limits were in the range between 5 mV and 2.0 V at a constant current of 0.1 C.) (From Su, F., Zhao, X.S., Wang, Y., Zeng, J., Zhou, Z., and Lee, J.Y. *J Phys Chem B* 109: 20200–20206, 2005. With permission.)

number of active sites for charge-transfer reactions because of the highly accessible surface area, (3) reasonable electrical conductivity of the carbon, (4) high ionic conductivity of the electrolyte within the 3-DOM carbon matrix, and (5) the absence of a binder or a conducting agent.

We fabricated graphitizable macroporous carbons and examined their electrochemical properties as negative electrode materials [256]. The specific capacity of the carbon (GMC1000) was measured to be about 326 mAh/g at a specific current of 40 mA/g. When current was increased to 600 mA/g, capacity decreased to 303 mAh/g. In comparison with a less ordered 3-D macroporous carbon having a similar porous structure [306], there was a substantial improvement in rate performance. Figure 2.51 compares the cycling performance of sample GMC1000 with that of SnO_2–GMC1000 composite materials having different SnO_2 contents. GMC1000 displays very stable cyclability. Upon coating SnO_2 nanoparticles, higher initial capacity was observed in the first cycle. Furthermore, the average capacity decrease of SnO_2/GMC1000 (with 18.6 wt% loading of SnO_2) was about 0.16% per cycle, better than that of previously reported 3-D macroporous carbon coated with SnO_2 nanoparticles [306], as well as that of OMC deposited with SnO_2 nanoparticles [304] and hollow carbon spheres encapsulated with Sn particles [307].

b. Supercapacitors

Electrochemical capacitors are important in supporting the voltage of a system during increased loads in everything from portable equipment to electric vehicles [299]. Carbon-based electrochemical capacitors are of two types, depending on the kind of accumulated energy, namely, EDLCs and supercapacitors [308]. For the former, only a pure electrostatic attraction between ions and carbon surface occurs, whereas for the latter the capacitance is based additionally on faradic (pseudocapacitance) reactions. The term *pseudo* originates from the fact that the double-layer capacitance arises from quick redox charge transfer reactions and not only from electrostatic charging. The key factors that dictate the selection of carbon materials for such electrodes are the following: very well-developed surface area, pore geometry, PSD, high conductivity, good wettability, and surface functional groups. In general, the capacitance is thought to be proportional to the "electrochemically active" surface area of the material and to the relative permittivity of the solution, and is reciprocally dependent on the double-layer thickness. Obviously, the higher the specific surface area of a porous carbon is, the higher its specific capacitance should be. The porous texture of carbon also affects the ionic conductivity, which relates to the mobility of ions inside the pores and, thus, the kinetic performance of an EDLC. It has been proved that the electronic conductivity of porous carbons strongly affects the formation of the electrical double layer [309]. Wettability of the carbon surface, largely determined by surface functionalities, also affects capacitance [310]. Among these factors, the most critical aspect appears to be a compromise between high surface area (to ensure high capacitance) and wide PSD (to permit easy access for the electrolyte) [299,311]. Templated porous carbons with a 3-D porous network and high surface area would therefore offer advantages over activated carbons [311–314].

Compared to activated carbon MSC-25, a commercially used negative electrode material in ELDCs, template-synthesized mesoporous carbons were demonstrated to exhibit superior electrochemical performance [63,196,315–317]. Another study showed that OMCs with a 2-D pore symmetry displayed superior capacitance behavior, power output, and high-frequency performance in EDLCs [318]. Microporous carbons replicated from zeolites Y, Beta, and ZSM-5 were observed to possess significant charge–discharge hysteresis and high irreversible capacity [72]. Because of enhanced electrical conductivity, OMCs containing a ferrocene derivative (Fc-MC) have higher capacitance than the parent pure mesoporous carbon [159]. An OMC material with 3.9-nm pores and 900 m^2/g exhibited an electrochemical double-layer capacitance of 60 to 90 F/g [319]. After 100 cycles, the capacity was decreased by only 20%. A detailed study showed that a pore packing defect (as a consequence of having an OMC with defective pore channels) is an important dynamic factor at high current intensity [320]. Colloidal-crystal-templated bimodal macroporous carbons were found to exhibit better capacitance per unit surface area than commercially available activated carbons [321].

The performance of a supercapacitor combines electrostatic attraction (as in EDLC capacitors) and faradic reactions similar to processes taking place in accumulators. Pseudocapacitance effects (e.g., electrosorption of H or metal ad-atoms and redox reactions of electroactive species) strongly depend on the chemical affinity of carbon for ions adsorbed on the electrode surface. Therefore, capacitance enhancement can be realized by surface modification of carbon [322], the formation of carbon–conducting polymer composites by polymerization [323], or insertion of electroactive particles of transition metal oxides (e.g., RuO_2, TiO_2, Cr_2O_3, MnO_2, or Co_2O_3) into carbon [324]. However, nowadays, few relevant reports on templated porous carbons as supercapacitors can be found. For example, MnO_2 nanoparticles have been embedded into pore walls of mesoporous carbon CMK-3 by a redox reaction between permanganate ions and carbon [325]. A large specific capacitance (>200 F/g) for the Mn–carbon composite, compared to 600 F/g for the MnO_2, was obtained; these materials have high electrochemical stability and high reversibility [326].

c. Fuel Cells

In recent decades, various electrode materials have been investigated to improve the performance of fuel cells [299]. A conventional low-temperature fuel cell electrode is composed of polytetrafluoroethylene, a high-surface-area carbon black loaded with a precious metal catalyst, and a current collector, as well as other minor components. The most challenging issue for electrode performance is the electrocatalyst [327]. Carbon has been established as the best catalyst support because of its good electrical conductivity, high surface area, surface hydrophobicity, and stability [328–331]. In the past few years, template-synthesized carbons with various structures have been tried as components of fuel cells.

Ryoo and coworkers observed a significant increase in oxygen reduction using Pt-loaded OMC compared to Pt-loaded carbon black [128]. In another study [332], highly porous carbons synthesized using USY zeolite and MCM-48 silica as templates have been reported to possess a higher electrochemically active surface area and to give a higher Pt dispersion than the commercial Vulcan XC-72R carbon black; the new electrocatalyst was more active and stable as an anodic material than Pt(50)Ru(50) in methanol electro-oxidation. The metal loading can also take place before, during, or after carbonization, resulting in different properties of metal nanoparticles and different carbon structures. Simultaneous pyrolysis of carbon precursor (sucrose) and decomposition of platinum precursor ($(NH_3)_4Pt(NO_3)_2$) in the mesoporous channels of SBA-15 yielded a Pt–C nanocomposite [333] in which small Pt clusters were studded in micropore walls of the OMC. This composite can be used as a methanol-tolerant cathode material for oxygen reduction in direct methanol fuel cells (DMFCs). The greatest advantage is that aggregation and mobility of Pt are hampered by polymerization and carbonization of the carbon precursor, which is thought to lead to high catalyst dispersion. A synthesized Pt/CMK-3 electrocatalyst was found to outperform a commercial electrocatalyst for oxygen reduction in a gas-diffusion electrode; the Pt–Ru/CMK-3 electrocatalyst was also not as effective for methanol oxidation [334]. A polymer–carbon composite prepared by impregnating an OMC with an organic monomer [335] exhibits the surface characteristics of the polymer and the electrical conductivity of the carbon, suggesting new applications as advanced electrode material. Mesoporous carbons prepared using commercial silica particles as a template were also employed as cathode Pt support in a polymer electrolyte fuel cell (PEMFC) [336]. It was found that the high surface area combined with carbon mesoporosity had a positive influence on metal dispersion and on distribution of ions, leading to enhanced cell performance.

A 3-D ordered macroporous carbon loaded with Pt–Ru alloy particles was evaluated for electrochemical performance in a DMFC [248,251]. The carbon displayed advantages over Vulcan XC-72 carbon black. Another study also showed that a Pt–Ru catalyst supported on ordered macroporous carbon exhibited higher specific activity for methanol oxidation than the commercial E-TEK catalyst [254]. We also demonstrated that the electrochemical properties of Pt catalysts, supported on microporous carbon with an "amorphous" carbon core/"graphitic" carbon shell structure [75,337], on an OMC [135], and on a 3-D macroporous carbon [256] improved the specific activity of methanol electrochemical oxidation at room temperature.

D. Materials Processing

Because of the uniquely ordered structure of template-synthesized porous carbons, they themselves can be used as templates to replicate other materials with an ordered porous structure that is difficult to make using traditional methods [121,338–348]. Thus, CMK-3 carbon was first demonstrated as a template to prepare mesoporous silica [121,338]. An OMC prepared from an MCM-48 template was used to prepare nanostructured silica [339]. Nanocasting of CMK-3 using ZSM-5 crystals yielded a mesoporous zeolite with both mesopores and micropores [340]. More recently, the preparation of a novel class of mesoporous aluminosilicate molecular sieves was described [341]. The preparation of mesoporous boron nitride (MBN) and mesoporous carbon nitride (MBCN) with very high surface area and pore volume was recently realized using a well-ordered hexagonal mesoporous carbon as a template and boron trioxide as a boron source [342]. Nonspherical silica "nanocases" with a hollow core and mesoporous shell were also produced [343].

With the assistance of template-synthesized porous carbon networks [250], Kim et al. [344] successfully synthesized TiO_2 nanospheres. Filling the carbon pores with a mixture of titanium ethoxide and chloroform, followed by drying and removal of the carbon template, yielded TiO_2 particles whose shape and size are determined by the pores in the carbon template. Synthesis of ordered mesoporous magnesium oxide using CMK-3 carbon by exotemplating was reported [345]. Hollow spheres and shells of crystalline porous metal oxides were fabricated using hollow spheres of mesoporous carbon as a hard template and alkoxides as metal precursors [346]. Other mesoporous materials such as boron nitride were similarly prepared [347,348].

E. Other Applications

Conductive carbon is known to be a good matrix for reproducible construction of sensors and biosensors [349–351]. The high conductivity of some carbon materials is ideal for electrochemical signal transduction, whereas the presence of high porosity facilitates the adsorption of large molecules, such as polyelectrolyte–enzyme complexes [352]. It has been shown that mesoporous carbons with oxygen-containing surface functional groups (preferably thermally stable ones such as lactones) are the most suitable electrochemical transducers [353]. The CMK-3 carbon was shown to have an adsorption capacity of 18.5 μmol/g for horse heart cytochrome C enzyme [354]. Adsorption of vitamin E onto mesoporous carbons CMK-1 and CMK-3 from solutions with different concentrations in nonpolar (*n*-heptane) and polar (*n*-butanol) solvent was studied [355]. It was found that vitamin E uptake on different adsorbents depends on both the solvent and the adsorbent's mesopore volume and surface area. Adsorption of lysozyme onto carbon molecular sieves with varying pore diameters was studied by Vinu et al. [356]; results demonstrated, as expected, that the amount adsorbed depends on solution pH as well as adsorbent porosity. Carbon has superior stability in water and is thus a more appropriate adsorbent for biomolecules than nanoporous silicates. Additionally, it was found that carbons CMK-3 and CMK-1 coated with about 10 wt% of poly(methyl methacrylate) (PMMA) exhibited higher adsorption capacity for vitamin B_{12} from water than their uncoated counterparts, because of formation of hydrogen bonding between PMMA and vitamin B_{12} molecules [171]. Adsorption of L-histidine on mesoporous carbon molecular sieves was also studied [357]. A glucose biosensor with high sensitivity and fast response was fabricated by immobilizing glucose oxidase (GOx) on a mesocellular carbon foam; owing to its unique structure, the mesocellular carbon afforded high enzyme loading and fast mass transfer, resulting in high catalytic efficiency [358].

The brief discussion presented suggests that template-synthesized porous carbons with carefully controlled pore structure may soon find applications in the immobilization of biomolecules (proteins, enzymes, and vitamins) and in bioconversion, reversible amperometric immunosensors, regeneration of enzyme electrodes, switchable biofuel cells, and artificial kidneys [359].

VIII. SUMMARY

In comparison with traditional porous carbon materials such as activated carbons and carbon blacks, template-synthesized carbons can more easily and controllably achieve prescribed physical, chemical, and morphological properties for a wide range of novel applications. Our review of this relatively new field has shown that templating on length scales ranging from nanometers to micrometers is readily achievable, enabling one to design and build a desired scaffold of porous carbon. Clearly, a number of issues and challenges must be addressed before such carbons find practical applications. The high cost (because of the involvement of templates such as SBA-15 and MCM-48) and environmental impact (because of the use of corrosive chemicals such as HF and NaOH) represent two key issues. Control of infiltration of carbon into template pores is also a challenge. Nevertheless, we are optimistic that such template-synthesized hierarchical porous carbon structures may soon find applications, especially in emerging high-tech areas such as nanotechnology, biotechnology, and photonics.

ACKNOWLEDGMENTS

We thank the National University of Singapore (NUS) for financial support. Thanks are also due to Professor Ljubisa Radovic for his invaluable suggestions.

REFERENCES

1. Davis, M.E. *Nature* 417: 813–821, 2002.
2. Lu, G.Q. and Zhao, X.S. Eds. *Nanoporous Materials: Science and Engineering*. London: Imperial College Press, 2004.
3. Loy, D.A. and Shea, K.J. *Chem Rev* 95: 1431–1442, 1995.
4. Aoyama, Y. *Top Curr Chem* 198: 131–161, 1998.
5. Barton, T.J., Bull, L.M., Klemperer, W.G., Loy, D.A., McEnaney, B., Misono, M., Monson, P.A., Pez, G., Scherer, G.W., Vartuli, J.C., and Yaghi, O.M. *Chem Mater* 11: 2633–2656, 1999.
6. Stein, A. *Adv Mater* 15: 763–775, 2003.
7. Yaghi, O.M., O'Keeffe, M., Ockwig, N.W., Chae, H.K., Eddaoudi, M., and Kim, J. *Nature* 423: 705–714, 2003.
8. Kitagawa, S., Kitaura, R., and Noro, S. *Angew Chem Int Ed* 43: 2334–2375, 2004.
9. Schüth, F. *Angew Chem Int Ed* 42: 3604–3622, 2003.
10. van Bommel, K.J.C., Friggeri, A., and Shinkai, S. *Angew Chem Int Ed* 42: 980–999, 2003.
11. Zhao, X.S., Su, F., Yan, Q., Guo, W., Bao, X.Y., Lv, L., and Zhou, Z. *J Mater Chem* 16: 637–648, 2006.
12. Beck, D.W. *Zeolite Molecular Sieves*. New York: Wiley, 1974, pp. 340–341.
13. Cundy, C.S. and Cox, P.A. *Chem Rev* 103: 663–702, 2003.
14. Zhao, X.S., Lu, G.Q., and Millar, G.J. *Int Eng Chem Res* 35: 2075–2090, 1996.
15. Ying, J.Y., Mehnert, C.P., and Wong, M.S. *Angew Chem Int Ed.* 38: 56–77, 1999.
16. Holmberg, K. *J Colloid Interface Sci* 274: 355–364, 2004.
17. Stein, A. *Micropor Mesopor Mater* 44–45: 227–239, 2001.
18. Kyotani, T. *Carbon* 38: 269–286, 2000.
19. Ryoo, R., Joo, S.H., Kruk, M., and Jaroniec, M. *Adv Mater* 13: 677–681, 2001.
20. Lee, J., Han, S., and Hyeon, T. *J Mater Chem* 14: 478–486, 2004.
21. Hentze, H.-P. and Antonietti, M. *Curr Opin Solid State Mater Sci* 5: 343–353, 2001.
22. Patrick, J.W. *Porosity in Carbons: Characterization and Applications*. London: Edward Arnold, 1995.
23. Marsh, H., Heintz, E.A., and Rodriguez-Reinoso, F. *Introduction to Carbon Technology*. Universidad de Alicante, Alicante, Spain, 1997.
24. Rodríguez-Reinoso, F. *Carbon* 36: 159–175, 1998.
25. Yang, R.T. *Adsorbents: Fundamentals and Applications*. John Wiley and Sons, New York, 2003.
26. Bansal, R.C., Donnet, J.-B., and Stoeckli, F. *Active Carbon*. New York: Marcel Dekker, 1988.
27. Linares-Solano, A., Lozano-Castelló, D., Lillo-Ródenas, M.A., and Cazorla-Amorós, D. In Radovic, L.R. Ed. *Chemistry and Physics of Carbon*, Vol. 30, 2007.
28. Kroto, H.W., Heath, J.R., O'Brien, S.C., Curl, R.F., and Smalley, R.E. *Nature* 318: 162–163, 1985.
29. Iijima, S. *Nature* 354: 56–58, 1991.
30. Krishnan, A., Dujardin, E., Treacy, M.M.J., Hugdahl, J., Lynum, S., and Ebbesen, T.W. *Nature* 388: 451–454, 1997.
31. Saito, Y. and Matsumoto, T. *Nature* 1998: 392: 237.

32. Che, G., Lakshmi, B.B., Fisher, E.R., and Martin, C.R. *Nature* 393: 346–349, 1998.
33. Ajayan, P.M., Nugent, J.M., Siegel, R.W., Wei, B., and Kohler-Redich, P.H. *Nature* 404: 243, 2000.
34. Zhang, G.Y., Jiang, X., and Wang, E.G. *Science* 300: 472–474, 2003.
35. Viculis, L.M., Mack, J.J., and Kaner, R.B. *Science* 299: 1361, 2003.
36. Helveg, S., López-Cartes, C., Sehested, J., Hansen, P.L., Clausen, B.S., Rostrup-Nielsen, J.R., Abild-Pedersen, F., and Nørskov, J.K. *Nature* 427: 426–429, 2004.
37. Subramoney, S. *Adv Mater* 10: 1157–1171, 1998.
38. Inagaki, M. and Radovic, L.R. *Carbon* 40: 2279–2282, 2002.
39. Inagaki, M., Kaneko, K., and Nishizawa, T. *Carbon* 42: 1401–1417, 2004.
40. Johnson, S.A., Ollivier, P.J., and Mallouk, T.E. *Science* 283: 963–965, 1999.
41. Yang, H. and Zhao, D. *J Mater Chem* 15: 1217–1231, 2005.
42. Sakintuna, B. and Yürüm, Y. *Int Eng Chem Res* 44: 2893–2902, 2005.
43. Gilbert, M.T., Knox, J.H., and Kaur, B. *Chromatographia* 16: 138–146, 1982.
44. Knox, J.H., Kaur, B., and Millward, G.R. *J Chromatogr* A 352: 3–25, 1986.
45. Sonobe, N., Kyotani, T., Hishiyama, Y., Shiraishi, M., and Tomita, A. *J Phys Chem* 92: 7029–7034, 1998.
46. Sonobe, N., Kyotani, T., and Tomita, A. *Carbon* 26: 573–578, 1998.
47. Sonobe, N., Kyotani, T., and Tomita, A. *Carbon* 28: 483–488, 1990.
48. Sonobe, N., Kyotani, T., and Tomita, A. *Carbon* 29: 61–67, 1991.
49. Kyotani, T., Yamada, H., Sonobe, N., and Tomita, A. *Carbon* 32: 627–635, 1994.
50. Kyotani, T., Tsai, L., and Tomita, A. *Chem Mater* 7: 1427–1428, 1995.
51. Kyotani, T., Tsai, L., and Tomita, A. *Chem Mater* 8: 2109–2113, 1996.
52. Hernadi, K., Fonseca, A., Nagy, J.B., Bernaerts, D., Fudala, A., and Lucas, A.A. *Zeolites* 17: 416–423, 1996.
53. Bandosz, T.J., Putyera, K., Jagiello, J., and Schwarz, J.A. *Micropor Mater* 1: 73–79, 1993.
54. Bandosz, T.J., Jagiello, J., Putyera, K., and Schwarz, J.A. *Carbon* 32: 659–664, 1994.
55. Bandosz, T.J., Gomez-Salazar, S., Putyera, K., and Schwarz, J.A. *Micropor Mater* 4: 177–184, 1994.
56. Bandosz, T.J., Jagiello, J., Putyera, K., and Schwarz, J.A. *Chem Mater* 8: 2023–2029, 1996.
57. Ma, Z., Kyotani, T., and Tomita, A. *Chem Commun* 2365–2366, 2000.
58. Ma, Z., Kyotani, T., Liu, Z., Terasaki, O., and Tomita, A. *Chem Mater* 13: 4413–4415, 2001.
59. Ma, Z., Kyotani, T., Liu, Z., Terasaki, O., and Tomita, A. *Carbon* 40: 2367–2374, 2002.
60. Kresge, C.T., Leonowicz, M.E., Roth, W.J., Vartuli, J.C., and Beck, J.S. *Nature* 359: 710–712, 1992.
61. Zhao, D., Feng, J., Huo, Q., Melosh, N., Fredrickson, G.H., Chmelka, B.F., and Stucky, G.D. *Science* 279: 548–552, 1998.
62. Ryoo, R., Joo, S.H., and Jun, S. *J Phys Chem* B 103: 7743–7746, 1999.
63. Lee, J., Yoon, S., Hyeon, T., Oh, S.M., and Kim, K.B. *Chem Commun* 2177–2178, 1999.
64. Zakhidov, A.A., Baughman, R., Iqbal, Z., Cui, C., Khayrullin, I., Dantas, S.O., Marti, J., and Ralchenko, V.G. *Science* 282: 897–901, 1998.
65. Han, S. and Hyeon, T. *Carbon* 37: 1645–1646, 1999.
66. Li, Z. and Jaroniec, M. *J Am Chem Soc* 123: 9208–9209, 2001.
67. Li, Z. and Jaroniec, M. *Chem Mater* 15: 1327–1333, 2003.
68. Li, Z. and Jaroniec, M. *J Phys Chem B* 108: 824–826, 2004.
69. Kyotani, T., Nagai, T., Inoue, S., and Tomita, A. *Chem Mater* 9: 609–615, 1997.
70. Kyotani, T., Ma, Z., and Tomita, A. *Carbon* 41: 1451–1459, 2003.

71. Rodriguez-Mirasol, J., Cordero, T., Radovic, L.R., and Rodriguez, J.J. *Chem Mater* 10: 550–558, 1998.
72. Meyers, C.J., Shah, S.D., Patel, S.C., Sneeringer, R.M., Bessel, C.A., Dollahon, N.R., Leising, R.A., and Takeuchi, E.S. *J Phys Chem B* 105: 2143–2152, 2001.
73. Barata-Rodrigues, P.M., Mays, T.J., and Moggridge, G.D. *Carbon* 41: 2231–2246, 2003.
74. Su, F., Zhao, X.S., Lv, L., and Zhou, Z. *Carbon* 42: 2821–2831, 2004.
75. Su, F., Zeng, J., Yu, Y., Lv, L., Lee, J.Y., and Zhao, X.S. *Carbon* 43: 2366–2373, 2005.
76. Su, F., Lv, L., and Zhao, X.S. *Int J Nanosci* 4: 261–268, 2005.
77. Su, F., Lv, L., and Zhao, X.S. *Stud Surf Sci Catal* 156: 557–565, 2005.
78. Matsuoka, K., Yamagishi, Y., Yamazaki, T., Setoyama, N., Tomita, A., and Kyotani, T. *Carbon* 43: 876–879, 2005.
79. Hou, P.X., Yamazaki, T., Orikasa, H., and Kyotani, T. *Carbon* 43: 2621–2627, 2005.
80. Tosheva, L., Parmentier, J., Valtchev, V., Vix-Guterl, C., and Patarin, J. *Carbon* 43: 2474–2480, 2005.
81. Kyotani, T., Nagai, T., and Tomita, A. *Proceedings of Carbon '92,* Essen, Germany, 1992, pp. 477–478.
82. Enzel, P. and Bein, T. *Chem Mater* 4: 819–824, 1992.
83. Enzel, P., Zoller, J.J., and Bein, T. *Chem Commun* 633–635, 1992.
84. Cordero, T., Thrower, P.A., and Radovic, L.R. *Carbon* 30: 365–374, 1992.
85. Johnson, S.A., Brigham, E.S., Ollivier, P.J., and Mallouk, T.E. *Chem Mater* 9: 2448–2458, 1997.
86. Matsuoka, K., Yamagishi, Y., Kyotani, T., and Tomita, A. *Carbon Conference 2003*, Oviedo, Spain, July 6–10, 2003.
87. Garsuch, A., Klepel, O., Sattler, R.R., Berger, C., Gläser, R., and Weitkamp, J. *Carbon* 44: 593–596, 2006.
88. Garsuch, A. and Klepel, O. *Carbon* 43: 2330–2337, 2005.
89. Gaslain, F.O.M., Parmentier, J., Valtchev, V.P., and Patarin, J. *Chem Commun* 991–993, 2006.
90. Fitzer, E. and Schafer, W. *Carbon* 8: 353–364, 1970.
91. Radovic, L.R., Moreno-Castilla, C., and Rivera-Utrilla, J. In Radovic, L.R. Ed. *Chemistry and Physics of Carbon*, Vol. 27, New York: Marcel Dekker, 2001, pp. 227–405.
92. Burg, P. and Cagniant, D. In Radovic, L.R. Ed. *Chemistry and Physics of Carbon*, Vol. 30, New York: Marcel Dekker, 2007.
93. Radovic, L.R. and Rodríguez-Reinoso, F. In Thrower, P.A. Ed. *Chemistry and Physics of Carbon*, Vol. 25, New York: Marcel Dekker, 1997, pp. 243–358.
94. Su, F., Lv, L., Tee, M.H., and Zhao, X.S. *Carbon* 43: 1156–1164, 2005.
95. Darmstadt, H., Roy, C., Kaliaguine, S., Choi, S.J., and Ryoo, R. *Carbon* 40: 2673–2683, 2002.
96. Hou, P.-X., Orikasa, H., Yamazaki, T., Matsuoka, K., Tomita, A., Setoyama, N., Fukushima, Y., and Kyotani, T. *Chem Mater* 17: 5187–5193, 2005.
97. Yang, Z., Xia, Y., and Mokaya, R. *Micropor Mesopor Mater* 86: 69–80, 2005.
98. Garsuch, A., Sattler, R.R., Witt, S., and Klepel, O. *Micropor Mesopor Mater* 89: 164–169, 2006.
99. On, D.T., Desplantier-Giscard, D., Danumah, C., and Kaliaguine, S. *Appl Catal A* 222: 299–357, 2001.
100. Schumacher, K., Ravikovitch, P.I., Du Chesne, A., Neimark, A.V., and Unger, K.K. *Langmuir* 16: 4648–4654, 2000.
101. Carlsson, A., Kaneda, M., Sakamoto, Y., Terasaki, O., Ryoo, R., and Joo, S.H. *J Electron Microsc* 48: 795–798, 1999.

102. Joo, S.H., Jun, S., and Ryoo, R. *Micropor Mesopor Mater* 44–45: 153–158, 2001.
103. Kruk, M., Jaroniec, M., Ryoo, R., and Joo, S.H. *J Phys Chem B* 104: 7960–7968, 2000.
104. Ryoo, R., Joo, S.H., Jun, S., Tsubakiyama, T., and Terasaki, O. *Stud Surf Sci Catal* 135: 150–158, 2001.
105. Solovyov, L.A., Zaikovskii, V.I., Shmakov, A.N., Belousov, O.V., and Ryoo, R. *J Phys Chem B* 106: 12198–12202, 2002.
106. Kaneda, M., Tsubakiyama, T., Carlsson, A., Sakamoto, Y., Ohsuna, T., Terasaki, O., Joo, S.H., and Ryoo, R. *J Phys Chem B* 106: 1256–1266, 2002.
107. Sakamoto, Y., Kim, T.-W., Ryoo, R., and Terasaki, O. *Angew Chem Int Ed* 43: 5231–5234, 2004.
108. Lee, J., Sohn, K., and Hyeon, T. *J Am Chem Soc* 123: 5146–5147, 2001.
109. Yoon, S.B., Kim, J.Y., and Yu, J.S. *Chem Commun* 559–560, 2001.
110. Yoon, S.B., Kim, J.Y., and Yu, J.S. *Chem Commun* 1536–1537, 2002.
111. Kim, S.-S., Lee, D.-K., Shah, J., and Pinnavaia, T.J. *Chem Commun* 1436–1437, 2003.
112. Mbileni, C.N., Prinsloo, F.F., Witcomb, M.J., and Coville, N.J. *Carbon* 44: 1476–1483, 2006.
113. Vix-Guterl, C., Boulard, S., Parmentier, J., Werckmann, J., and Patarin, J. *Chem Lett* 10: 1062–1063, 2002.
114. Li, Z., Del Cul, G.D., Yan, W., Liang, C., and Dai, S. *J Am Chem Soc* 126: 12782–12783, 2004.
115. Kruk, M., Jaroniec, M., Joo, S.H., and Ryoo, R. *J Phys Chem B* 107: 2205–2213, 2003.
116. Kruk, M., Jaroniec, M., Ko, C.H., and Ryoo, R. *Chem Mater* 12: 1961–1968, 2000.
117. Joo, S.H., Ryoo, R., Kruk, M., and Jaroniec, M. *J Phys Chem B* 106: 4640–4646, 2002.
118. Lettow, J.S., Han, Y.J., Schmidt-Winkel, P., Yang, P.D., Zhao, D., Stucky, G.D., and Ying, J.Y. *Langmuir* 16: 8291–8295, 2000.
119. Che, S., Lund, K., Tatsumi, T., Iijima, S., Joo, S.H., Ryoo, R., and Terasaki, O. *Angew Chem Int Ed* 42: 2182–2185, 2003.
120. Jun, S., Joo, S.H., Ryoo, R., Kruk, M., Jaroniec, M., Liu, Z., Ohsuna, T., and Terasaki, O. *J Am Chem Soc* 122: 10712–10713, 2000.
121. Kang, M., Yi, S.H., Lee, H.I., Yie, J.E., and Kim, J.M. *Chem Commun* 1944–1945, 2002.
122. Solovyov, L.A., Shmakov, A.N., Zaikovskii, V.I., Joo, S.H., and Ryoo, R. *Carbon* 40: 2477–2481, 2002.
123. Fuertes, A.B. *Micropor Mesopor Mater* 67: 273–281, 2004.
124. Shin, H.J., Ryoo, R., Kruk, M., and Jaroniec, M. *Chem Commun* 349–350, 2001.
125. Lee, J.S., Joo, S.H., and Ryoo, R. *J Am Chem Soc* 124: 1156–1157, 2002.
126. Darmstadt, H., Roy, C., Kaliaguine, S., Joo, S.H., and Ryoo, R. *Micropor Mesopor Mater* 60: 139–149, 2003.
127. Kruk, M., Jaroniec, M., Kim, T.W., and Ryoo, R. *Chem Mater* 15: 2815–2823, 2003.
128. Joo, S.H., Choi, S.J., Oh, I., Kwak, J., Liu, Z., Terasaki, O., and Ryoo, R. *Nature* 412: 169–172, 2001.
129. Darmstadt, H., Roy, C., Kaliaguine, S., Kim, T.W., and Ryoo, R. *Chem Mater* 15: 3300–3307, 2003.
130. Lu, A., Schmidt, W., Spliethoff, B., and Schüth, F. *Adv Mater* 15: 1602–1606, 2003.
131. Lu, A.-H., Li, W.-C., Schmidt, W., and Schüth, F. *Micropor Mesopor Mater* 80: 117–128, 2005.

132. Lu, A.-H., Li, W.-C., Schmidt, W., Kiefer, W., and Schüth, F. *Carbon* 42: 2939–2948, 2004.
133. Liu, X., Chang, F., Xu, L., Yang, Y., He, Y., and Liu, Z. *Carbon* 44: 184–187, 2006.
134. Solovyov, L.A., Kim, T.-W., Kleitz, F., Terasaki, O., and Ryoo, R. *Chem Mater* 16: 2274–2281, 2004.
135. Su, F., Zeng, J., Bao, X.Y., Yu, Y., Lee, J.Y., and Zhao, X.S. *Chem Mater* 17: 3960–3967, 2005.
136. Yu, C., Fan, J., Tian, B., Zhao, D., and Stucky, G.D. *Adv Mater* 14: 1742–1745, 2002.
137. Xia, Y. and Mokaya, R. *Adv Mater* 16: 886–891, 2004.
138. Fuertes, A.B. *J Mater Chem* 13: 3085–3088, 2003.
139. Fuertes, A.B. and Nevskaia, D.M. *J Mater Chem* 13: 1843–1846, 2003.
140. Fuertes, A.B. and Nevskaia, D.M. *Micropor Mesopor Mater* 62: 177–190, 2003.
141. Kruk, M., Dufour, B., Celer, E.B., Kowalewski, T., Jaroniec, M., and Matyjaszewski, K. *J Phys Chem B* 109: 9216–9225, 2005.
142. Lu, A., Kiefer, A., Schmidt, W., and Schüth, F. *Chem Mater* 16: 100–103, 2004.
143. Kim, T.-W., Park, I.-S., and Ryoo, R. *Angew Chem Int Ed* 42: 4375–4379, 2003.
144. Vix-Guterl, C., Saadallah, S., Vidal, L., Reda, M., Parmentier, J., and Patarin, J. *J Mater Chem* 13: 2535–2539, 2003.
145. Yang, H., Yan, Y., Liu, Y., Zhang, F., Zhang, R., Meng, Y., Li, M., Xie, S., Tu, B., and Zhao, D. *J Phys Chem B* 108: 17320–17328, 2004.
146. Fuertes, A.B. and Alvarez, S. *Carbon* 42: 3049–3055, 2004.
147. Kim, C.H., Lee, D.-K., and Pinnavaia, T.J. *Langmuir* 20: 5157–5159, 2004.
148. Yang, C.-M., Weidenthaler, C., Spliethoff, B., Mayanna, M., and Schüth, F. *Chem Mater* 17: 355–358, 2005.
149. Fuertes, A.B. and Centeno, T.A. *J Mater Chem* 15: 1079–1083, 2005.
150. Sevilla, M. and Fuertes, A.B. *Carbon* 44: 468–474, 2006.
151. Qiao, W.M., Song, Y., Hong, S.H., Lim, S.Y., Yoon, S.-H., Korai, Y., and Mochida, I. *Langmuir* 22: 3791–3797, 2006.
152. Ehrburger-Dolle, F., Morfin, I., Geissler, E., Bley, F., Livet, F., Parmentier, J., Reda, M., Patarin, J., Iliescu, M., and Werckmann, J. *Langmuir* 19: 4303–4308, 2003.
153. Zhang, F., Yan, Y., Yang, H., Meng, Y., Yu, C., Tu, B., and Zhao, D. *J Phys Chem B* 109: 8723–8732, 2005.
154. Parmentier, J., Vix-Guterl, C., Gibot, P., Reda, M., Ilescu, M., Werckmann, J., and Patarin, J. *Micropor Mesopor Mater* 62: 87–96, 2003.
155. Kawabuchi, Y., Kishino, M., Kawano, S., Whitehurst, D.D., and Mochida, I. *Langmuir* 12: 4281–4285, 1996.
156. Zheng, G.-B., Kouda, K., Sano, H., Uchiyama, Y., Shi, Y.-F., and Quan, H.J. *Carbon* 42: 635–640, 2004.
157. Hata, K., Futaba, D.N., Mizuno, K., Namai, T., Yumura, M., and Iijima, S. *Science* 306: 1362–1364, 2004.
158. Zhang, W.H., Liang, C., Sun, H., Shen, Z., Guan, Y., Ying, P., and Li, C. *Adv Mater* 14: 1776–1778, 2002.
159. Furukawa, H., Hibino, M., Zhou, H.S., and Honma, I. *Chem Lett* 32: 132–133, 2003.
160. Xia, Y. and Mokaya, R. *Adv Mater* 16: 1553–1558, 2004.
161. Xia, Y., Yang, Z., and Mokaya, R. *J Phys Chem* B 108: 19293–19298, 2004.
162. Xia, Y. and Mokaya, R. *Chem Mater* 17: 1553–1560, 2005.
163. Xia, Y., Yang, Z., and Mokaya, R. *Chem Mater* 18: 140–148, 2006.
164. Fitzer, E., Köchling, K.-H., Boehm, H.P., and Marsh, H. *Pure Appl Chem* 67: 473–506, 1995.

165. Fischbach, D.B. In Walker, P.L., Jr. Ed. *Chemistry and Physics of Carbon*, New York: Marcel Dekker, 1971, chap. 1.
166. Su, F. and Zhao, X.S. Unpublished work, 2006a.
167. Bokros, J.C. In Walker, P.L., Jr. Ed. *Chemistry and Physics of Carbon*, Vol. 5, New York: Marcel Dekker, 1969, chap. 1.
168. Kotlensky, W.V. In Walker, P.L., Jr. and Thrower, P.A. Eds. *Chemistry and Physics of Carbon*, Vol. 9, New York: Marcel Dekker, 1973, chap 3.
169. Kawabuchi, Y., Oka, H., Kawano, S., Mochida, I., and Yoshizawa, N. *Carbon* 36: 377–382, 1998.
170. Kim, D.J., Lee, H.I., Yie, J.E., Kim, S.-J., and Kim, J.M. *Carbon* 43: 1868–1873, 2005.
171. Guo, Z., Zhu, G., Gao, B., Zhang, D., Tian, G., Chen, Y., Zhang, W., and Qiu, S. *Carbon* 43: 2344–2351, 2005.
172. Li, Z. and Dai, S. *Chem Mater* 17: 1717–1721, 2005.
173. Li, Z., Yan, W., and Dai, S. *Langmuir* 21: 11999–12006, 2005.
174. Vinu, A., Ariga, K., Mori, T., Nakanishi, T., Hishita, S., Golberg, D., and Bando, Y. *Adv Mater* 17: 1648–1652, 2005.
175. Wang, L., Zhao, Y., Lin, K., Zhao, X., Shan, Z., Di, Y., Sun, Z., Cao, X., Zou, Y., Jiang, D., Jiang, L., and Xiao, F.-S. *Carbon* 44: 1336–1339, 2006.
176. Huwe, H. and Fröba, M. *Micropor Mesopor Mater* 60: 151–158, 2003.
177. Lu, A.-H., Li, W.-C., Kiefer, A., Schmidt, W., Bill, E., Fink, G., and Schüth, F. *J Am Chem Soc* 126: 8616–8617, 2004.
178. Lu, A.-H., Schmidt, W., Matoussevitch, N., Bönnemann, H., Spliethoff, B., Tesche, B., Bill, E., Kiefer, W., and Schüth, F. *Angew Chem Int Ed* 43: 4303–4306, 2004.
179. Lee, J., Jin, S., Hwang, Y., Park, J.-G., Park, H.M., and Hyeon, T. *Carbon* 43: 2536–2543, 2005.
180. Lee, J., Lee, D., Oh, E., Kim, J., Kim, Y., Jin, S., Kim, H.-S., Hwang, Y., Kwak, J.H., Park, J.-G., Shin, C.-H., Kim, J., and Hyeon, T. *Angew Chem Int Ed* 44: 7427–7432, 2005.
181. Li, H., Zhu, S., Xi, H., and Wang, R. *Micropor Mesopor Mater* 89: 196–203, 2006.
182. Holmes, S.M., Foran, P., Roberts, E.P.L., and Newton, J.M. *Chem Commun* 1912–1913, 2005.
183. Su, F., Lee, F.Y., Lv, L., Liu, J., Tian, X.N., and Zhao, X.S. *Adv Funct Mater* 17: 1926–31, 2007.
184. Kim, J.Y., Yoon, S.B., Kooli, F., and Yu, J.S. *J Mater Chem* 11: 2912–2914, 2001.
185. Yoon, S.B., Kim, J.Y., Kooli, F., Lee, C.W., and Yu, J.S. *Chem Commun* 1740–1741, 2003.
186. Wang, L., Lin, S., Lin, K., Yin, C., Liang, D., Di, Y., Fan, P., Jiang, D., and Xiao, F.-S. *Micropor Mesopor Mater* 85: 136–142, 2005.
187. Su, F., Li, X., Lv, L., and Zhao, X.S. *Carbon* 44: 801–803, 2006.
188. Su, F., Zhao, X.S., Wang, Y., and Lee, J.Y. *Micropor Mesopor Mater* 98: 323–329, 2007.
189. Sakamoto, Y., Kaneda, M., Terasaki, O., Zhao, D.Y., Kim, J.M., Stucky, G., Shin, H.J., and Ryoo, R. *Nature* 408: 449–453, 2000.
190. Kim, M.J. and Ryoo, R. *Chem Mater* 11: 487–491, 1999.
191. Kim, T.-W., Ryoo, R., Gierszal, K.P., Jaroniec, M., Solovyov, L.A., Sakamoto, Y., and Terasaki, O. *J Mater Chem* 15: 1560–1571, 2005.
192. Guo, W., Su, F., and Zhao, X.S. *Carbon* 43: 2423–2426, 2005.
193. Guo, W. and Zhao, X.S. *Stud Surf Sci Catal* 156: 551–556, 2005.
194. Tanev, P.T. and Pinnavaia, T.J. *Science* 267: 865–867, 1995.
195. Zhang, W.Z., Pauly, R.T., and Pinnavaia, T.J. *Chem Mater* 9: 2491–2498, 1997.

196. Lee, J., Yoon, S., Oh, S.M., Shin, C.H., and Hyeon, T. *Adv Mater* 12: 359–362, 2000.
197. Sevilla, M., Alvarez, S., and Fuertes, A.B. *Micropor Mesopor Mater* 74: 49–58, 2004.
198. Kim, S.-S., Pauly, T.R., and Pinnavaia, T.J. *Chem Commun* 1661–1662, 2000.
199. Kim, S.-S. and Pinnavaia, T.J. *Chem Commun* 2418–2419, 2001.
200. Lee, J., Sohn, K., and Hyeon, T. *Chem Commun* 2674–2675, 2002.
201. Lee, J., Kim, J., and Hyeon, T. *Chem Commun* 1138–1139, 2003.
202. Alvarez, S. and Fuertes, A.B. *Carbon* 42: 433–436, 2004.
203. Ania, C.O. and Bandosz, T.J. *Micropor Mesopor Mater* 89: 315–324, 2006.
204. Prouzet, E. and Pinnavaia, T.J. *Angew Chem Int Ed* 36: 516–518, 1997.
205. Fuertes, A.B. *Chem Mater* 16: 449–455, 2004.
206. Shi, Z.-G., Feng, Y.-Q., Xu, L., Da, S.-L., and Zhang, M. *Carbon* 42: 1677–1682, 2004.
207. Yang, H., Shi, Q., Liu, X., Xie, S., Jiang, D., Zhang, F., Yu, C., Tu, B., and Zhao, D. *Chem Commun* 2842–2843, 2002.
208. Taguchi, A., Smatt, J.H., and Linden, M. *Adv Mater* 15: 1209–1211, 2003.
209. Lu, A.-H., Smått, J.-H., and Lindén, M. *Adv Funct Mater* 15: 865–871, 2005.
210. Böhme, K., Einicke, W.-D., and Klepel, O. *Carbon* 43: 1918–1925, 2005.
211. Li, W.-C., Lu, A.-H., and Schüth, F. *Chem Mater* 17: 3620–3626, 2005.
212. Hampsey, J.E., Hu, Q., Rice, L., Pang, J., Wu, Z., and Lu, Y. *Chem Commun* 3606–3608, 2005
213. Hampsey, J.E., Hu, Q., Wu, Z., Rice, L., Pang, J., and Lu, Y. *Carbon* 43: 2977–2982, 2005.
214. Gierszal, K.P., Kim, T.-W., Ryoo, R., and Jaroniec, M. *J Phys Chem* B 109: 23263–23268, 2005.
215. Zhou, L., Li, H., Yu, C., Zhou, X., Tang, J., Meng, Y., Xia, Y., and Zhao, D. *Carbon* 44: 1601–1604, 2006.
216. Vinu, A., Miyahara, M., Sivamurugan, V., Morib, T., and Ariga, K. *J Mater Chem* 15: 5122–5127, 2005.
217. Lee, H.I., Pak, C., Shin, C.-H., Chang, H., Seung, D., Yied, J.E., and Kim, J.M. *Chem Commun* 6035–6037, 2005.
218. Kruk, M., Dufour, B., Celer, E.B., Kowalewski, T., Jaroniec, M., and Matyjaszewski, K. *Chem Mater* 18: 1417–1424, 2006.
219. Cott, D.J., Petkov, N., Morris, M.A., Platschek, B., Bein, T., and Holmes, J.D. *J Am Chem Soc* 128: 3920–3921, 2006.
220. Chae, W.-S., An, M.-J., Lee, S.-W., Son, M.-S., Yoo, K.-H., and Kim, Y.-R. *J Phys Chem* B 110: 6447–6450, 2006.
221. Hu, Q., Pang, J., Wu, Z., and Lu, Y. *Carbon* 44: 1298–1352, 2006.
222. Kim, M., Sohn, K., Kim, J., and Hyeon, T. *Chem Commun* 652–653, 2003.
223. Antonietti, M., Berton, B., Göltner, C., and Hentz, H.P. *Adv Mater* 10: 154–159, 1998.
224. Han, S. and Hyeon, T. *Chem Commun* 1955–1956, 1999.
225. Han, S., Sohn, K., and Hyeon, T. *Chem Mater* 12: 3337–3341, 2000.
226. Han, S., Kim, S., Lim, H., Choi, W., Park, H., Yoon, J., and Hyeon, T. *Micropor Mesopor Mater* 58: 131–135, 2003.
227. Han, S., Lee, K.T., Oh, S.M., and Hyeon, T. *Carbon* 41: 1049–1056, 2003.
228. Jang, J., Lim, B., and Choi, M. *Chem Commun* 4214–4216, 2005.
229. Li, Z. and Jaroniec, M. *Carbon* 39: 2077–2088, 2001.
230. Li, Z., Jaroniec, M., Lee, Y.J., and Radovic, L.R. *Chem Commun* 1346–1347, 2002.
231. Lukens, W.W. and Stucky, G.D. *Chem Mater* 14: 1665–1670, 2002.
232. Su, F. and Zhao, X.S. Unpublished work, 2006b.
233. Pierre, A.C. *Introduction to Sol-Gel Processing*. Boston, Kluwer, 1998.

234. Kawashima, D., Aihara, T., Kobayashi, Y., Kyotani, T., and Tomita, A. *Chem Mater* 14: 3397–3401, 2000.
235. Han, S., Kim, M., and Hyeon, T. *Carbon* 41: 1525–1532, 2003.
236. Pang, J., Li, X., Wang, D., Wu, Z., John, V.T., Yang, Z., and Lu, Y. *Adv Mater* 16: 884–886, 2004.
237. Pang, J., Hu, Q., Wu, Z., Hampsey, J.E., He, J., and Lu, Y. *Micropor Mesopor Mater* 74: 73–78, 2004.
238. Lee, J., Kim, J., Lee, Y., Yoon, S., Oh, S.M., and Hyeon, T. *Chem Mater* 16: 3323–3330, 2004.
239. Moriguchi, I., Koga, Y., Matsukura, R., Teraoka, Y., and Kodama, M. *Chem Commun* 1844–1845, 2002.
240. Han, B.-H., Zhu, W., and Sayari, A. *J Am Chem Soc* 125: 3444–3445, 2003.
241. Kim, J., Lee, J., and Hyeon, T. *Carbon* 42: 2711–2719, 2004.
242. Su, F. and Zhao, X.S. Unpublished work, 2006c.
243. Rouquerol, F., Rouquerol, J., and Sing, K. *Adsorption by Powders and Porous Solids*. San Diego, CA: Academic Press, pp. 288–311, 1999.
244. Iler, R.K. *The Chemistry of Silica: Solubility, Polymerization, Colloid and Surface Properties, and Biochemistry*. New York: John Wiley and Sons, 1979.
245. Attia, Y.A. *Sol-Gel Processing and Applications*. New York: Plenum Press, 1994, pp. 237–256.
246. Brinker, C.J. and Scherer, G.W. *Sol-Gel Science: The Physics and Chemistry of Sol-Gel Processing*. Boston, MA: Academic Press, 1990.
247. Gundiah, G., Govindaraj, A., and Rao, C.N.R. *Mater Res Bull* 36: 1751–1757, 2001.
248. Chai, G.S., Yoon, S.B., Yu, J.-S., Choi, J.-H., and Sung, Y.-E. *J Phys Chem B* 108: 7074–7079, 2004.
249. Yu, J.S., Yoon, S.B., and Chai, G.S. *Carbon* 39: 1442–1446, 2001.
250. Lei, Z., Zhang, Y., Wang, H., Ke, Y., Li, J., Li, F., and Xing, J. *J Mater Chem* 11: 1975–1977, 2001.
251. Yu, J.S., Kang, S., Yoon, S.B., and Chai, G. *J Am Chem Soc* 124: 9382–9383, 2002.
252. Perpall, M.W., Perera, K.P.U., DiMaio, J., Ballato, J., Foulger, S.H., and Smith, D.W., Jr. *Langmuir* 19: 7153–7156, 2003.
253. Kang, S., Yu, J.S., Kruk, M., and Jaroniec, M. *Chem Commun* 1670–1671, 2002.
254. Chai, G.S., Shin, I.S., and Yu, J.-S. *Adv Mater* 16: 2057–2061, 2004.
255. Kim, P., Joo, J.B., Kim, W., Kang, S.K., Song, I.K., and Yi, J. *Carbon* 44: 381–392, 2006.
256. Su, F., Zhao, X.S., Wang, Y., Zeng, J., Zhou, Z., and Lee, J.Y. *J Phys Chem B* 109: 20200–20206, 2005.
257. Shim, S.-E., Cha, Y.-J., Byun, J.-M., and Choe, S. *J Appl Polym Sci* 71: 2259–2269, 1999.
258. Zhou, Z. and Zhao, X.S. *Langmuir* 20: 1524–1526, 2004.
259. Yoon, S.B., Chai, G.S., Kang, S.K., Yu, J.-S., Gierszal, K.P., and Jaroniec, M. *J Am Chem Soc* 127: 4188–4189, 2005.
260. Wang, Z., Ergang, N.S., Al-Daous, M.A., and Stein, A. *Chem Mater* 17: 6805–6813, 2005.
261. Baumann, T.F. and Satcher, J.H., Jr. *Chem Mater* 15: 3745–3747, 2003.
262. Kim, P., Joo, J.B., Kim, W., Kim, H., Song, I.K., and Yi, J. *Carbon* 43: 2409–2412, 2005.
263. Collins, P.G., Zettl, A., Bando, H., Thess, A., and Smalley, R.E. *Science* 278: 100–102, 1997.
264. Mochida, I., Ku, C.H., and Korai, Y. *Carbon* 39: 399–410, 2001.

265. Forgács, E. and Cserháti, T. *J Pharm Biomed Anal* 18: 15–20, 1998.
266. Jang, L. and Lim, B. *Adv Mater* 14: 1390–1393, 2002.
267. Liu, J., Shao, M., Tang, Q., Chen, X., Liu, Z., and Qian, Y. *Carbon* 41: 1682–1685, 2003.
268. Niwase, K., Homae, T., Nakamura, K.G., and Kondo, K. *Chem Phys Lett* 362: 47–50, 2002.
269. Jang, J., Li, X.L., and Oh, J.H. *Chem Commun* 794–795, 2004.
270. Kim, M., Yoon, S.B., Sohn, K., Kim, J.Y., Shin, C.H., Hyeon, T., and Yu, J.S. *Micropor Mesopor Mater* 63: 1–9, 2003.
271. Stöber, W., Fink, A., and Bohn, E. *J Colloid Interface Sci* 26: 62–69, 1968.
272. Tamai, H., Sumi, T., and Yasuda, H. *J Colloid Interface Sci* 8: 454–462, 1996.
273. Kim, J.Y., Yoon, S.B., and Yu, J.S. *Chem Commun* 790–791, 2003.
274. Yoon, S.B., Sohn, K., Kim, J.Y., Shin, C.H., Yu, J.S., and Hyeon, T. *Adv Mater* 14: 19–21, 2002.
275. Kim, M., Sohn, K., Bin, H.B., and Hyeon, T. *Nano Lett* 2: 1383–1387, 2002.
276. Oda, Y., Fukuyama, K., Nishikawa, K., Namba, S., Yoshitake, H., and Tatsumi, T. *Chem Mater*, 16: 3860–3866, 2004.
277. Wang, Y., Su, F., Lee, J.Y., and Zhao, X.S. *Chem Mater* 18: 1347–1353, 2006.
278. Su, F., Zhao, X.S., Wang, Y., Wang, L., and Lee, J.Y. *J Mater Chem 16: 4413–4419,* 2006.
279. Jiang, P., Betone, J.F., and Colvin, V.L. *Science* 291: 453–457, 2001.
280. Zhou, Z., Yan, Q., Su, F., and Zhao, X.S. *J Mater Chem* 15: 2569–2574, 2005.
281. Zhao, X.S., Zhou, Z., Chong, A.S.M., Bao, X.Y., Gu, C., Su, F., Lv, L., Guo, W., and Yan, Q. *J Mater Sci Technol* 21: 20–24, 2005.
282. Tanaka, S., Nishiyama, N., Egashira, Y., and Ueyama, K. *Chem Commun* 2125–2127, 2005.
283. Zhang, F., Meng, Y., Gu, D., Yan, Y., Yu, C., Tu, B., and Zhao, D. *J Am Chem Soc* 127: 13508–13509, 2005.
284. Meng, Y., Gu, D., Zhang, F., Shi, Y., Yang, H., Li, Z., Yu, C., Tu, B., and Zhao, D. *Angew Chem Int Ed* 44: 7053–7059, 2005.
285. Liang, C., Hong, K., Guiochon, G.A., Mays, J.W., and Dai, S. *Angew Chem Int Ed.* 43: 5785–5789, 2004.
286. Liang, C. and Dai, S. *J Am Chem Soc* 128: 5316–5317, 2006.
287. Wang, H., Lam, F.L.Y., Hu, X., and Ng, K.M. *Langmuir* 22: 4583–4588, 2006.
288. Ohkubo, T., Miyawaki, J., Kaneko, K., Ryoo, R., and Seaton, N.A. *J Phys Chem B* 106: 6523–6528, 2002.
289. Liu, X., Zhou, L., Li, J., Sun, Y., Su, W., and Zhou, Y. *Carbon* 44: 1386–1392, 2006.
290. Pang, J., Hampsey, J.E., Wu, Z., Hu, Q., and Lu, Y. *Appl Phys Lett* 85: 4887–4889, 2004.
291. Gadiou, R., Saadallah, S.-E., Piquero, T., David, P., Parmentier, J., and Vix-Guterl, C. *Micropor Mesopor Mater* 79: 121–128, 2005.
292. Fang, B., Zhou, H., and Honma, I. *J Phys Chem B* 110: 4875–4880, 2006.
293. Vix-Guterl, C., Frackowiak, E., Jurewicz, K., Friebe, M., Parmentier, J., and Béguin, F. *Carbon* 43: 1293–1302, 2005.
294. Ahn, W.S., Min, K.I., Chung, Y.M., Rhee, H.K., Joo, S.H., and Ryoo, R. *Stud Surf Sci Catal* 135: 313–320, 2001.
295. Lapkin, A., Bozkaya, B., Mays, T., Borello, L., Edler, K., and Crittenden, B. *Catal Today* 81: 611–621, 2003.
296. Hu, Q., Pang, J., Jiang, N., Hampsey, J.E., and Lu, Y. *Micropor Mesopor Mater* 81: 149–154, 2005.

297. Banks, C.E., Davies, T.J., Wildgoose, G.G., and Compton, R.G. *Chem Commun* 829–841, 2005.
298. Biniak, S., Swiatkowski, A., and Pakula, M. In L.R. Radovic, Ed. *Chemistry and Physics of Carbon*, Vol. 27, New York: Marcel Dekker, 2001, pp. 125–225.
299. Aricò, A.S., Bruce, P., Scrosati, B., Tarascon, J.-M., and Schalkwijk, W.V. *Nat Mater* 4: 366–377, 2005.
300. Tirado, J.L. *Mater Sci Eng* R 40: 103–136, 2003.
301. Flandrois, S. and Simon, B. *Carbon* 37: 165–180, 1999.
302. Zhou, H., Zhu, S., Hibino, M., Honma, I., and Ichihara, M. *Adv Mater* 15: 2107–2111, 2003.
303. Wang, T., Liu, X., Zhao, D., and Jiang, Z. *Chem Phys Lett* 389: 327–331, 2004.
304. Fan, J., Wang, T., Yu, C., Tu, B., Jiang, Z., and Zhao, D. *Adv Mater* 16: 1432–1436, 2004.
305. Grigoriants, I., Sominski, L., Li, H., Ifargan, I., Aurbach, D., and Gedanken, A. *Chem Commun* 921–923, 2005.
306. Lee, K.T., Lytle, J.C., Ergang, N.S., Oh, S.M., and Stein, A. *Adv Funct Mater* 15: 547–556, 2005.
307. Lee, K.T., Jung, Y.S., and Oh, S.M. *J Am Chem Soc* 125: 5652–5653, 2003.
308. Frackowiak, E. and Béguin, F. *Carbon* 39: 937–950, 2001.
309. Wu, N.L. and Wang, S.W. *J Power Sources* 110: 233–236, 2002.
310. Wei, Y.-Z., Fang, B., Iwasa, S., and Kumagai, M. *J Power Sources* 141: 386–391, 2005.
311. Lozano-Castello, D., Cazorla-Amoros, D., Linares-Solano, A., Shiraishi, S., Kurihara, H., and Oya, A. *Carbon* 41: 1765–1775, 2003.
312. Lin, Y.R. and Teng, H.A. *Carbon* 41: 2865–2871, 2003.
313. Biniak, S., Dzielendziak, B., and Siedlewski, J. *Carbon* 33: 1255–63, 1995.
314. Hsieh, C.-T. and Teng, H. *Carbon* 40: 667–674, 2002.
315. Yoon, S., Lee, J., Hyeon, T., and Oh, S.M. *J Electrochem Soc* 147: 2507–2512, 2000.
316. Jurewicz, K., Vix-Guterl, C., Frackowiak, E., Saadallah, S., Reda, M., Parmentier, J., Patarin, J., and Béguin, F. *J Phys Chem Solids* 65: 287–293, 2004.
317. Vix-Guterl, C., Saadallah, S., Jurewicz, K., Frackowiak, E., Reda, M., Parmentier, J., Patarin, J., and Beguin, F. *Mater Sci Eng B* 108: 148–155, 2004.
318. Xing, W., Qiao, S.Z., Ding, R.G., Li, R., Lu, G.Q., Yan, Z.F., and Cheng, H.M. *Carbon* 44: 216–224, 2006.
319. Zhou, H., Zhu, S., Hibino, M., and Honma, I. *J Power Sources* 122: 219–223, 2003.
320. Wang, D.-W., Li, F., Fang, H.-T., Liu, M., Lu, G.-Q., and Cheng, H.-Mi. *J Phys Chem B* 110: 8570–8575, 2006.
321. Moriguchi, I., Nakahara, F., Furukawa, H., Yamada, H., and Kudo, T. *Electrochem Solid-State Lett* 7(8): A221–A223, 2004.
322. Momma, T., Liu, X., Osaka, T., Ushio, Y., and Sawada, Y. *J Power Sources* 60: 249–253, 1996.
323. Ingram, M.D., Pappin, A.J., Delalande, F., Poupard, D., and Terzulli, G. *Electrochim Acta* 43: 1601–1605, 1998.
324. Miller, J.M. and Dunn, B. *Langmuir* 15: 799–806, 1999.
325. Dong, X., Shen, W., Gu, J., Xiong, L., Zhu, Y., Li, H., and Shi, J. *Micropor Mesopor Mater* 91: 120–127, 2006.
326. Dong, X., Shen, W., Gu, J., Xiong, L., Zhu, Y., Li, H., and Shi, J. *J Phys Chem B* 110: 6015–6019, 2006.
327. Chan, K.-Y., Ding, J., Ren, J., Cheng, S., and Tsang, K.Y. *J Mater Chem* 14: 505–516, 2004.
328. Che, G., Lakshmi, B.B., Fisher, E.R., and Martin, C.R. *Nature* 393: 346–349, 1998.

329. Anderson, M.L., Stroud, R.M., and Rolison, D.R. *Nano Lett* 2: 235–240, 2002.
330. Rolison, D.R. *Science* 299: 1698–1701, 2003.
331. Ye, S. and Vijh, A.K. *Electrochem Commun* 5: 272–275, 2003.
332. Choi, W.C., Kim, Y.J., Woo, S.I., and Hong, W.H. *Stud Surf Sci Catal* 145: 395–398, 2003.
333. Choi, W.C., Woo, S.I., Jeon, M.K., Sohn, J.M., Kim, M.R., and Jeon, H.J. *Adv Mater* 17: 446–451, 2005.
334. Ding, J., Chana, K.-Y., Rena, J., and Xiao, F. *Electrochim Acta* 50: 3131–3141, 2005.
335. Choi, M. and Ryoo, R. *Nat Mater* 2: 473–476, 2003.
336. Joo, J.B., Kim, P., Kim, W., Kim, J., and Yi, J. *Catal Today* 111: 171–175, 2006.
337. Zeng, J., Su, F., Lee, J.Y., Zhou, W., and Zhao, X.S. *Carbon* 44: 1713–1717, 2006.
338. Wang, L., Lin, K., Di, Y., Zhang, D., Li, C., Yang, Q., Yin, C., Sun, Z., Jiang, D., and Xiao, F.-S. *Micropor Mesopor Mater* 86: 81–88, 2005
339. Kim, J.Y., Yoon, S.B., and Yu, J.S. *Chem Mater* 15: 1932–1934, 2003.
340. Yang, Z.X., Xia, X.D., and Mokaya, R. *Adv Mater* 16: 727–732, 2004.
341. Sakthivel, A., Huang, S.-J., Chen, W.-H., Lan, Z.-H., Chen, K.-H., Kim, T.-W., Ryoo, R., Chiang, A.S.T., and Liu, S.-B. *Chem Mater* 16: 3168–3175, 2004.
342. Vinu, A., Terrones, M., Golberg, D., Hishita, S., Ariga, K., and Mori, T. *Chem Mater* 17: 5887–5890, 2005.
343. Yu, J.-S., Yoon, S.B., Lee, Y.J., and Yoon, K.B. *J Phys Chem B* 109: 7040–7045, 2005.
344. Kim, J.C., Kim, Y.N., Chi, E.O., Hur, N.H., Yoon, S.B., and Yu, J.S. *J Mater Res* 18: 780–783, 2003.
345. Roggenbuck, J. and Tiemann, M. *J Am Chem Soc* 127: 1096–1097, 2005.
346. Xia, Y. and Mokaya, R. *J Mater Chem* 15: 3126–3131, 2005.
347. Dibandjo, P., Bois, L., Chassagneux, F., Cornu, D., Letoffe, J.-M., Toury, B., Babonneau, F., and Miele, P. *Adv Mater* 17: 571–574, 2005.
348. Dibandjo, P., Chassagneux, F., Bois, L., Sigala, C., and Miele, P. *J Mater Chem* 15: 1917–1923, 2005.
349. Gavalas, V.G., Chaniotakis, N.A., and Gibson, T. *Biosens Bioelectron* 13: 1205–1211, 1998.
350. Gavalas, V.G. and Chaniotakis, N.A. *Anal Chim Acta* 409: 131–135, 2000.
351. Gavalas, V.G. and Chaniotakis, N.A. *Mikrochim Acta* 136: 211–215, 2001.
352. Gavalas, V.G. and Chaniotakis, N.A. *Anal Chim Acta* 404: 67–73, 2000.
353. Sotiropoulou, S., Gavalas, V., Vamvakaki, V., and Chaniotakis, N.A. *Biosens Bioelectron* 18: 211–215, 2003.
354. Vinu, A., Streb, C., Murugesan, V., and Martin, H. *J Phys Chem* B 107: 8297–8299, 2003.
355. Hartmann, M., Vinu, A., and Chandrasekar, G. *Chem Mater* 17: 829–833, 2005.
356. Vinu, A., Miyahara, M., and Ariga, K. *J Phys Chem B* 109: 6436–6441, 2005.
357. Vinu, A., Hossain, K.Z., Kumar, G.S., and Ariga, K. *Carbon* 44: 530–536, 2006.
358. Lee, D., Lee, J., Kim, J., Kim, J., Na, H.B., Kim, B., Shin, C.-H., Kwak, J.H., Dohnalkova, A., Grate, J.W., Hyeon, T., and Kim, H.-S. *Adv Mater* 17: 2828–2833, 2005.
359. Botella, J., Ghezzi, P.M., and Sanz-Moreno, C. *Kidney Int* 56, Suppl. 76: S60–S65, 2000.

3 Characterization of Carbon Surface Chemistry

P. Burg and D. Cagniant

CONTENTS

I. INTRODUCTION: CARBONACEOUS MATERIALS

The term *carbonaceous materials* covers a large set of products with different degrees of functionalization. Their applications are diverse and, for many of them, are based on a good knowledge of their surface properties. These can be described in different ways, according to miscellaneous concepts that will be detailed in this chapter. Investigations of the surface of such materials require physicochemical, thermodynamic, and spectroscopic approaches with the aim of improving understanding of their behavior. The importance of the surface chemistry of carbonaceous materials, besides their surface area and porosity, to explain many of their properties in numerous applications began to be widely recognized only in the 1980s [1]. In particular, the amphoteric character of carbons, with both acidic and basic centers coexisting on their surface, was pointed out [1,2]. The presence of various functional groups on the carbon surface depends on the nature of the base material and on the activation technique employed in the manufacturing process. Furthermore, special posttreatments can modify the surface chemistry, resulting in new useful properties.

Over the last few decades, many studies have been devoted to the characterization and evaluation of these centers by many analytical methods of varying degrees of sophistication. It is not excessive to assert that during the last 20 years an “avalanche” of analytical methods have been applied to characterize the surface chemistry of carbons, chiefly in the main application areas, as adsorbents, catalyst supports, and catalysts. Taking into account the structural complexity of carbonaceous materials, their heterogeneity, as well as the diversity of their application areas, one method alone is generally considered insufficient and the trend over the last decade has been to use several complementary methods.

Here, we review the main methods used for characterization of the surface chemistry of carbonaceous materials. We point out the diversity of methods and their ease or difficulty of use, linked to required laboratory equipment. According to the specific type of carbonaceous material and keeping in mind the desired field of its utilization, one is then faced with the problem of choice of the most appropriate analytical method as well as the complementary ones. This problem is illustrated by numerous examples. Our chief aim is to present the most useful methods available, pointing out the pros and cons of each as well as their complementarities. In doing so, we have also strived to provide an update on this important and rapidly evolving topic, which has been reviewed recently in this series [1,2].

Because the choice of analytical methods depends on the type of carbonaceous material studied, a brief overview of materials of interest is necessary as a first step. Thus, we present here some significant examples of typical functional groups encountered, naturally or after treatment, on the surface of representative carbonaceous materials. These functional groups are constituted by heteroelements (oxygen, nitrogen, sulfur, chlorine, and hydrogen) that can profoundly affect their sorption properties.

The importance of surface chemistry in different areas is described hereafter either in terms of applications or in terms of surface chemical modifications. A summary of different applications of carbonaceous materials and their chemical modifications is given in Table 3.1. For example, the sorption properties of active carbons are known to be mostly influenced by the presence of surface oxygen groups. Thus, many of the surface characterization methods described hereafter concern not only the raw materials but also their products of oxidation by various reagents.

A. Carbon Materials Used in Heterogeneous Catalysis

The role of carbonaceous materials in heterogeneous catalysis was recently reviewed by Radovic and Rodriguez-Reinoso [2], as well as by Rodriguez-Reinoso [3]. Although they have been used for a long time, it was only recently that research was directed at understanding all aspects of their physical and chemical characteristics. Future growth of their use will depend on better knowledge of the carbon surface chemistry, and not only of surface area [3] and porosity [3] as was thought to be the case not many years ago.

The list of carbon materials used to prepare carbon-supported catalysts is impressive: graphites, carbon blacks, active carbons and active carbon fibers, glassy carbons, pyrolytic carbons, fullerenes, and nanotubes. Among them, the high-surface-area active carbons and carbon blacks have often been the materials of choice for most catalyst supports, but this may soon change. Depending to some degree on their origin, a high level of functional groups can characterize the surface chemistry of any carbon material, as has been demonstrated recently for functionalized carbon nanotubes [4].

Besides their role as supports, there are several industrial reactions using carbon molecular sieves, carbon blacks, or active carbons as catalysts (e.g., in hydrogenation, oxidation, reduction, chlorination, and polymerization). In these applications too, although they are not as well developed as for carbon-supported catalysts, the key to their future growth is linked to knowledge of their surface chemical properties. The oxygen surface groups have attracted much attention recently owing to their role in catalyst preparation [1,3] and not only in carbon gasification [5–8]. Besides direct reaction with oxygen (e.g., oxygen chemisorption on active sites), their origin may be either the raw material itself, the activation process, or the posttreatment (e.g., reaction with oxidizing gases such as ozone, nitrous oxide, or carbon dioxide and with oxidizing solutions such as nitric acid, sodium hypochlorite, ammonium peroxydisulfate, or hydrogen peroxide).

The changes in surface chemistry of active carbons, mainly after treatment with HNO_3 [9–12] and $(NH_4)_2S_2O_8$ [11–13], were followed recently by several methods described later in this chapter: temperature-programmed desorption (TPD), Fourier transform infrared spectroscopy (FTIR), Boehm titration, and point of zero charge (pH_{PZC}) measurements. The impact on porosity of severe oxidation with HNO_3 [10,12] demonstrated its erosive effect on the pore walls. Indeed, when carried out under severe acidic conditions, oxidation leads to the fixation of a large amount of surface oxygen functionalities with simultaneous partial destruction of the porous structure. In a comparative study of the oxidation of

AU: Table title is missing.

TABLE 3.1
Treatments and Characterization Methods of Some Carbon Materials

Samples	Treatments[a]	Methods[b]
Nonactivated Carbons		
Coals	Py	XPS [124,128–130,133,134,136–138,140], XANES [135,137–139], ^{13}C NMR [140,152], ^{15}N NMR [152,153], FT/IR(ATR) [106]
	DM–Ox (1)	XPS [143]
	R	XPS [136]
	N (2)	XPS [37], FT/IR [37]
Lignites	Py	XPS [127,138,140], XANES [138], ^{13}C NMR [140]
	Chemical	^{13}C NMR [151]
	N (1)	XPS [38]
	N (2)	XPS [37], FT/IR [37]
Synthetic chars		XPS [28,31,125,127,132,134], XANES [126,130,135,139]
Model compounds		
Sulfur derivatives		XPS [134], XANES [135,139]
Nitrogen derivatives		XANES [125,126]
Cellulose	N (1)	XPS [39], FT/IR [39]
Green River oil shale		XPS [154], ^{15}N NMR [154]
Carbon blacks	Oxidation	Boehm [48], PZC [66], electrophoresis [63], mass titration [63]
Graphites	$(NH_4)_2S_2O_8$ [40,48]	ASA [40–43,79,83], IGC [40–43,97], TPD [63], calorimetry [78]

Furnace carbon blacks	HNO_3 [48,63,97,142]	Flow microcalorimetry [89–94,97,99], microcalorimetry [66,97]
	Air [40], H_2O_2 [63]	XPS [41–43,48,97,99,142,146], FT/IR [7,48,99,142], micro ATR [119]
	Halogenation [40–43]	^{13}C NMR [40,142]
	Alkylation [40]	
Microporous carbons	Oxidation	Mass titration [45,64], TPD [45], calorimetry [78], enthalpy of neutralization [45], LTPD [75], Boehm [75]
	HNO_3 [45,64,76]	
	Air [75], H_2O_2 [63]	
Soot		IGC LSER [161]

Activated Carbons

Samples	**Treatments**[a]	**Methods**[b]
Granular & Fibrous	Oxidation	Boehm [9,13,47,48,51,52,53–55,72,73,75,96,100,104,108]
	HNO_3 [8,9,12,25,30,54,55,76,100,104,108]	Potentiometric titration [9,51,54–59,108]
	$(NH_4)_2S_2O_8$ [12,13,48,54,72–74,96,101,108]	PZC [13,66–72,104]
	H_2O_2 [25,54,55,74,102,108]	IEP [30,67,69,104]
	Air [48,69]	ASA [28,31,52,76]
	O_2 [25]	TPD [13,25,30,54,67,72–74,75,76,80,96,104]
	Nitrogenation	Calorimetry [78]
	NH_3 [9,29,104]	Flow microcalorimetry [52,89–96]
	HCN [29]	Microcalorimetry [67,68,78,104]
	Ammoxidation [30]	Enthalpy of immersion [72,75,78,100,101]
	Thermal treatments	Enthalpy of neutralization [72,75]
	HTT [8,9,66,67,75,76,86,100,118]	Gas adsorption calorimetry [78,103]
	Nitrogen enriched [111]	FT/IR [8,9,13,25,30,48,54,76,100,108,111,118,157]

[a] Pyrolysis (Py), demineralization (DM), oxidation (Ox) (1-peroxyacetic acid), reductive and nonreductive alkylation (R), chemical ($SiCl_4$-NaI), nitrogenation (N)(1 ammoxidation, 2-urea).

[b] Attenuated total reflectance (ATR).

active carbons (from olive stones) with HNO_3 and $(NH_4)_2S_2O_8$, Moreno-Castilla and coworkers [11,13] showed that the highest catalytic activity of the oxidized samples in dehydration of methanol was obtained after peroxydisulfate treatment, in spite of the fact that this treatment fixed the smallest number of both oxygen and surface acidic groups. This could be due to the fixation of carboxyl groups close to other groups that by both negative inductive and resonance effects enhanced their acidity.

Several types of oxygen functional groups—such as carboxylic acid and anhydride, phenol, lactone, quinone [1]—are polar and make carbon more hydrophilic and acidic. Their presence can be exploited in many aspects of preparation, dispersion, and activity of carbon-supported catalysts. This topic was exhaustively discussed [2,3], and it was concluded that there has been an avalanche of studies devoted particularly to noble-metal catalysts. Thus, for example, in the study of Prado-Burguete et al. [14] it was shown that oxygen surface groups play a major role in the final metal dispersion (e.g., Pt/C catalysts): the more acidic ones, those giving rise to CO_2 by TPD (see Section II.D) and introduced by treatment with H_2O_2, decrease the hydrophobicity of carbon, making the surface more accessible to aqueous solution of metal precursor during impregnation; the less acidic ones (those giving rise to CO by TPD) somehow increase the interaction between the metal and the carbon support, thus decreasing the sintering propensity of the catalyst. The presence of different amounts and types of oxygen surface groups gives rise to both negatively and positively charged surface sites in aqueous solution depending on its pH: at the isoelectric point (IEP) (see Section II.C) the net overall surface charge will be zero, at $pH > pH_{IEP}$ the carbon surface will attract cations from solution, and at $pH < pH_{IEP}$ it will attract anions. This point affects the chemical surface accessibility for the catalyst precursor. In this field, Radovic and coworkers have carried out extensive work to understand the amphoteric character of the carbon surface and to optimize the preparation of some catalysts [2,3].

B. Carbon Materials Used as Adsorbents

This topic is of interest in two main applications: stationary phases in chromatography and materials for removal or preconcentration of organic pollutants (e.g., volatile organic compounds) from air and water streams.

The chromatographic column applications of carbons were well reviewed by Leboda and coworkers [15,16], both in gas chromatography [15] (GC) and high-performance liquid chromatography [16] (HPLC).

Different types of carbons have been employed more or less successfully in gas chromatography. Active carbons have been used the longest, practically since GC was invented. Their surface character is determined by the kind and the amount of the various heteroatom-containing functionalities and may be very complex (i.e., phenols, carboxyls, carbonyls, ethers, quinones, and lactones). To adjust these properties for efficient and effective chromatographic separation, special attention has been paid to carbonization and activation processes and to various posttreatments (e.g., removal of mineral impurities, removal of specific oxygen surface groups, and modification of adsorptive properties by chemical reaction with nitrogen derivatives [16]).

Another group of carbon adsorbents used in GC is that of microporous carbons having properties similar to molecular sieves (CMSs). This group includes, for example, Saran-based carbons and, more generally, carbons formed by controlled pyrolysis of suitable polymeric materials or petroleum pitches. CMSs have been used for GC since the late 1960s; nevertheless, the introduction of new types of such adsorbents has continued to this date owing to their unique performance in the separation of polar compounds (water, formaldehyde, and methanol). For example [15,17], the literature during the past decades offers many more examples of preparation of carbon molecular sieves from various raw materials. One well-known example is graphitized carbon black (commercially known as Carbopack) that is obtained by high-temperature heat treatment, e.g., at 3000°C in vacuum.

Adsorption on the carbon black surface depends mainly on the geometry of the adsorbate molecules and their polarizability. Thus, graphitized carbon black is an ideal adsorbent for separation of isomers of similar physical properties but of different geometrical structures, because of its surface homogeneity [18]. Chemical modification of carbon black, on the other hand, enhances specific adsorption affinity by virtue of the presence of different functional groups on its surface [17,19].

The review by Leboda and coworkers [16] describes the requirements for carbon adsorbents for HPLC and the methods of their preparation. Preliminary attempts to use general-purpose carbons were basically unsuccessful until the use of porous graphitized carbon prepared using the template-replication method (Hypercarb commercial product [20–22]), resistant to both strongly acidic and basic solutions.

The use of carbons as sorbents for preconcentration of pollutants, an application that has much in common with the preceding one, was recently reviewed by Matisova and Skrabakova [23], who investigated the applicability of carbon sorbents (active carbon, graphitized carbon black, molecular sieves, and porous carbon) for preconcentration of organic pollutants in environmental samples and their analysis by GC and HPLC. Another example of an application of carbonaceous materials in which surface chemistry is involved is the reduction of pollutant emissions from aqueous and gaseous media [24].

In summary, it is possible to modify the surface chemistry of activated carbons by introducing oxygenated groups while essentially maintaining their original porous texture. Gas-phase oxidation increases mainly the concentration of hydroxyl and carbonyl surface groups, whereas oxidation in the liquid phase increases especially the concentration of carboxylic acids [25]. A thorough knowledge of surface chemistry enables the preparation of adsorbents with appropriate characteristics for specific applications. This point was illustrated in recent studies of the effect of surface chemistry on the removal of dyes from textile effluents [26,27]: for reactive and acid dyes, a close relationship between the surface basicity of the adsorbents and dye adsorption was shown, and for a basic dye the acidic oxygen-containing surface groups had a positive effect on adsorption.

The usefulness of nitrogen in increasing the sorption capacity toward anions [9] and the thermal stability of catalysts [28] are emphasized by the development of methods of preparation of nitrogen-enriched materials, mainly active carbons. This topic has received great attention since the 1990s, as briefly summarized in later text.

A first group of methods starts from already made active carbons subjected to a chemical posttreatment: NH_3 or HCN [29], amination or ammoxidation [30]. A second group starts from raw materials with a high content of nitrogen, such as vinyl pyridine–divinyl benzene copolymer [28,31].

The influence of ammonia surface treatment of a viscose-based activated carbon cloth on oxidative retention of H_2S and SO_2 at 25°C has been assessed by x-ray photoelectron spectroscopy (XPS), x-ray absorption near edge structure (XANES), and TPD [32]. Also, active carbon fibers with basic properties were obtained by pyrolysis of pyridine over the fibers at 725°C. The pyridine provided basic surface functionalities, and the catalytic SO_x removal was very much enhanced [33]. This result was confirmed by recent studies [34,35] demonstrating that surface nitrogen favors SO_2 retention because it enhances the ability of the material to effect oxidation of SO_2 into SO_3. Starting from a series of nitrogenated materials, using active carbons and active carbon fibers as precursors and various nitrogenated compounds, Raymundo-Piñero et al. [34] were able to demonstrate the role of different N-containing groups in SO_2 adsorption. Pyridinic functional groups, located at graphene edges, appear to contribute mostly to the enhancement of catalytic activity of porous carbon materials. Good H_2S removal capacity of coal-based activated carbons modified with nitrogen (by impregnation with melanine followed by heat treatment at 850°C or with melanine + urea and heat-treated at 650 to 850°C) was recently pointed out as well [35].

More recent studies were carried out with coals (lignite and subbituminous), enriched with nitrogen, before carbonization and activation, by treatment with ammonia and its derivatives, mainly urea [36,37]. The ammoxidation reaction (reaction with an ammonia–air mixture) is another route to obtaining nitrogen-enriched precursors of active carbons from various carbonaceous materials, such as pine wood, peat, lignite, subbituminous coal [38], and cellulose [39].

Finally, a series of chemical treatments, including particularly halogenation [40–43] (bromination [41], chlorination [42], and fluorination [43]), but also oxidation and alkylation [40], were investigated starting from carbon black samples of different origins. The variations of surface properties upon chemical modification were studied [40] by various techniques such as XPS (see Section V.B), active surface area determination (see Section III), and inverse gas chromatography (see Section VI.B). The nature of surface acid sites obtained by chlorination of an active carbon prepared from olive stones was studied by a battery of analytical methods [44]. The results—increase in Lewis acidity and decrease in Brönsted acidity—were explained by resonance effects introduced into the graphene layers by the chlorine atoms covalently bound to their edges.

To sum up this section, the carbonaceous materials studied in the papers reviewed here are presented in Table 3.1, which includes the origin of the material, its treatment (Ox = oxidation, N = nitrogenation, Hal = halogenation, HTT = heat treatment temperature), and the analytical methods used for the characterization of its surface chemistry.

II. CHARACTERIZATION OF SURFACE CHEMISTRY IN TERMS OF FUNCTIONAL GROUPS

Carbon surfaces can display both acidic and basic properties. This amphoteric character influences the behavior of carbonaceous materials to a great extent in most of their applications. This fact accounts for the impressive literature devoted to the analytical methods of characterization of acidic and basic surface groups developed for many years, especially in the last decade. The important recent studies will be reviewed here, which will emphasize the current trend of using a battery of complementary methods rather than only one of them. This is justified on the one hand by the complexity of carbonaceous raw materials and their products (obtained by thermal or chemical processes such as oxidation or functionalization) and on the other hand by the great variety of applications.

A. Carbon Acidity

There is general consensus that acidic character is related to the oxide structures that are part of the chemisorbed oxygen found on all carbon surfaces that have been exposed to air or other oxidizing agents [45]. The existence of surface functional groups such as carboxyls, phenols, lactones, and anhydrides is now well established as the source of surface acidity, as reviewed in several studies [46–48].

For example, confirmation of the influence of surface oxides on the adsorptive properties of activated carbons can be found in recent papers [49,50]. It was shown that acidic surface functional groups hinder the ability of active carbons to adsorb phenolic compounds by reducing the effectiveness of oxygen-containing basic groups to promote adsorption via oxidative coupling reactions. Thus, catalytic activity may be greatly enhanced by eliminating the acidic surface groups and promoting the formation of (e.g., pyrone-type) basic groups, for example, by outgassing at 900°C. In the same way, it has been increasingly realized since the mid-1970s that neither the support surface area nor its pore structure is sufficient to explain many of the properties of carbon-supported catalysts, and it was only in the late 1980s that the importance of carbon surface chemistry began to be analyzed in depth. The role of oxygen surface groups, especially the acidic ones, in the preparation (e.g., by ion exchange), dispersion, and activity of such catalysts was thus demonstrated [2]. As a consequence, the identification and quantification of the surface functional groups are of prime importance.

The well-known Boehm method [47,48], based on the titration of acidic centers using several basic solutions of increasing strengths (EtONa, NaOH, Na_2CO_3, and $NaHCO_3$) has become very popular (see Table 3.1) owing to its simplicity: its protocol is exhaustively repeated in the literature (see, for example, Reference

[51]). It should be understood that it gives only a semiquantitative measure of the surface functionalities and that quantitative agreement is often poor, which is the case when the amount of irreversibly adsorbed molecules is more than twice the amount of surface functional groups determined by the Boehm method, as for example reported by Lahaye [52]. This method is not practical when dealing with small samples, and it may not account for a large proportion (as high as 50%) of the total surface oxygen content [25]. It is not surprising that, in recent publications, the Boehm method is used as a complement to others (see Table 3.1). For example, when combined with the results of Boehm titration, adsorption of dyes (such as methylene blue and metanil yellow) also provides information about the surface distribution of acidic functional groups in the case of carbon fibers [53].

The acidic groups have different pK_a values depending on their location on the carbon surface relative to the location of nonacidic groups that can exert an inductive effect on them [46]. Then, by a potentiometric titration method it is assumed that the system under study consists of acidic sites characterized by their acidity constants K_a. The site population can thus be described by a continuous pK_a distribution function $f(pK_a)$ [54–57]. The experimental titration data are thus transformed into a proton-binding curve from which the distribution of acidity constants is obtained by using, for example, the splines-based numerical procedure SAIEUS suggested by Jagiello [58].

The groups with $pK_a < 8$ are classified as carboxyls, and those with $pK_a > 8$ are classified as phenols and quinones [54]. The results obtained are in agreement with those obtained from the Boehm titration. Such potentiometric titration was recently applied [59] to study the mechanisms involved during the activation of wood by phosphoric acid.

The Boehm and potentiometric titration methods provide useful information about surface species that behave as acids in aqueous solutions. Nevertheless, a significant number of other oxygenated functionalities—such as carbonyls, esters, and ethers—are not taken into consideration even though they can play a significant role in adsorption processes by virtue of their polar properties and hydrogen-bonding abilities. Other methods described below (e.g., TPD, XPS, FTIR, as well as linear solvation energy relationship approaches) are therefore useful complements.

B. Carbon Basicity

Although there is general agreement in attributing acidic groups to oxide complexes present on carbon surfaces, the types of sites responsible for basic properties of carbons are not so clear, and this has given rise to occasional debates for many years. Clearly, much more work is needed to resolve the many outstanding issues [60]. Several hypotheses were put forward [48,61] involving either oxygen-free Lewis basic sites or oxygen-containing surface groups such as γ-pyrone-like structures (in which two nonneighboring oxygen atoms, preferably located in two different rings of a graphene layer, presumably constitute one basic site). Pyrones were postulated to be formed by room-temperature air reexposure of

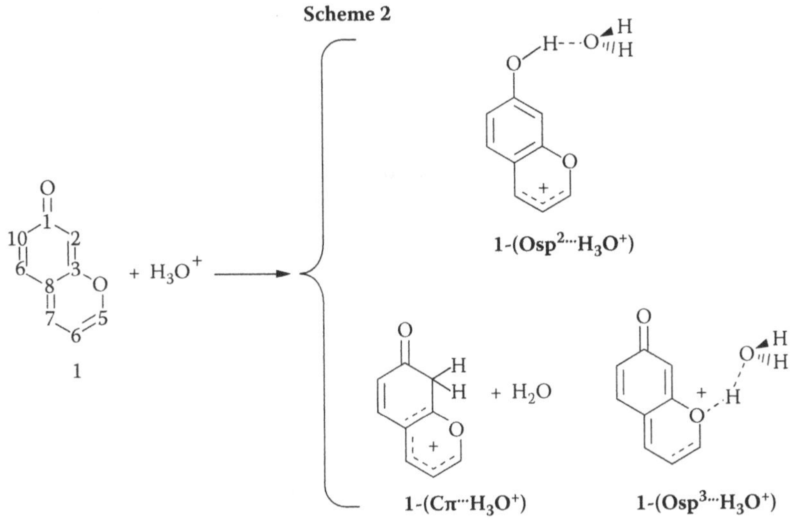

Scheme 2

1 + H_3O^+ →

$1\text{-}(Osp^{2} \cdots H_3O^+)$

$1\text{-}(C\pi \cdots H_3O^+)$ + H_2O

$1\text{-}(Osp^{3} \cdots H_3O^+)$

FIGURE 3.1 Examples of carbon basicity from ab initio computational chemistry study of pyrone-type structures by Suarez et al. (From Suarez, D., Menendez, J.A., Fuente, E., and Montes-Moran, M.A. *Langmuir* 15: 3897–3904, 1999. With permission.)

heat-treated carbons [61]. Such structures seem more likely than the chromene types [48].

More recently, the contribution to carbon basicity of several pyrone structures was examined by carrying out ab initio molecular simulations of various model clusters [62], as illustrated in Figure 3.1. It was argued that gas-phase interactions of such basic structures with H_3O^+ cations lead preferentially to protonation of the quinone group, with reaction energies of about –100 kcal/mol, which can be compared to quinoline (–76 kcal/mol) [62]. Such pyrone structures could be clearly considered basic centers according to the Brönsted definition. In the case of model clusters involving two nonneighboring oxygen atoms separated by two or three rings, a dramatic increase in reaction energy is predicted. If a sufficient number of surface pyrones are present, they could indeed be responsible for the overall carbon basicity.

Another hypothesis attributes the basic properties to oxygen-free basic sites according to the following equation:

$$C_\pi + H_3O^+ \rightarrow C_\pi\text{----}H_3O^+ \tag{3.1}$$

C_π is a basal plane site characterized by the presence of delocalized π electrons, which thus acts as a Lewis base center. Such sites are probably located in π-electron-rich regions within the basal planes of carbon crystallites (graphenes), away from the edges [63]. Evidence favoring their electron donor–acceptor (EDA) interactions, such as Reaction (3.1), was presented [63] by investigating a polymer-derived microporous carbon and a furnace carbon black, using HCl adsorption isotherms, stepped TPD, electrophoresis, and mass titration. The basicity of carbon surfaces was maximized by inert gas heat treatments in the range 1023 to 1123 K.

As in the case of pyrones, the contribution to carbon basicity of π-cation interactions in model aromatic systems was investigated by carrying out ab initio calculations [64]. The energies for benzene–H_3O^+, pyrene–benzene–H_3O^+, and coronene–H_3O^+ interactions were considered in an attempt to understand the basicity contributions of the basal plane. These theoretical results did support the experimental data concerning the ability of basal planes to contribute to carbon basicity and suggested that π–cation interactions may play an important role in the surface chemistry of carbon materials.

Considering these two hypotheses, Darmstadt and Roy [65] used XPS (see Section V.B) to study the basic sites on furnace carbon blacks with a low surface oxygen concentration and insignificant amounts of nitrogen. Under these conditions, it can be expected that oxygen-containing groups represent only a small proportion of the basic sites. A comparison of the concentration of basic sites determined by titration with HCl with the concentration of oxygen atoms determined by XPS showed that the concentration of basic sites is much higher than the maximum concentration of basic sites associated with oxygen-containing groups; most basic sites were thus not associated with oxygen. Furthermore, these authors obtained a very good correlation between the concentration of basic sites determined by titration and the full width at half maximum (FWHM) of the graphite peak determined by XPS: the highest concentration of basic sites was found for samples with the narrowest peaks. Therefore, at least for the carbon black sample studied, the basic sites are mainly associated with the basal planes of the graphene layers.

An application of these concepts is found in the pursuit of active carbons with stable basic properties by modification of their surface properties [66–68]. Generally, high-temperature heat treatment (950°C) under H_2 or N_2 is applied [67]. The nature and the number of active surface sites can be manipulated by adjusting these two parameters (temperature and atmosphere). Interestingly, highly basic carbons can also be produced by treating mixtures of carbon and platinum in hydrogen at 500°C or at even lower temperatures. Platinum produces atoms of hydrogen that spill over onto the carbon surface and stabilize it against oxidation [68].

C. Point of Zero Charge (PZC), Isoelectric Point (IEP), and Electrochemical Methods

The role of electrochemical surface properties in the use of carbons as catalyst supports and adsorbents was often pointed out in the literature [1,63,66–70]. The surface charge distribution of carbons was most often assessed by two methods, mass titration and electrophoresis.

Mass titration is presumed to vary in response to the net total (external and internal) surface charge of particles immersed in solution, as it measures the overall pH resulting from the maximum possible transfer of protons between the particle surface and the bulk solution [63]. According to the procedure described by Noh and Schwarz [70,71], weighed amounts of carbon are added sequentially to a given volume of aqueous 0.1 N NaCl until the pH of the solution does not change with further addition of carbon. This value (PZC) corresponds to the H^+ ion concentration after all acid groups present on the surface reach their dissociative–associative equilibria. At this limiting pH (PZC), it is supposed [45] that the hydrogen ion concentration in solution is high enough that surface ionization of even the strongest acidic species is largely suppressed.

The PZC values have been shown to change systematically with the extent of oxidation: for example, the more oxidized the carbon, the lower its PZC [12] value. Other ways to determine the PZC are the "pH drift method" used by Lopez-Ramon et al. [72], as well as the reverse mass titration, a modification of the Noh and Schwarz method [63,68].

By electrophoresis, one determines the isoelectric point (IEP), that is, the pH value below which the (external) surface of particles will be, on average, positively charged. The difference PZC – IEP can be taken as a measure of surface charge distribution in a porous carbon: a value of 0 should hold for homogeneously charged carbon surfaces [63,69]. These measurements were applied with success to a wide variety of carbons subjected to various pretreatments. For example, by selecting the oxidation conditions it is possible to prepare active carbons with different acidity levels and oxygen surface group distributions. In this vein, the influence of progressive surface oxidation (e.g., by hot air or nitric acid) on their physicochemical properties has been assessed by electrophoretic mobility measurements and potentiometric titration [73]. By choosing a suitable chemical treatment, the concentration of functional groups may be controlled, and it is possible to prepare carbons possessing "tailor-made" surface chemical properties.

D. Temperature-Programmed Desorption (TPD)

As a complement to acid-base titration, temperature-programmed desorption (TPD) has become a rather popular method over the last decade, in an attempt to identify the oxygenated groups either in raw materials or after their oxidation [11,25,54,73–77]. It has been applied mainly to active carbons of various origins (Table 3.1) as these materials are well known to have a surface chemistry with high contents of oxygenated species.

Surface oxygen groups decompose upon heating in inert gas by releasing CO or CO_2 at different temperatures depending on their stability. The TPD profiles are obtained in various types of reactors, at various flow rates of carrier gas, and at various heating rates. The amounts of CO and CO_2 desorbed are monitored typically by a mass spectrometer [13,25,72–74], by a gas chromatograph [75,76], by GC-MS [76], by nondispersive infrared analyzers [77], or by thermal conductivity detectors [54]. In this last case, the plots obtained combine all of the desorbed species, and they can be deconvoluted by using standard procedures, for example, an arbitrary choice of four peaks [54]: peaks 1 (570 K) and 2 (750 K) represented CO_2 desorption, and peaks 3 (970 K) and 4 (1150 K) represented CO desorption. Generally speaking, a CO_2 peak results from carboxylic acids at

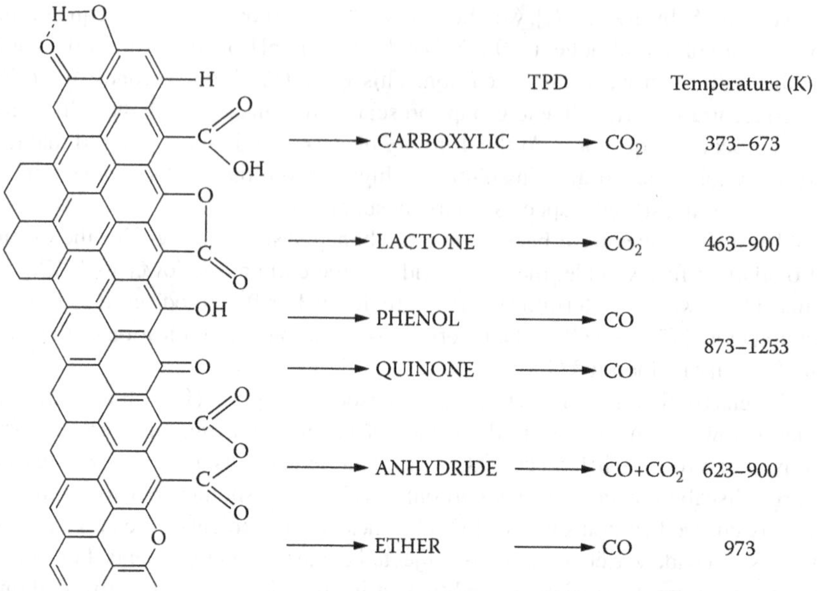

	TPD		Temperature (K)
	CARBOXYLIC	CO_2	373–673
	LACTONE	CO_2	463–900
	PHENOL	CO	873–1253
	QUINONE	CO	
	ANHYDRIDE	$CO+CO_2$	623–900
	ETHER	CO	973

FIGURE 3.2 Trends of decomposition temperatures by TPD of surface groups on carbons. (From Figueiredo, J.L., Pereira, M.F.R., Freitas, M.M.A., and Orfao, J.J.M. *Carbon* 37: 1379–1389, 1999. With permission.)

lower temperature or lactones at higher temperature. Carboxylic anhydrides produce both CO and CO_2 peaks. Phenols, ethers, and quinones produce a CO peak. As noted by Figueiredo et al. [25], the peak temperatures and intensities may be affected by spurious effects (e.g., heating rate and geometry of the experimental system, or porous texture of the material). Nevertheless, some general trends can be summarized, as shown in Figure 3.2. Thus, TPD profiles often show composite CO and CO_2 peaks that must be deconvoluted before the surface composition can be estimated [25]. These results are generally discussed in conjunction with DRIFTS, XPS (see Section V), and titration methods. For example, Dandekar et al. [8] combined the CO and CO_2 TPD evolution profiles with DRIFTS results, before and after pretreatments in H_2, to characterize an active carbon, graphitized carbon fibers, and diamond powder. An insight into the details of surface group stability was also obtained using intermittent TPD (ITPD), by determining the apparent desorption activation energy as a function of surface coverage [74].

E. Immersion Calorimetry

In addition to the aforementioned methods, the characterization of acidic and basic sites on carbon surfaces has been carried out by using immersion calorimetry [72,78]. As will be analyzed further in Section IV, calorimetry has not been employed widely for carbon characterization and particularly not for the study of acid/base site distribution [78]. Nevertheless, in recent papers, Stoeckli and coworkers [72,73] used the enthalpies of immersion of oxidized active carbon samples in benzene and water [73] and the corresponding enthalpy of immersion into 2 N NaOH [72] determined with a calorimeter of the Tian–Calvet type. They found that the net neutralization enthalpy, ΔH(NaOH)net = ΔH(NaOH)-ΔH(H_2O), was proportional to the amount of acid present on the surface. In the same way, the calorimetric technique was applied to a variety of activated carbons [72] in the case of NaOH, HCl, and $NaHCO_3$ neutralization. Good correlations were found between the net neutralization enthalpies by 2 N NaOH, 2 N HCl, and 1 N $NaHCO_3$ and, respectively, the number of milliequivalents obtained from direct titration with NaOH, HCl, and $NaHCO_3$ (Boehm method).

The surface chemistry of carbonaceous materials in terms of functional groups can thus be assessed by a large set of complementary methods.

III. SURFACE CHARACTERIZATION IN TERMS OF ACTIVE, POLAR, AND NONPOLAR SITES

It is generally assumed that a carbon surface is mainly composed of two types of planes: basal and edge. The edge planes include all types of defects that may be present in the structure (mainly potentially "unsaturated" sp2-hybridized carbon atoms bound to only two other carbon atoms) and thus constitute the so-called active sites or active surface area (ASA). During the last four decades, the concept of active sites has been very useful in the study of carbon reactivity [52,79–81] as well as to the gasification kinetics of coal chars [82–85].

The ASA is often determined on the basis of the method originally proposed by Laine et al. [83], involving oxygen chemisorption (at a temperature that represents a compromise between the adsorption rate and the absence of concomitant gasification) followed by desorption of the surface complexes as CO and CO_2. Indeed, oxygenated functionalities are formed on active sites in the conditions specified by the authors.

It is now understood [6,86,87] that the extent of oxygen chemisorption depends on experimental conditions and, consequently, the ASA depends also on the method used. For example, in the gravimetric method, the weight increase assigned to chemisorbed oxygen may overestimate the active site concentration owing to the contribution of physisorbed oxygen. In the TPD method, the formation of C–O complexes that are stable even at 1273 K (and are thus typically not included in the quantification of the CO and CO_2 desorbed) underestimates the ASA to some extent. A third method, the use of oxygen chemisorption isotherms (OCI) [86], based on the determinations of two isotherms, the first corresponding to chemical and physical and the second only to physical adsorption of oxygen (a technique commonly used in heterogeneous catalysis), avoids the preceding errors but gives very low values of ASA in comparison with the other two methods, which deserves further explanation. Figure 3.3 illustrates this issue by presenting a comparison of the ASA values obtained from the three methods (gravimetric, TPD, and OCI): The gravimetric method gives the highest ASA values, whereas the OCI method gives the lowest. However, the three series of ASA

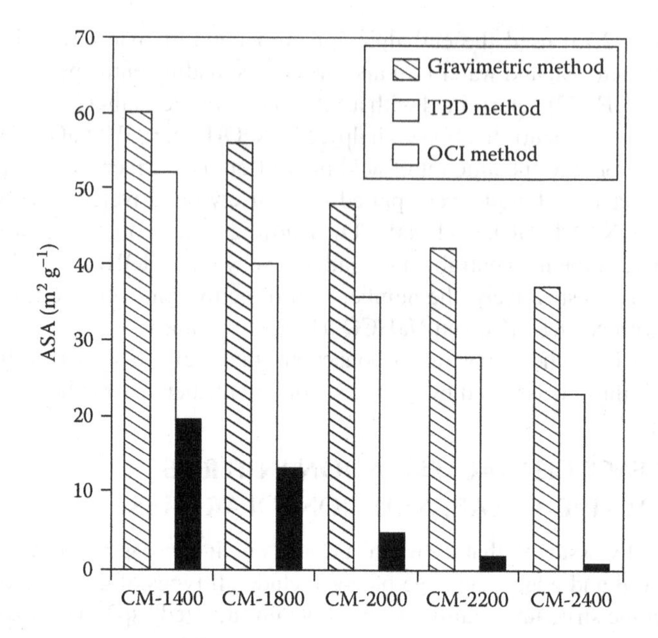

FIGURE 3.3 Comparison between the gravimetric, TPD, and oxygen chemisorption isotherms methods for evaluation of "surface active area" in carbonaceous materials. (From Arenillas, A., Rubiera, F., Barra, J.B., and Pils, J.J. *Carbon* 40: 1381–1383, 2002. With permission.)

values follow the same trend. Thus, the results are comparable if obtained by the same method and allow relative "rankings" of samples.

Other approaches acknowledge the existence of a distribution of site activities and go further in the titration of the surface that is reactive in a specific process, such as gasification. These studies suggest that the surface oxygen complexes formed should be divided at least into reactive and temporarily stable entities [5,6]. The reactive part can be titrated by the transient kinetics (TK) technique directly at reaction conditions [88].

The active sites computed from CO/CO_2 evolution (e.g., by the TPD method) correspond to a polar surface. Thus, it is interesting to compare these results with the polar and nonpolar areas as determined, for example, by the flow immersion calorimetry method developed by Groszek (its principle will be detailed in Section IV). Since the 1960s this author has been describing its use, mainly for graphites [89], but also for carbon blacks [90], activated carbons [91–93] and chars [94], as well as for coals [92] and cokes [95]. The method also identifies and quantifies the hydrophilic and hydrophobic sites and their relationship to porosity [95]. The agreement between ASA and polar surface determination obtained by this method in the case of a nonporous graphitized carbon black was excellent [90], in spite of the fact that the ASA method depends on oxygen chemisorption and the calorimetric method depends on physical adsorption of *n*-butanol. The same method was recently used to study the dynamic adsorption of ammonia on activated carbons

[96]. The authors pointed out the importance of the accessibility to the functional groups for the understanding of their behavior in adsorption and catalysis. The mere evaluation of the chemical functionalities (e.g., by using the Boehm method) and surface areas is not always sufficient to arrive at reliable conclusions.

IV. APPLICATION OF CALORIMETRIC METHODS TO CHARACTERIZATION OF SURFACE CHEMISTRY

Recently, Menéndez [78] reviewed the use of calorimetric techniques to assess the chemical properties of carbons (e.g., the nature of surface groups, hydrophobic/hydrophilic character, acidic/basic behavior). Here we summarize the key issues.

A. Immersion and Flow Adsorption Calorimetry

The measured parameter here is the adsorption heat released by a sample immersed either in a liquid or in a stream of a carrier liquid and a probe compound as a function of the amount adsorbed. These techniques have been found useful in the study of the hydrophobic/hydrophilic nature of carbons and the understanding of their surface acidity-basicity. For example, Groszek and coworkers [89–96] used flow immersion calorimetry to determine the heats of adsorption of *n*-butanol and *n*-dotriacontane (*n*-C32) solutions in *n*-heptane. The latter value was interpreted as the basal plane surface area of graphites and a graphitized carbon black according to the following relation 1 J/g = 19.3 m^2/g. The former was taken to be the polar surface area according to the relation 1 J/g = 6.7 m^2/g. Thus, for example, in case of "graphitic" sites located in micropores, such as in active carbons [92], adsorption of *n*-C32 is obviously not suitable and the titration of basal plane sites is based on the uptake of *n*-butanol from dilute aqueous solution. Another example can be given for a microporous active carbon [95]; the following results were obtained (m^2/g): total surface area (BET), 1000; hydrophilic sites, 53.6; hydrophobic sites, 1151 (with butanol/water); hydrophobic sites, only 83 (with *n*-C32/*n*-heptane).

Flow microcalorimetry (FMC) has been used in combination with thermogravimetric analysis (TGA) and XPS for the characterization of oxidized carbon blacks (HNO_3, peroxydisulfate) [97]. By using either the heat of adsorption of *n*-butanol from *n*-heptane or the heats of neutralization of acidic and basic surface sites, the surface polarity of the samples and the acidic and basic surface sites were characterized before and after oxidation [89,90,92,97,98]. As shown by XPS, for example [97], for two different powdered carbon blacks (CB1 and CB2) the strength of the oxidative conditions (i.e., in particular, the increase of the concentration of HNO_3) influences the quantity of surface oxygen groups by increasing the percentages of C=O (see Table 3.2, component O1, assigned to carboxylic groups) and C-O functionalities (see Table 3.2, component O2, assigned to ether/hydroxyl groups). A general trend is apparent: increasing oxidative strength for both carbon blacks leads to increasing heats of adsorption of *n*-butanol (see Figure 3.4) and of neutralization of acidic sites (see Figure 3.5).

TABLE 3.2
Compositions of functional groups on CB1 and CB2 samples derived from XPS analysis

Carbon Sample	Functional Group Type Percentage (wt.)			
	O1 (C=O)	O2 (C–O)	O3 (quinone)	Total Percentage
CB1 base	0.81	1.19		2.0
CB1 HNO_3 (69%)	6.61	9.09		15.7
CB1 H_2O_2 (30%)	2.99	4.31		7.3
CB1 $(NH_4)_2\ S_2O_8$ in H_2SO_4	3.13	5.26		8.39
CB2 base	—	1.10	—	1.1
CB2 HNO_3 (17.25%)	0.25	1.54	—	1.8
CB2 HNO_3 (69%)	0.33	2.00	0.07	2.4
CB2 H_2O_2 (30%)	1.45	1.45	—	2.9
CB2 $(NH_4)_2\ S_2O_8$ in H_2SO_4	1.38	4.12	—	5.5

Source: From Beck, N.V., Meech, S.E., Norman, P.R., and Pears, L.A. *Carbon* 40: 531–540, 2002. With permission.

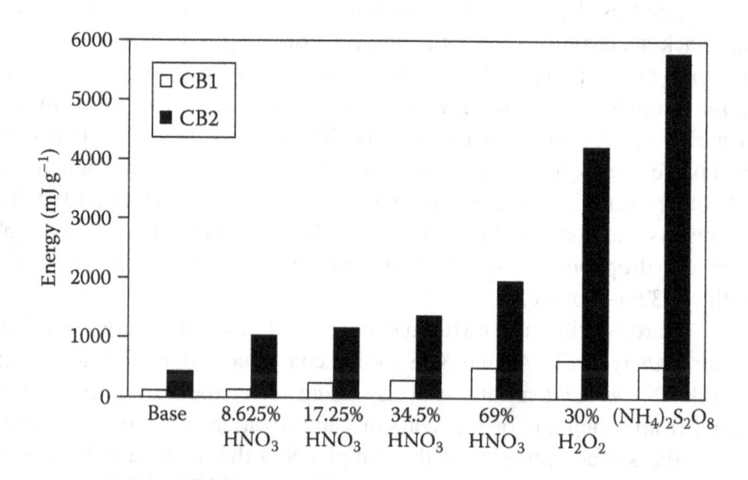

FIGURE 3.4 Heats of adsorption of *n*-butanol from *n*-heptane for base and oxidized carbon blacks. (From Beck, N.V., Meech, S.E., Norman, P.R., and Pears, L.A. *Carbon* 40: 531–540, 2002. With permission.)

More recently, the surface activity of different types of carbon black with adsorbed secondary antioxidants was examined using FMC [99]. This technique seems to have become an established method for characterizing filler surfaces in terms of their chemistry and their interactions upon surface treatment, combined with FTIR and XPS. It enables heats of adsorption and desorption to be measured

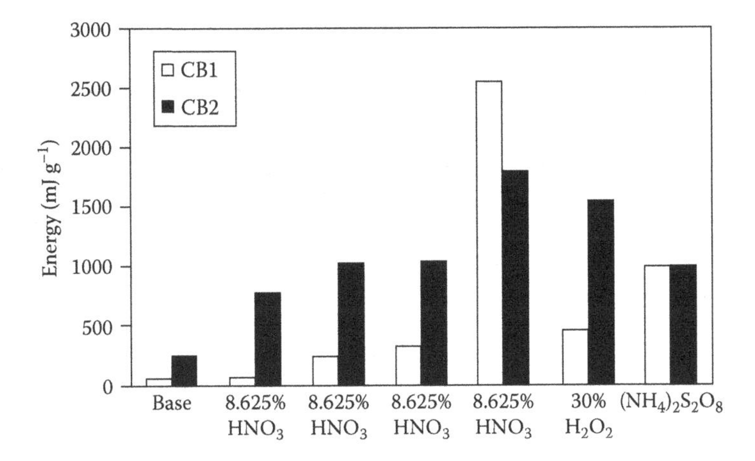

FIGURE 3.5 Acidic site heats of neutralization for base and oxidized carbon blacks. (From Beck, N.V., Meech, S.E., Norman, P.R., and Pears, L.A. *Carbon* 40: 531–540, 2002. With permission.)

as well as the amounts adsorbed and desorbed. Thus, for example, Peña et al. [99] have studied several formulations containing carbon black and various stabilizers. The quantity of stabilizer remaining on the carbon surface was determined by subtraction of desorption from adsorption values. The results were also compared to those obtained by other methods, that is, a gravimetric method and UV spectroscopy, as presented in Table 3.3, and allowed classification of samples in terms of adsorption power. It is obvious, as commented by the authors, that the results obtained by FMC and UV spectroscopy are convergent, contrary to those obtained by TGA, where thermal decomposition often occurs during analysis.

B. Immersion Calorimetry in Water

This technique has also been used for the determination of carbon surface polarity [100]. It was assumed that the water molecules interact only with oxygen surface complexes located on the polar sites at graphene edges. Starting from active carbons with different surface chemistry, the authors determined the enthalpies of immersion into water (298 K) with an isothermal calorimeter of the Tian–Calvet type, in conjunction with FTIR, XPS, and Boehm titration. The enthalpy increased linearly with the sum of acidic and basic sites.

A variety of active carbons were also oxidized to different degrees with $(NH_4)_2S_2O_8$, HNO_3, and H_2O_2, and these results [101] revealed a simple correlation (see Figure 3.6) between the enthalpy of immersion and (i) the oxygen content of the surface obtained by TPD; (ii) the number of basic groups characterized by their HCl equivalents; and (iii) the amount of water filling the micropores (N_{am}) and covering the external surface (S_e) with a monolayer (obtained from water adsorption isotherms). In case of a series of ozone-oxidized nonporous carbon blacks, the enthalpy of immersion (ΔH_i) was found to correlate directly with the

TABLE 3.3
Quantity of residual amount of stabilizers adsorbed on carbon black determined by UVA, TG, and FMC analyses

	Remaining Amount by UVA Analysis (% w/w of CB)	Remaining Amount by TG Analysis (% w/w of CB)	Remaining Amount by FMC Analysis (% w/w of CB)
C 944/CB-A	3.33	2.66	3.69
C 944/CB-B	2.38	3.92	4.08
C 944/CB-C	3.39	4.23	2.16
C 944/CB-D	4.26	4.7	4.97
U 299/CB-A	0.90	0.22	0.18
U 299/CB-B	0.41	1.77	1.07
U 299/CB-C	6.36	6.32	2.33
U 299/CB-D	8.59	1.93	1.80
ATMP/CB-A	0.09	0.21	0.18
ATMP/CB-B	–0.01	0.33	0.39
ATMP/CB-C	1.68	0.36	0.70
ATMP/CB-D	2.34	4.2	8.99
I P-EPQ/CB-A	3.12	0.63	0.57
I P-EPQ/CB-B	2.43	2.12	1.58
I P-EPQ/CB-C	3.36		3.38
I P-EPQ/CB-D	3.79	3.8	3.77
C 1790/CB-A	5.77	5.11	4.30
C 1790/CB-B	5.57	3.75	3.19
C 1790/CB-C	6.53	3.85	3.91
C 1790/CB-D	7.23	2.80	1.80

Source: From Peña, J.M., Allen, N.A., Edge, M., Liauw, C.M., and Valange, B. *Polym. Degrad. Stabil.* 74: 1–24, 2001. With permission.

total surface oxygen content [O_t] measured by XPS [102]. A correlation between the point of zero charge of the carbons and ΔH_i or [O_t] was also presented [102].

The use of immersion calorimetry for the determination of enthalpies of neutralization into aqueous solutions of NaOH, $NaHCO_3$, and HCl was previously cited as a useful tool for the characterization of acidic and basic surface sites [72].

C. Gas Adsorption Calorimetry

Here the heat generated by the adsorption of measured doses of a gas or vapor is determined, and it can be used to obtain information regarding the types and concentration of active sites on carbon surfaces. In addition, the energy of interaction

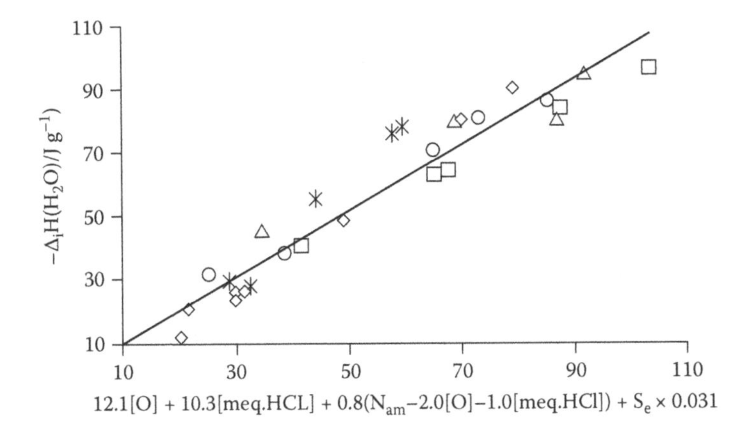

FIGURE 3.6 Correlation between the enthalpy of immersion of active carbons into water, the total surface oxygen, the basic groups titrated with HCl, the micropore filling, and the wetting of the nonporous surface area S_e. (From Lopez-Ramon, M.V., Stoeckli, F., Moreno-Castilla, C., and Carraco-Marin, F. *Carbon* 38: 825–829, 2000. With permission.)

of such sites with probe molecules (mainly O_2, but also NH_3, NO, SO_2, and CH_4) can be evaluated by this technique [78].

For example, microcalorimetry was used to study the surface chemistry of activated carbons subjected to several pretreatment procedures. By adsorption of oxygen in a laboratory-designed calorimeter on activated carbons pretreated at high temperature (950°C) in nitrogen or in hydrogen, it was shown [67,78,103] that the chemistry of the surface is very much dependent on the pretreatment procedure: upon treatment at 950°C in nitrogen the carbon adsorbed a great deal of oxygen at 25°C with a high heat of adsorption (125 kcal/g mol O_2), but after treatment at high temperature in hydrogen, the same carbon adsorbed virtually no oxygen at 25°C.

In the same way, calorimetric studies of ammonia adsorption on acid- and ammonia-pretreated active carbons [104] demonstrated that the technique is capable of titrating acid sites according to their relative strengths, and it was shown, as expected, that ammonia pretreatments create a surface whose chemistry is completely different from that of the original carbon. The authors [104] concluded that calorimetry can be considered a valuable addition to the current set of techniques normally employed for the study of acid carbon sites on carbon (TPD, Boehm titration, PZC and IEP, FTIR and XPS).

In summary, calorimetric methods can have important applications in the determination of active sites in relation with their polarity and their acid/base character, but the results need to be discussed in conjunction with other methods such as TPD, PZC, and IEP, as well as water vapor isotherm measurements [67]. Still, there is a perception that the greatest impediment to increased application of calorimetric techniques is the lack of instrumentation, which should be inexpensive and relatively easy to employ.

V. SPECTROSCOPIC CHARACTERIZATION OF CARBON SURFACE CHEMISTRY

Another way to characterize carbon surfaces is the nondestructive approach by spectroscopy. Some techniques, such as infrared or nuclear magnetic resonance spectroscopy, are well known from classical organic chemistry. The solid state and the complexity of carbon materials require new, delicate developments of classical spectroscopic methods or call for the use of new nontraditional ones (e.g., x-ray photoelectron spectroscopy).

Classical spectroscopic methods, and more particularly their new developments in the study of carbon surfaces, as well as new techniques, are described in this section.

A. Infrared Spectroscopy

Transmission infrared spectroscopy has been extensively applied since the 1950s to the characterization of functional groups in carbonaceous materials and to their changes after different treatments [105–107]. Many such studies have been conducted on various forms of carbon, including coals, chars, charcoals, carbon films, graphites, pitches, activated carbons, and soot, but with various degrees of success depending on the carbon system examined. For example, the major limitation in applying traditional infrared methods to coals is the overlap and superposition of the absorption bands of such complex carbonaceous materials. Furthermore, many carbons act as very effective black body absorbers, diminishing or completely attenuating beam energy throughput [12].

The literature is very abundant, and it is not within the scope of this paper to present an exhaustive review. The state of the applicability of FTIR to the characterization of coal structure, including quantitative results, was assessed by Painter et al. [105]. These authors also pointed out the controversies concerning band assignments, particularly in the range 1000 to 1350 cm^{-1} and near 1600 cm^{-1}.

Here we focus our attention on some new techniques applied during the last decade with the aim of overcoming these difficulties. They are mainly used to study the formation of surface groups on carbon by oxidation or other chemical treatments (e.g., introduction of nitrogenated functions).

Among them, diffuse reflectance FTIR spectroscopy (DRIFTS) has been very popular, as is the case in the field of heterogeneous catalysis in general. Changes in surface chemistry of activated carbons, oxidized with different agents (HNO_3, H_2O_2, and ammonium persulfate), were studied by DRIFTS in addition to Boehm titration, potentiometric titration, and water adsorption measurements [108]. The FTIR characterization of coals and their *N*-methyl-2-pyrrolidinone extraction products were described by application of Fourier self-deconvolution [109]; deconvoluted spectra were obtained using a standard software package and four spectral regions were selected for the identification of functional groups either in coals or their extracts: 3100 to 2800 cm^{-1} (νC-H arom, νC-H aliph), 1750 to 1600 cm^{-1} (νC = Carom, νC=O), 1260 to 1034 cm^{-1} (νC–O), and 900 to 650

cm^{-1} (γC-H arom). Under the same conditions, coal samples modified by chemical treatments (polyanions or the potassium coal adducts), for example, protonation, alkylation, oxidation, and butylation, were studied [110]. When the same technique is applied to materials such as carbon blacks or polymer-derived chars, it is extremely difficult to characterize them [7]. The same result was found for nitrogen-enriched active carbons with a high level of burn-off [111].

A successful and straightforward detection of functional groups by infrared spectroscopy is apparently dependent on incomplete carbonization of the source material, whereas complete carbonization leads to the disappearance of distinguishable spectral features [112]. Furthermore, oxidation of carbonized chars is claimed to produce no detectable IR bands [112]. Nevertheless, despite the experimental challenges, Fanning and Vannice [7] were able to use DRIFTS to detect and monitor surface functionalities on carbon surfaces that develop under oxidizing conditions. The use of CaF_2 as a diluent apparently yields a very low signal-to-noise ratio [7,8], increasing the performance of this type of analysis in terms of spectra resolution.

The band assignments have been made according to those of model compounds, and expectations for the types of functional groups present were based mainly on identification by chemical methods. These assignments are summarized in Table 3.4.

In situ FTIR spectroscopy (DRIFTS) has been used [113] for semiquantitative determination of surface group evolution during oxidation: the sample studied, a mixture of crushed coke with KBr (5/95%), was introduced in the sample cup of a Graseby–Specac environmental chamber, which permits sample exposure to an atmosphere controlled in pressure (vacuum to 500 psi) and temperature (up to 500°C) [114]. For example, the surface group evolution during oxidative stabilization of mesocarbon microbeads was studied under the following conditions [115]: the spectra were acquired with a high-sensitivity liquid-nitrogen-cooled MCT detector using a Graseby–Specac selector accessory with optical off-axis configuration for elimination of specular reflectance (sample to KBr mass ratio 5/9). The Kubelka–Munk function was applied to obtain semiquantitative data. During the oxidation step a spectrum is recorded every 2°C. Each spectrum results from the coaddition of 500 scans at a resolution of 2 cm^{-1} in the 4000- to 700-cm^{-1} range. The obtained results gave some evidence for kinetics and the mechanism of oxygen functional groups formation.

Infrared microspectroscopy has been applied to the study of macerals (organic fractions) from a wide range of coals [106,116] and oxidized coal [117]. It may be performed either in transmission or reflection [106]. In the former case, sample preparation is tedious, and transmission IR microspectroscopy is seldom used. The preparation of samples for reflection measurements is simpler; for example, the technique of attenuated total reflectance (ATR) has been applied to the study of coal [106]. Indeed, if the standard ATR accessories (which have been available for many years) were not suitable for block coal samples, recently, an ATR lens, equipped with a silicon (or germanium) internal reflection element, has become available for use with IR microscopes, and the technique was successfully applied

TABLE 3.4
Proposed principal functional groups on carbon surfaces and their corresponding infrared assignments

Group or functionality	Assignment regions (cm^{-1}) 1000–1500	1500–2050	2050–3700
C-O stretch of ethers	1000–1300		
Ether bridge between rings	1230–1250		
Cyclic ethers containing COCOC groups[a]	1025–1141		
Alcohols	1049–1276		3200–3640
Phenolic groups:			
C-OH stretch	1000–1220		
O-H bend/stretch	1160–1200		2500–3620
Carbonates: carboxyl-carbonates	1100–1500	1590–1600	
Aromatic C=C stretching		1585–1600	
Quinones		1550–1680	
Carboxylic acids (COOH)	1120–1200	1665–1760	2500–3300
Lactones	1160–1370	1675–1790	
Anhydrides	980–1300	1740–1880	
Ketenes (C=C=O)			2080–2200
C-H stretch			2600–3000

[a] For example,

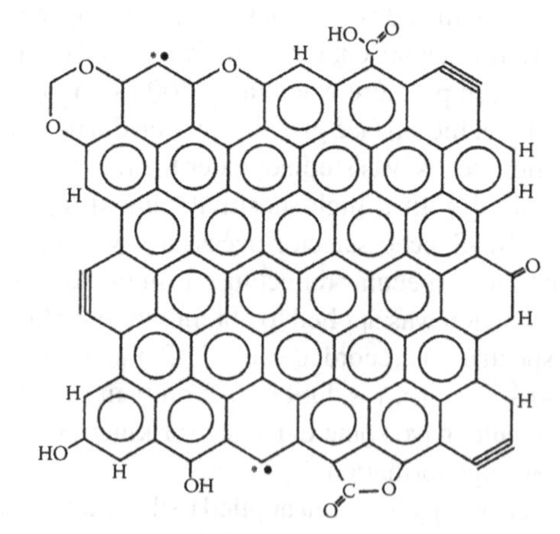

Source: From Fanning, P.E. and Vannice, M.A. *Carbon* 31: 721–731, 1993. With permission.

to seven coal samples [106]. The spectra are similar to absorption spectra, though band intensities are different because of the relationship between penetration depth and wavelength that exists in ATR spectra.

The micro-ATR technique was also applied to investigate surface functional groups of pitch-based active carbon fibers and their changes as a result of heat treatment [118]. To produce good optical contact between the prism (germanium in this case) and the material, a very thin KBr layer was introduced between the prism and the carbon fiber sample. The same technique has been shown [119] to also be a useful method for obtaining IR spectra of a wide range of carbon-black-filled rubbers. These materials are extremely difficult to characterize because of their very high absorptivity and their propensity to scatter IR light. Using a silicon internal reflection element, good-quality spectra were collected in a matter of minutes [119].

B. X-Ray Photoelectron Spectroscopy (XPS)

The usefulness of XPS in the study of the surface chemistry of carbonaceous materials is well established [120–123]. It is significant, however, that in 1994, while reviewing some aspects of the surface chemistry of carbon blacks and other carbons (in particular, active carbons), Boehm [48] devoted only a few lines to XPS and argued that XPS is not very useful for quantitative determinations. The selected examples included here are meant to give the reader an idea of the use of XPS in the last decade, often in combination with other methods such as FTIR and TPD, and more recently with solid-state nuclear magnetic resonance (NMR), for both qualitative and quantitative characterization of the surface chemistry (i.e., functional groups) either of raw carbonaceous materials or after their treatment (e.g., activation, thermal or chemical treatment, and pyrolysis).

Two main topics are selected: (1) study of the surface chemistry of raw carbonaceous materials, their behavior during thermal treatment (pyrolysis) and oxidation; (2) study of the modifications of surface chemistry after chemical treatments, such as nitrogenation or halogenation.

During coal combustion, fuel nitrogen is a significant major source of NOx pollution. This makes the fate of nitrogen during pyrolysis an important area of study for the design and implementation of NO_x control strategies [124]. In this field, XPS has been a successful nondestructive technique to study nitrogen functionalities in coals and chars. Besides XPS, x-ray absorption near edge structure (XANES) appears to be the only reasonable alternative, though developed to a lesser extent [125,126].

From the work of Pels et al. [127], XPS studies on nitrogen functionalities in coal give strong indications that the major constituents are present in pyrrolic and pyridone (binding energy of 400.5 ± 0.3 eV), named N_5, and pyridinic (binding energy 398.6 ± 0.3 eV), named N_6, forms. A component with binding energy of about 401.4 eV must be included to achieve an acceptable fit in some cases. It corresponds to "quaternary" nitrogen (N_Q). These three peaks are generally used for the fitting of the N_{1s} envelope, besides a fourth peak with binding energy of 402 to 405 eV (N-X), assigned to some form of oxidized nitrogen. The appearance of N_Q in fresh lower-rank coal and the partial retention of this form after mild pyrolysis imply that it represents a significant class of strong noncovalent interactions such as a pyridinic or other basic nitrogen associated with adjacent or nearby hydroxyl groups from phenols or carboxylic acids [128]. Increasing the pyrolysis temperature results in a dramatic decrease in the quantity of pyrrolic nitrogen and a dramatic N_Q increase; the level of pyridinic nitrogen remains in a fairly narrow range. The N_Q produced by high-temperature pyrolysis is different from the N_Q present in the parent coal, and it has been assigned to nitrogen incorporation within the graphene structure [113,114].

An exhaustive study of the quantification of nitrogen forms in coal samples and the identification and quantification of changes that these nitrogen forms present in tars and chars undergo during pyrolysis was carried out by Kelemen and coworkers [128,129]. A similar determination of the fate of nitrogen was made by Thomas and coworkers [125,126] using XANES and comparing these results with XPS spectra; carbons used in their study were prepared by carbonization of carbazole, acridine, and polyacrylonitrile.

XANES has been shown to be a very useful technique for the identification of nitrogen, sulfur, and oxygen functional groups in carbonaceous materials [130]. It allows, in particular, the distinction between pyrrolic and pyridonic structures, which is impossible to make by XPS. However, it does suffer from difficulties in the quantification of the species present, and it is of course less accessible and more costly [126] as it needs a synchrotron. It must be kept in mind that although XPS permits analysis of the sample surface only, XANES allows analysis of the bulk of a sample.

An NO_x index to predict NO_x emissions from pulverized coal combustion was proposed [131] based on a correlation between functional forms of coal nitrogen and the yield of N-containing species evaluated by XPS. A schematic representation of the burn-off of nitrogen-containing graphene during gasification in CO_2 and O_2 was presented by Kapteijn et al. [132]. In the same vein, the fate of nitrogen functionalities in coal was studied [133] following the changes occurring in the N_6, N_5, and N_Q forms, according to the severity of thermal treatment.

An XPS characterization of surface composition and chemistry of a vinylpyridine–divinyl benzene copolymer and its chars obtained under different conditions (of temperature and atmosphere) was carried out by Lahaye and coworkers [28,31]. Separation and identification of different types of heterocyclic nitrogen was carried out by fitting the N_{1s} envelope, and at least six different forms of pyridinic nitrogen were proposed to exist. This study offers a good example of a methodology that can be applied to address such complex analytical challenges.

Besides the characterization of nitrogen functionalities in carbonaceous materials, XPS and XANES were applied to the problem of speciating and quantifying organically bound forms of sulfur in these materials. Kelemen et al. [134] pointed out the difficulties presented by divalent organic sulfur forms and, based on the determination of organic sulfur in Illinois n°6 coal, they concluded that XPS is a viable characterization technique. Kelemen and coworkers [135] also investigated the application of XANES for the same purpose in the case of sulfide and thiophenic sulfur in coals. Usually, the S_{2p} signal can be resolved into peaks corresponding to pyrite, sulfides, thiophenes, sulfoxides, sulfones, and sulfonates/sulfates [134,136–138]. An example is shown in Figure 3.7 of a typical S_{2p} XPS spectrum of a coal (Leonardite), appearing as a single peak (binding energy 166 eV); in the same figure, the XANES spectrum is presented below. The

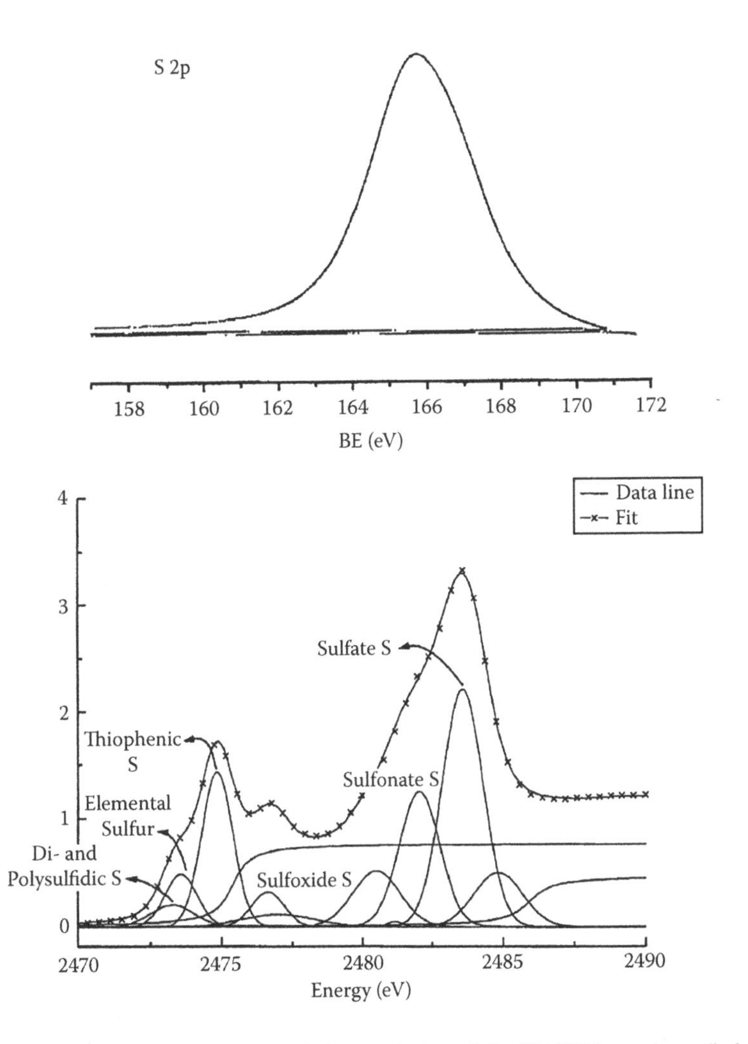

FIGURE 3.7 XPS S-2p spectrum and deconvolution of the XANES spectrum (below) for a Spanish leonardite coal. The intensities are given in arbitrary units. (From Olivella, M.A., del Rio, J.C., Palacios, J., Vairavamurthy, M.A., and de las Heras, F.X.C. *J. Anal. Appl. Pyrol.* 63: 59–68, 2002. With permission.)

comparison of both spectra reveals that this method can go further in the characterization of sulfur functionalities, evidencing two distinguishable regions after deconvolution: sulfates and sulfonates, and also sulfoxide, thiophenic, elemental, and di- and polysulfidic sulfur.

Sulfur characterization in coal by XANES was also carried out by Kasrai and coworkers [139], using as fingerprints the spectra of model compounds: alkyl and aryl sulfides and disulfides and heterocyclic sulfur species.

Organic oxygen functionalities in peats, lignites, and higher-rank coals were also determined by analyzing oxygen's effect on the XPS carbon 1s [140] signal.

Three peaks were used to resolve the XPS C_{1s} band representing, respectively, carbon bound to one oxygen by a single bond (286.3 eV), carbon bound to oxygen by two oxygen bonds (287.5 eV), and carbon bound to oxygen by three bonds (289.0 eV) [141]. This study was combined with ^{13}C NMR (see Section V.C). The characterization of oxygen groups has also been carried out for a series of graphite oxides [142], considering the C_{1s} and O_{1s} bands, in addition to FTIR and ^{13}C NMR spectra; in a series of coals with different degrees of coalification and different amounts of sulfur, subjected to oxidation [143]; in polyacrylonitrile-derived carbon fibers [144]; and in vitrain of a high volatile bituminous coal [145].

The usefulness of XPS for analysis of complex carbonaceous materials is illustrated well by the case of surface analysis (oxygen content and acidity) of carbon black waste materials from tire residues [146]; here, XPS revealed both information about major functional groups with C-O, C=O, and O-C=O bonds and elemental identification [146].

C. XPS Combined with Solid-State NMR Spectroscopy

It is obvious that solid-state ^{13}C NMR techniques have been well developed in the past for the characterization of carbonaceous materials [147,148]. More particularly, in the field of identification of organically bound oxygen forms in these materials, recent papers were published on the use of ^{13}C NMR, in combination with traditional chemical methods or XPS [149,150]. Interestingly, the use of ^{15}N NMR spectroscopy applied to coal samples has also witnessed an increase in the last decade.

The distribution of oxygen functional groups (alcohol, phenol, carboxyl, quinone, and ether linkages) in brown coals has been evaluated by the combined use of chemical analysis and ^{13}C NMR [151]. Evaluation of ether groups was investigated by treating the samples with $SiCl_4$–NaI, which is able to cleave alkylaryl and dialkyl ethers. The characterization of organic oxygen species in peats, pyrolyzed peats, lignites, and other coal samples was performed by Kelemen et al. [140], taking advantage of the fact that the carbon 1s XPS peak is quite sensitive to chemical shifts caused by its bonding to oxygen; comparison between the data obtained by the two methods was discussed and related with the pathways (thermal decarboxylation/decarbonylation and demethoxylation) that can be followed by natural coalification processes.

Although solid-state ^{13}C NMR is certainly one of the most powerful tools for the analysis of carbon structure in coals, evaluation of nitrogen forms in coal by ^{15}N NMR appears to suffer from severe sensitivity problems, as argued by Knicker and coworkers [152]: a low natural abundance of ^{15}N (0.36%), a negative gyromagnetic ratio, and low nitrogen concentration (generally less than 2%) in most coals are among the disadvantages. Nevertheless, a recent paper suggested the contrary, and the first spectra of bituminous and subbituminous coals were reported [152] in 1995. The spectra displayed the main signal intensities in the region of pyrrole and indole derivatives, but no signal could be discerned from the noise for pyridine derivatives. This result contradicts those obtained with XPS and XANES. The authors suggested that pyridinic nitrogen content is rank dependent and that the technique needs further optimization. Subsequently, Solum et al. [153] explained these discrepancies using model heterocyclic nitrogen compounds as well as representative coal samples. Only evidence of the presence of protonated nitrogen in five-membered rings was found (–230 to –235 ppm in pyrrole, –246 to –254 ppm in indole, and –260 to –264 ppm in carbazole). The absence of detectable NMR signals in the region expected for pyridinic nitrogen is in agreement with the results of Knicker et al. [152]. However, after treatment with *p*-toluene-sulfonic acid, the ^{15}N spectra were similar to those of untreated samples, except for an increase in intensity in the region between –150 and –200 ppm, where pyridinic nitrogens are protonated. Acid treatment is expected to protonate the available basic nitrogen forms, thus making ^{1}H-^{15}N cross-polarization to these nitrogen forms more favorable. These results are in agreement with XPS data [152] that about 25% of the nitrogen is present as pyridinic (6-n) type.

In conclusion, the nitrogen signals are very small (with poor signal-to-noise ratio), the ^{15}N NMR studies in coal are very time consuming (accumulation of 10^5 scans is necessary), and for this technique to become a useful tool in coal and carbon surface analysis, it is imperative to develop methodologies to improve sensitivity [153].

More recently, Kelemen et al. [154] discussed the pros and cons of XPS, XANES, and ^{15}N NMR for characterizing and identifying the chemical forms of nitrogen in complex carbonaceous systems. They used both XPS and ^{15}N NMR quantitatively to study kerogen obtained by demineralization of a Green River oil shale and of a peat sample, as well as chars obtained by pyrolysis and isoquinoline- and quinoline-derived chars. The inherent advantage of using a combination of these methods has thus been demonstrated.

XPS was used to study the modifications of the surface chemistry of carbonaceous materials after chemical treatments [111]. For example, recent studies were devoted to the preparation of active carbons modified by incorporation of nitrogen aiming at the development of new solid catalysts [33,34], new catalyst support materials [155], and new competitive adsorbents [111].

Jansen and van Bekkum studied the amination and ammoxidation of active carbons [30]. Amination was carried out by reaction of ammonia with carbons preoxidized with nitric acid and was expected to take place at carboxylic acid sites formed by oxidation. Ammoxidation is a direct reaction with ammonia–air gas mixtures and was expected to take place only at aliphatic substituents of the aromatic rings. On the basis of FTIR spectroscopy, amides, lactams, and imides were suggested as the main functional groups whose heat treatment may lead to pyrroles and pyridines. The products of amination and ammoxidation were studied by XPS [156]. All the N_{1s} binding energies were in the range of 399.7 to 399.9 eV and assigned to amides, lactams, imides, and tertiary amides. After heat treatment (>400°C), pyridine and pyrrole types were the most abundant, principally owing to the conversion of amides and *N*-alkyl imides. Several mechanisms were suggested [37].

Stöhr and Boehm [29] studied the enhancement of oxidative catalytic activity of active carbons treated with ammonia or cyanhydric acid at 600 to 900°C.

Determination of the nitrogen functionalities by XPS suggested the presence of amide and amine groups (binding energy of 400 to 401 eV) and of pyridine and nitrile groups (binding energy 398 eV).

In these studies, the nitrogen was introduced in the already activated carbons. Another way to prepare such active carbons is to start from coals enriched in nitrogen before the step of pyrolysis or activation [36,37]. Two coals were selected: a Polish subbituminous coal, preoxidized with performic acid, and a Russian lignite. Ammonia and its derivatives (ammonium carbonate, hydrazine, hydroxylamine, and urea) were applied as N-reagents [36,37]. The N-rich-activated carbons showed good oxidative removal activity for traces of hydrogen sulfide and for its oxidized by-products (elemental sulfur and sulfur dioxide) [36].

The same treatment with urea was also applied to a Polish lignite [24,157]. The nitrogen-enriched material was carbonized and activated at 800°C under steam, with increasing percentages of burn-off (25, 50, and 70%). Surface characterization was made by IR spectroscopy and XPS. Changes in behavior were linked to the chemical modifications and were tested against their selectivity toward two pairs of volatile organic compounds (VOCs) using also a linear soluation energy relationship (LSER) approach (see Section VI). The XPS data showed that all the nitrogen in the activated materials was in pyridinic form.

The ammoxidation reaction was also recently applied to various carbonaceous materials, for example, pine wood, peat, lignite, and subbituminous coal [38]. Taking cellulose as a precursor, the characterization of functional groups introduced by ammoxidation and their gradual change in the course of heat treatment was followed by IR and XPS analyses [39]. Analysis of the N1 bands allowed the identification and quantification mainly of imine –C=NH (binding energy 399 ± 0.1 eV), besides amides, lactams, and nitriles (binding energy 400.2 ± 0.1 eV). After reaction (4 h at 250°C), the total amount of nitrogen was about 24.5%, and some 60% were imines. A general reaction pathway scheme was proposed, which explained the formation of all nitrogenated derivatives from conjugated imines.

Another field of treatment of carbonaceous materials exhaustively studied by XPS is related to chlorination, bromination, and fluorination. A series of papers concerning halogenation, in both gaseous and aqueous phase, of various samples of carbon blacks were published by Papirer and coworkers [40–43]. The objective was to discriminate between different forms of halogen surface functionalities according to halogenation procedure and sample origin. XPS was the characterization method of choice. A quantitative interpretation of the spectra was made possible by careful and critical analysis of the relevant peaks with the help of model compounds [41,42]. In particular, a critical literature review of XPS studies of fluorinated carbons was offered and guidelines for identification and assignment of the different surface groups, individualized by fitting the C_{1s} and F_{1s} peaks, were provided [40]. It was shown that the fluorination reaction is very complex, the products obtained are of polyphasic nature, including intercalated zones in which fluorine is apparently semi-ionically bound to the sp2 carbons in the graphene layers, as well as fluorinated layers in which fluorine is covalently bound to surface and edge C atoms, and perfluorinated clusters whose structure should be identical to that of fluorographites [40]. These studies included furnace black samples, a thermal black, and an electrically conducting black. They also showed that the total amount of fluorine fixed by a given carbon black is a function of surface reactivity, which in turn depends on the extent of graphitic character.

In summary, the spectroscopic methods reviewed in Section V, by presenting various examples reviewed (although nonexhaustively), illustrate the fact that these nondestructive methods are of fundamental importance, either used alone or in conjunction with other analytical methods. The XPS and XANES methods are particularly useful for the identification of sulfur and nitrogen functionalities.

VI. CHARACTERIZATION OF SURFACE CHEMISTRY IN TERMS OF MOLECULAR INTERACTIONS

In addition to the methods reviewed in the preceding sections, a new aspect of the characterization of carbon surface chemistry has now been developed. It concerns thermodynamic surface properties and, more precisely, phenomena based on molecular interactions between the carbon surface and some specific target molecules, mainly acidic and basic probes. Important applications can be found in the recent literature:

- Adsorbent properties of active carbons, e.g., VOCs [24]
- Fiber-matrix interfacial interactions governing the performance of carbon-fiber-reinforced composites [158–160]
- Interaction of soot with adhering soluble organic fraction (SOF), present in diesel exhaust [161]

In all of them the acid–base interactions are fundamental. Their evaluations can be based on thermodynamic surface properties such as the work of adhesion/pH diagrams and inverse gas chromatography.

A. Work of Adhesion

This concept has been used to characterize the acidity and basicity of carbon fiber surfaces [162]. It is useful to recall that, according to Fowkes [163,164], the total work of adhesion, W_{SL}, is dependent on the pH of an aqueous solution, according to its two components:

- W_{SL}^{L}, the work of adhesion due to nonspecific dispersion interactions (L), which is independent of pH. Its value is about 70 mJ/m^2 and equal to the surface tension of water [165].
- W_{SL}^{AB}, the work of adhesion due to acid–base (AB) interactions, which is a function of pH and includes the effects of any hydrogen bonding.

Values of W_{SL} can be directly obtained from contact angle measurement (CAM) [162]. In case of a negligible acid–base interaction, $W_{SL} = W_{SL}^{L}$, and W_{SL}^{L} is obtained

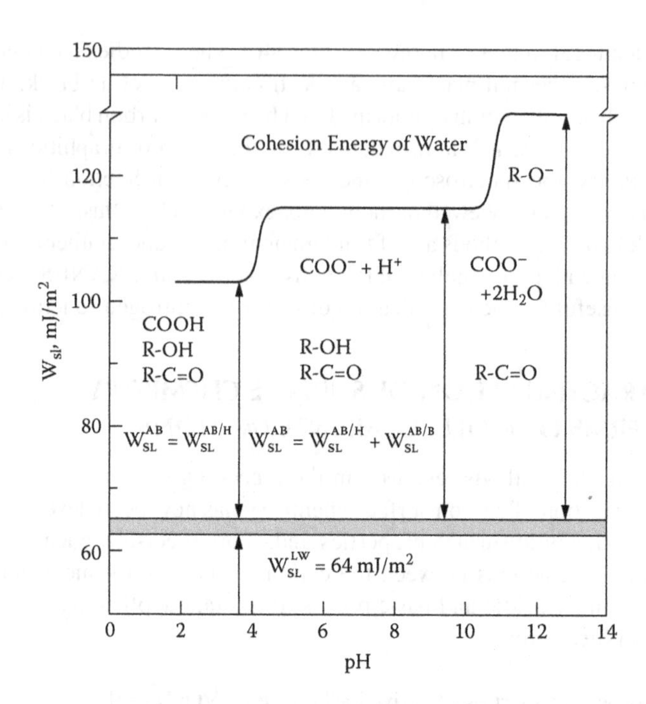

FIGURE 3.8 Schematic representation of the work of adhesion versus pH for surface-oxidized carbon fibers. W_{SL} is the total work of adhesion, $W_{SL}{}^{L}$ is the work of adhesion due to van der Waals/Lifshitz dispersion interactions, $W_{SL}{}^{AB}$ is the work of adhesion due to acid–base interactions, $W_{SL}{}^{AB/B}$ is the work of adhesion due to Brönsted acid–base complexes, and $W_{SL}{}^{AB/H}$ is the work of adhesion due to hydrogen bonds. (From Zielke, U., Hüttinger, K.J., and Hoffman, W.P. *Carbon* 34: 1007–1013, 1996. With permission.)

from the geometric mean of the dispersion components of surface free energies of the carbon and the liquid (water or acidic/basic aqueous solution). The value of W_{SL}^{AB} is obtained from the surface population of acid–base pairs and the enthalpy of the acid–base interactions [164].

Characterization of carbon fiber surfaces, after oxidation, by the work of adhesion/pH diagram was proposed by Zielke et al. [165]. It was shown that the W_{SL}/pH curves have two steps. The first, at low pH (2.5 to 3), disappears after thermal treatment at 250°C and after chemical modification of carboxyl groups (reduction by $LiAlH_4$ and esterification); it was assigned to Brönsted acid–base complexes. A second step was found at pH ≥ 8; it disappeared after heat treatment at 750°C, the temperature at which hydroxyl groups should be destroyed, but not after reduction of quinones by $NaBH_4$ and $LiAlH_4$. At pH values between 3 and 8, the carboxyl groups can form Brönsted acid–base complexes causing an increase in W_{SL}. A schematic representation of W_{SL} versus pH is given in Figure 3.8.

A comparison of the results obtained with the various surface characterization methods used [166] (CAM, XPS, and TPD) showed that neither TPD of functional groups nor quantitative XPS data can be regarded as an absolute measure of the exterior surface functionalities in a fiber. This is more appropriately reflected by CAM or the work of adhesion behavior. Indeed, XPS is a valuable method for identifying functional groups in an approximately 5-nm surface layer, but not in the outermost surface layer, which is more relevant for adhesion. By TPD, of course, the functional groups existing on the total accessible surface of the material are detected [167].

B. Inverse Gas Chromatography (IGC)

The IGC method analyzes a material placed in a column by injections of known pure volatile compounds, referred to as molecular probes or solutes. The retention times obtained for such molecules are controlled by their physicochemical properties and those of the column material under study. Thus, by knowing the solute characteristics, it is possible to evaluate those of the investigated condensed phase. The applications of IGC are numerous and concern both liquids and solids [168–170].

This analytical method is attractive because of its simplicity of implementation using a conventional gas chromatograph. The extensive literature concerning this subject shows different approaches to characterizing a material in terms of its molecular interactions. Among the most popular ones are those developed by Papirer [169] and Dong et al. [171]. These authors investigated carbonaceous adsorbents through IGC by splitting the free energy of adsorption (ΔG_a) of gaseous organic compounds into two contributions according to Equation 3.2:

$$\Delta G_a = \Delta G_a^D + \Delta G_a^{SP} \tag{3.2}$$

Here, ΔG_a^D and ΔG_a^{SP} are the free energy contributions corresponding to dispersive and specific interactions. The free energy of adsorption is obtained from chromatographic data such as the net volume of retention (V_N) through Equation 3.3:

$$\Delta G_a = -RT\ln(V_N) \tag{3.3}$$

where T is the column temperature.

The dispersive component is determined by injections of nonpolar probes such as *n*-alkanes. Generally, there is a linear relationship between the free energy of adsorption and the number of methylene groups for a homologous series of *n*-alkanes, as shown in Figure 3.9. This nonspecific component (also denoted γ_s^D) can be obtained (see Figure 3.9) according to the method of Dorris and Gray [172]. It is thus possible to determine the γ_s^D value of a carbonaceous material knowing the area of a methylene group through Equation 3.4,

$$\gamma_S^D = [\Delta G_{a\,(CH2)} / 2N.a_{CH2}]^2 \,.\, (1/\gamma_{CH2}) \tag{3.4}$$

where N is the Avogadro number, a_{CH2} is the area of a methylene group (6Å^2) [172], γCH_2 is the surface energy of a methylene group, for example, in polyethylene,

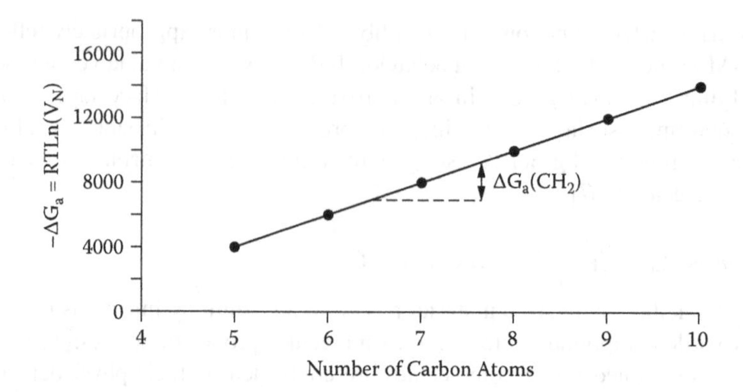

FIGURE 3.9 Free energy variation of a homologous series of *n*-alkanes versus the number of carbon atoms. V_N is the net volume of retention of the alkane, ΔG_a is the free energy of adsorption on the solid surface, and $\Delta G_a(CH_2)$ is the contribution of a methylene group to the free energy of adsorption.

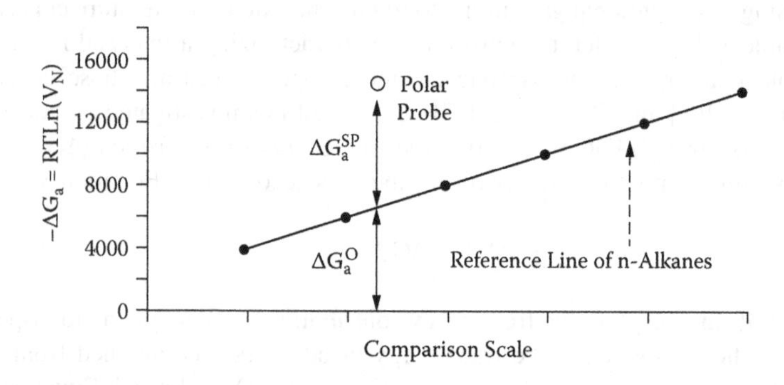

FIGURE 3.10 Illustration of the comparison scale method for deconvolution of total free energy of adsorption for a polar probe on a solid surface. $\Delta G_a{}^{SP}$ is the specific contribution, and $\Delta G_a{}^{D}$ is the dispersive contribution.

and $\Delta G_{a\,(CH2)}$ is determined from the slope of the line presented in Figure 3.9. This method is based on the assumption of interaction between a methylene group and a flat surface of homogeneous energy, and thus it is suitable for nonporous carbons. For porous materials, another method was proposed by Derouane et al. [173], by introducing a supplementary parameter in Equation 3.4 and thus taking into account the size of the adsorbate molecule and of the pores.

The specific interaction contribution of a polar probe (i.e., ΔG_a^{SP}) is calculated by subtracting the dispersive component of an *n*-alkane (i.e., ΔG_a^{D}) having some similar properties, from the total free energy (ΔG_a) as described by the general equation (Equation 3.2). This type of approach leads to a graph such as that presented in Figure 3.10, in which the free energy of adsorption of a polar probe on a given solid is plotted versus the selected comparison scale (see Table 3.5). The free energy of adsorption of a homologous series of *n*-alkanes is also shown to

TABLE 3.5
Different comparison scale approaches for determination of specific interactions of, for example, graphites [171,174,175] and carbon fibers [171,175,178]

Comparison Scale	Explanations	Authors
$\log(P_0)$	P_0 is the saturated vapor pressure of the probe	Saint-Flour and Papirer [174]
$\alpha(\gamma_L^D)^{1/2}$	α is the area of the adsorbed molecule and γ_L^D is the dispersive component of the surface energy of the molecule in the liquid state	Schultz et al. [175]
$\alpha_0(h\nu)^{1/2}$ 10 [49]	α_0 is the polarizability of the molecule, ν is the proper frequency of the molecule	Donnet and Dong [171]
χ_T	χ_T is the topology index of the molecule according to the method of Wiener modified by Barysz [179,180]	Papirer and Brendlé [178]

define the reference line that represents the limit between nonpolar and polar contributions of the total adsorption free energy of a polar probe molecule.

It is obvious that such separation is easy from a theoretical point of view, but in practice it is difficult to find a comparison scale between *n*-alkanes and polar probes. All these approaches are summarized in Table 3.5 and include the definition of parameters constituting each comparison term. Additional explanations concerning the different comparison scales are given in the following text.

In an early study, Saint-Flour and Papirer [174] used the logarithm of the saturated vapor pressure—$\log(P_0)$—as a comparison scale. Its advantage is the ready availability of pressure data in the literature and ease of calculation at various temperatures. On the other hand, this method compares macroscopic data of *n*-alkanes and polar probes, but for the latter the specific interactions lead to a diminished P_0; thus, it is often observed that for some polar probes the representative points for $RT\ln V_N$ values lie above the *n*-alkane reference line.

Schultz et al. [175] used $a(\gamma_L^D)^{1/2}$ as a comparison scale. The parameter a is the area of the adsorbed probe molecule [176]. It is obtained from IGC measurements on neutral reference solids such as polyethylene or PTFE. Some examples are *n*-hexane, 51.5Å^2; ether, 47Å^2; chloroform, 44Å^2; and acetone, 42.5Å^2. In standard fashion, γ_L^D is determined from contact angle measurements on reference solids [176]. Once again, some probes fall under the reference line of the *n*-alkanes, and the data are difficult to interpret. This constitutes a limiting factor of this method.

Dong et al. [171] developed a scale using the polarizability of probe molecules. An advantage of this method is that polarizability is not a macroscopic property of the molecules in a condensed state. In this case, no probe molecule lies under the reference line. Nevertheless, the polarizability values are not easy to find in the literature, and their determination requires the use of Debye's equation, which in turn requires macroscopic data such as refractive index, molecular weight, and density of the probe [177].

The calculation method of Papirer and Brendlé [178] is based on molecular topology and on calculation of Wiener's indices [179] as modified by Barysz et al. [180]. This method is based on the determination of an index that considers the molecular structure of the probe molecule. The calculation leads to a parameter χ_T that represents a non–whole number of carbon atoms of a hypothetical alkane that would interact with a solid surface with the same intensity as the molecule that it represents.

Some authors go further in the characterization of specific interactions by splitting them into two terms, for example, of basicity and acidity, using Gutmann's parameters DN (donor number) and AN (acceptor number) [181,182]. Some examples of DN and AN values can be given here for some probes [170,176,182]: *n*-hexane, DN = 0, AN = 0; ether, DN = 19.2, AN = 3.9; chloroform, DN = 0, AN = 23.1; acetone, DN = 17.0, AN = 12.5; and ethanol, DN = 20.5, AN = 37.3. Obviously, *n*-alkanes cannot interact with a solid surface through basic or acidic interactions, ethers are mainly basic, chlorinated compounds are considered only acidic, and ketones and alcohols are amphoteric.

This approach allows evaluation of the acidity and basicity of a solid as K_A and K_D, respectively. Thus, ΔG_a^{SP} is replaced by ΔG_{AB} and K_D or K_A are calculated through Equations 3.5 and 3.6.

$$-\Delta G_{AB} = K_A.DN + K_D.AN \quad (3.5)$$

$$-\Delta G_{AB}/AN = (K_A.DN/AN) + K_D \quad (3.6)$$

For a series of probes, $-\Delta G_{AB}/AN$ versus DN/AN gives a linear relationship, with K_A as the slope and K_D the intercept. As discussed elsewhere [170], the DN is obtained from the complexation of probes with $SbCl_5$ in dichloroethane solution; thus, it is a property of an isolated molecule. However, the AN is obtained from the influence of bulk solvents on the ^{31}P shift of triphenylphosphine oxide, and so it is a solvent parameter and not a characteristic parameter of monomeric probe molecules (the form in which they exist in IGC near zero concentration). Thus, the molecule is not well described in terms of acidity, and the determination of the basicity of solid surfaces can be considered with some reservation.

Characterization of the acidity of carbonaceous materials was also undertaken by Sidqi et al. [183] using an alkene as a basic probe (electron donor) and by comparing its chromatographic behavior with that of a corresponding alkane (i.e., with the same number of carbon atoms).

Other approaches, well known in the chromatographic area, are that of Kovats [184,185] and the use of system constants of McReynolds [186]. The specific interactions are estimated by injecting polar probes and making the difference for each of them equal to that between their retention index (I_X) and the index obtained at identical conditions for a nonpolar reference material (I_{ref}) such as solid polyethylene or Apiezon L.

All these approaches compare polar probes to hypothetical alkanes having the same properties. Most of them use properties of compounds in a condensed state, whereas for characterizing solid surfaces these molecules will behave in an isolated state (condition of infinite dilution) [170]. Moreover, some of these parameters (e.g., polarizability) are quite difficult to obtain. These are the main disadvantages of such methods.

Another approach decomposes the free energy of phase transfer into several contributions characterizing at the same time the dispersive and polar interactions, without a comparison between alkanes and polar molecules. Such methods can be summarized as linear free energy relationship (LFER) equations. With the aim of characterizing adsorptive properties of graphitized carbon blacks in particular, Larionov et al. [187] described a method of determining molecular interactions in gas adsorption chromatography using a set of 30 solutes. This work yielded four contributing terms defining dispersion, orientation, and donor–acceptor interactions. Another method that goes further into the characterization of molecular interactions involved in gas-solid adsorption is the linear solvation energy relationship (LSER) of Abraham et al. [188]. The history of this method and more details are given elsewhere [170,189]. It allows one to decompose the free energy of adsorption of a VOC on a given sorbent into five contributions. Knowing the properties of several injected solutes and using a statistical procedure such as a multiple linear regression analysis (MLRA), it is possible to extract five quantitative microscopic constants characterizing the material under study at a given temperature. The LSER equation is described as follows:

$$SP = c + rR_2 + s\pi_2^H + a\Sigma\alpha_2^H + b\Sigma\beta_2^H + l\log_{10}L^{16} \tag{3.7}$$

Or, more recently [190,191], as

$$SP = c + eE + sS + aA + bB + lL \tag{3.8}$$

Here, SP is a solute property in a given system that is a free energy parameter obtained by IGC measurements, and c is the regression constant. The solute parameters or descriptors are R_2 (or E), the molar excess refraction [192]; π_2^H (or S), the dipolarity/polarizability [193]; $\Sigma\alpha_2^H$ (or A), the hydrogen-bond acidity [193]; $\Sigma\beta_2^H$ (or B), the hydrogen-bond basicity [193,194]; and L^{16} (or L), the gas-to-hexadecane partition coefficient at 298 K [195]. The power of the method is that all these descriptors are free energy parameters that can be determined by physicochemical and thermodynamic measurements. Moreover, there is no comparison with alkanes, and all the descriptors are isolated molecule values. The relevant data can be found in the numerous articles of Abraham and coworkers, and come from a data bank of about 3000 compounds [196]. However, these parameters are not dependent on temperature. They are all measured at 298 K, except for R_2, which is determined at 293 K.

Thus, for each injected molecular probe, an LSER equation can be established. Then, interaction constants characterizing the studied sorbent are determined by MLRA. The r-coefficient (or e) represents the ability of the material to interact with solutes through π-(delocalized electrons such as in aromatic structures) and n-electron (nonbonding electrons such as in halogen atoms) pairs, the s-coefficient represents the tendency of the material to interact through dipole-type interactions, the a-coefficient is its hydrogen-bond basicity (i.e., ability to interact with acid solutes), the b-coefficient its hydrogen-bond acidity (i.e., ability to interact with basic solutes), and the l-coefficient is a measure of dispersion interactions. The details of the statistical procedures, as well as some precautions, are given elsewhere [197].

This method allows the description, classification, and even predictions of the behavior of various systems, most notably adsorbents [170,197]. Some recent applications in the characterization of carbonaceous materials are given in Table 3.6. It is obvious that such an approach is a complementary analytical tool to chemical surface characterization and classification of carbonaceous materials and is useful to distinguish different forms of carbon [161,170,197]. It is also a supplementary means of monitoring modifications of the carbon surface when they are chemically treated [24,111,157,170,198,199]. Moreover, the LSER coefficients are quantitative data that can be used to extrapolate the free energies of adsorption for organic solutes not used to set up the LSER equations; this is important, for example, in the selectivity of carbonaceous adsorbents toward pairs of VOCs [24,111,157,170]. The selectivity (α) toward a given pair of VOCs A/B was defined [200] as the ratio of the gas-solid partition coefficient of A (K_A) to the gas-solid partition coefficient of B (K_B). The LSER method allowed prediction of K_A and K_B, and thus α for several pairs of VOCs, successfully for mesoporous materials and, with some discrepancies, for microporous carbons [24]. For example, when studying the selectivity of a series of activated lignites toward methanol/dichloromethane, the LSER approach provided an explanation of why an activated lignite without chemical treatment preferentially retains the chlorinated VOC whereas the corresponding nitrogen-enriched material (that is, a basic adsorbent) had an increasing preference for adsorption of methanol (that is, an acidic VOC or donor of hydrogen bonds through OH groups) [24].

VII. CONCLUSIONS

The "dry methods" of carbon surface chemistry characterization are all based on spectroscopy. They can be applied either directly to the starting materials (such as coals, lignites, and chars) or after treatments for specific purposes (e.g., oxidation and nitrogenation). They are preferable for determining structural linkages between carbon and other elements, mainly oxygen, nitrogen, and sulfur. In this sense, x-ray photoelectron spectroscopy (XPS) is particularly useful for both the nitrogen functionalities, a problem linked to nitrogen oxides pollution, and organic sulfur forms. However, XPS is known to suffer from several potential drawbacks. Being a surface technique, it provides the chemical composition of the few uppermost layers of the sample (about 5 nm); furthermore, to determine the nature of the functional groups, the C_{1s}, O_{1s}, and N_{1s} bands must be deconvoluted, and this procedure is not so straightforward. The same is true for the less accessible and more costly XANES technique.

TABLE 3.6
Some examples of carbonaceous materials studied by the LSER approach

Nonactivated Carbonaceous Materials							
	c	r	s	a	b	l	Applications
Graphite [170]	−1.55	−0.26	0.99	1.11	0	0.59	Surface characterization
Carbotrap [170]	−4.82	0	0	0	0	2.41	Surface characterization
Carbosieve [170]	−0.40	0	0	0	0	2.59	Surface characterization
Charcoal cloth [170]	−1.17	0	0	0	0	3.00	Surface characterization
Carbon fibers [170]	−7.62	0	1.20	0	0	5.82	Surface characterization
Fullerene [170]	−1.58	−0.24	0.72	1.04	0	0.48	Surface characterization
Soot [161]	−1.61	−0.52	0.43	1.02	0.38	0.72	Adsorptive properties
Extracted soot [161]	0	0	0.37	0.81	0	0.36	Adsorptive properties
Coal tar pitch/PAN [198]	−2.62	0	0	1.59	1.06	1.23	Surface characterization
Nondemineralized lignite [199]	−2.82	0	1.27	1.15	0	0.531	Surface characterization
Strong acid demineralized lignite [199]	−2.76	−0.80	0.55	0.93	2.70	0.62	Surface characterization
Activated Carbonaceous Materials							
	c	r	s	a	b	l	Applications
Commercial AC sample [24,200]	−2.69	−1.58	0	2.47	1.83	1.63	Surface characterization Adsorptive properties Selectivity
Lignite (AcL50) [24,111]	−1.08	−1.82	0	0	0	1.43	Surface characterization Adsorptive properties Selectivity
Nitrogen-enriched lignite (AcLU50) [24,111,157]	0.42	0	0	0.83	0	0.38	Surface characterization Adsorptive properties Selectivity

Note: c is the regression constant, r is the ability of the solid to interact with gaseous compounds through n- and π-electron pairs, s is the dipolar character of the surface, a the hydrogen-bonding basicity, b the hydrogen-bonding acidity, and l its ability to interact through dispersive interactions.

Even though it is extensively used for the determination of the structure of organic materials in terms of their functional groups, infrared spectroscopy is quite difficult, from an experimental point of view, when applied to fully carbonized materials, owing to their high IR absorbance. After different pretreatments (e.g., heat treatment or oxidation), the band assignment is another difficult challenge and varying proposals have been made on the basis of model compounds, though not without some disagreements.

Besides old and classical determinations by ^{13}C NMR, not reviewed here, joint use of XPS and solid-state ^{13}C NMR, or XPS and ^{15}N NMR, deserve special attention, although the latter technique is only at its early stage of development.

In the case of practically important activated carbons, many analytical methods are available, and often several of them are used in a complementary fashion. Their judicious choice is of crucial importance.

It is obvious that the Boehm titration method is the most popular one for the determination of various types of acidic (and basic) surface functionalities in carbon materials. From 1966 until 2002, when Boehm himself published a critical assessment of the analysis of surface oxides on carbon [201], an exhaustive utilization of this method has been described by many authors. They underlined its simplicity, but pointed out also the need for using other complementary methods such as potentiometric titration, temperature-programmed desorption (TPD), spectroscopic methods (mainly XPS and FTIR), and thermodynamic approaches such as calorimetry. The case of TPD is of special interest, to identify oxygenated functionalities. However, the CO and CO_2 peaks must certainly be deconvoluted before the surface composition can be estimated. Thus, a quantitative TPD analysis of surface functional groups is sensitive to the deconvolution method and to experimental conditions. The results are generally discussed in relation to those of DRIFTS and XPS analysis, as can be seen from the references listed in Table 3.1.

For functional group analysis, infrared spectroscopy can be successfully applied in conjunction with Boehm titrations and TPD, particularly in the DRIFTS mode. The spectra are generally complicated owing to overlap of the absorption bands, and their interpretation prompted the various authors to propose their own assignments applicable to each specific study. The same can be said, to a lesser extent, of the use of XPS.

Another group of characterization methods explores a global aspect of surface chemistry in terms of molecular interactions, namely, adhesion and adsorption measurements. Such characterization, using the work of adhesion and inverse gas chromatography, allows one to obtain quantitative parameters that can be used to classify and predict behavior such as adsorptive properties of carbons. In effect, these methods constitute a quantitative structure behavior relationship (QSBR) and are complementary to the other methods. It is an interesting means of characterizing complex materials when chemical composition data are difficult or impossible to obtain. Moreover, molecular interaction data can often be determined with quite simple apparatus. Among these methods, the LSER approach goes farthest in the separation of specific interactions and thus offers promise for a detailed understanding, and even predictions, of the adsorption of, for example, numerous VOCs whose properties (descriptors) are easy to obtain.

ACKNOWLEDGMENTS

Professor Denise Cagniant, coauthor of this chapter, died on May 1, 2005.

She studied at the Chemistry School of Strasbourg and defended her PhD thesis in 1963, under the direction of Prof. Guy Ourisson. She was promoted to the Scientific College of Metz the same year. She developed organic chemistry in the young University of Metz and left her imprint on the subject, particularly in the area of the synthesis of chalcogen-derived heterocycles.

In the 1980s, she initiated, with Pr. Henri Charcosset from the University of Lyon, national and international networks, supported by CNRS, aiming at the characterization and valorization of coal.

She was author of 250 articles, and was a national authority in organic chemistry and in the area of carbonaceous materials.

REFERENCES

1. Leon y Leon, C.A. and Radovic, L.R. *Chem. Phys. Carbon* 24: 213–310, 1994.
2. Radovic, L.R. and Rodriguez-Reinoso, F. *Chem. Phys. Carbon* 25: 243–358, 1997.
3. Rodriguez-Reinoso, F. *Carbon* 36: 159–175, 1998.
4. Xia, W., Schlüter, O.F.K., Liang, C., van den Berg, M.W.E., Guraya, M., and Muhler, M. *Catalysis Today* 102–103: 34–39, 2005.
5. Zhuang, Q.L., Kyotani, T., and Tomita, A. *Energy Fuels* 8, 714–718, 1994.
6. Lizzio, A.A., Jiang, H., and Radovic, L.R. *Carbon* 28, 7–19, 1990.
7. Fanning, P.E. and Vannice, M.A. *Carbon* 31, 721–731, 1993.
8. Dandekar, A., Baker, R.T.K., and Vannice, M.A. *Carbon* 36: 1821–1831, 1998.
9. Biniak, S., Szymanski, G., Siedlewski, J., and Swiatkowski, A. *Carbon* 35, 1799–1810, 1997.
10. El-Hendawy, A.N.A. *Carbon* 41, 713–722, 2003.
11. Moreno-Castilla, C., Lopez-Ramon, M.V., and Carrasco-Marin, F. *Carbon* 38: 1995–2001, 2000.
12. Ania, C.O., Parra, J.B., and Pis, J.J., *Fuel Process. Technol.* 79: 265–271, 2002.
13. Moreno-Castilla, C., Carrasco-Marin, F., and Mueden, A. *Carbon* 35: 1619–1626, 1997.
14. Prado-Burguete, C., Linares-Solano, A., Rodriguez-Reinoso, F., and Salinas-Martinez de Lecea, C. *J. Catal.* 115: 98–106, 1989.
15. Leboda, R., Lodyga, A., and Gierak, A. *Mater. Chem. Phys.* 51: 216–232, 1997.
16. Leboda, R., Lodyga, A., and Charmas, B. *Mater. Chem. Phys.* 55: 1–29, 1998.
17. Gomez de Salazar, C., Sepulveda-Escribano, A., and Rodriguez-Reinoso, F. *Carbon* 38: 1879–1902, 2000.
18. Poole, S.K. and Poole, C.F. *Anal. Comm.* 34: 247–251, 1997.
19. Centeno, T.A. and Fuertes, A.B. *Carbon* 38: 1067–1073, 2000.
20. Knox, J.H., Kaur, B., and Millward, G.R. *J. Chromatogr.* 352: 3–25, 1986.
21. Knox, J.H. and Ross, P. *Adv. Chromatogr.* 37: 73–119, 1997.
22. Kamegawa, K. and Yishida, H. *Carbon* 35: 631–639, 1997.
23. Matisova, E. and Skrabakova, S. *J. Chromatogr.* 707, 145–179, 1995.
24. Burg, P., Fydrych, P., Bimer, J., Salbut, P.D., and Jankowska, A. *Carbon* 40: 73–80, 2002.
25. Figueiredo, J.L., Pereira, M.F.R., Freitas, M.M.A., and Orfao, J.J.M. *Carbon* 37: 1379–1389, 1999.
26. Pereira, M.F.R., Soares, S.F., Orfao, J.J.M., and Figueiredo, J.L. *Carbon* 41: 811–821, 2003.

27. Faria, P.C.C., Orfao, J.J.M., and Pereira, M.F.R. *Water Res.* 38: 2043–2052, 2004.
28. Lahaye, J., Nansé, G., Bagreev, A., and Strelko, V. *Carbon* 37: 585–590, 1999.
29. Stöhr, B. and Boehm, H.P. *Carbon* 29: 707–720, 1991.
30. Jansen, R.J.J. and van Bekkum, H. *Carbon* 32: 1507–1516, 1994.
31. Lahaye, J., Nansé, G., Fioux, P., Bagreev, A., Broshnik, A., and Strelko, V. *Appl. Surf. Sci.* 147: 153–174, 1999.
32. Boudou, J.P., Chehimi, M., Broniek, E., Siemieniewska, T., and Bimer, J. *Carbon* 41: 1999–2007, 2003.
33. Kawabuchi, Y., Sotowa, C., Kishino, M., Kawano, S., Whitehurst, D.D., and Mochida, I. *Langmuir* 13: 2314–2317, 1997.
34. Raymundo-Piñero, E., Cazorla-Amorós, D., and Linares-Solano, A. *Carbon* 41: 1925–1932, 2003.
35. Bagreev, A., Menendez, J.A., Dukhno, I., Tarasenko, Y., and Bandosz, T.J. *Carbon* 42: 469–476, 2004.
36. Bimer, J., Salbut, P.D., Berlozecki, S., Boudou, J.P., Broniek, E., and Siemieniewska, M.T. *Fuel* 77: 519–525, 1998.
37. Cagniant, D., Gruber, R., Boudou, J.P., Bilem, C., Bimer, J., and Salbut, P.D. *Energy Fuels* 12: 672–681, 1998.
38. Bimer, J., Berlozecki, S., Salbut, P.D., Bao Quing, L., and Zhu Yu, L. Eds. Prospects for Coal Science in the 21 Century. In *Proc. 10th Int. Conf. Coal Sci.,* Vol. 2, Shanxi Science and Technology Press, Taiyuan, 1999, pp. 957–962.
39. Cagniant, D., Magri, P., Gruber, R., Berlozecki, S., Salbut, P.D., Bimer, J., and Nansé, G. *J. Anal. Appl. Pyrol.* 65: 1–23, 2002.
40. Papirer, E., Lacroix, R., and Donnet, J.B. *Carbon* 34: 1521–1529, 1996.
41. Papirer, E., Lacroix, R., Donnet, J.B., Nansé, G., and Fioux, P. *Carbon* 32: 1341–1358, 1994.
42. Papirer, E., Lacroix, R., Donnet, J.B., Nanse, G., and Fioux, P. *Carbon* 33: 63–72, 1995.
43. Nanse, G., Papirer, E., Fioux, P., Moguet, F., and Tressaud, A. *Carbon* 35: 175–194, 1997.
44. Perez-Cadenas, A.F., Maldonado-Hodar, F.J., and Moreno-Castilla, C. *Carbon* 41: 473–478, 2003.
45. Barton, S.S., Evans, M.J.B., Halliop, E., and MacDonald, J.A.E. *Carbon* 35: 1361–1366, 1997.
46. Bansal, R.C., Donnet, J.B., and Stoeckli, H.F. *Active Carbon*, Marcel Dekker, New York, 1988, pp. 27–118.
47. Boehm, H.P. In *Advances in Catalysis,* Vol. 16, Eley, D.D., Pines, H., and Weisz, P.B. Eds., Academic Press, New York, 1966, pp. 179–275.
48. Boehm, H.P. *Carbon* 32: 759–769, 1994.
49. Uranowski, L.J., Tessmer, C.H., and Vidic, R.D. *Water Res.* 32: 1841–1851, 1998.
50. Vidic, R.D., Tessmer, C.H., and Uranowski, L.J. *Carbon* 35: 1349–1359, 1997.
51. Bandosz, T.J. *Carbon* 37: 483–491, 1999.
52. Lahaye, J. *Fuel* 77: 543–547, 1998.
53. Pittman, C.U., Jr., He, G.R., Wu, B., and Gardner, S.D. *Carbon* 35: 317–331, 1997.
54. Salame, I. and Bandosz, T.J. *J. Colloid Interface Sci.* 240: 252–258, 2001.
55. Contescu, A., Contescu, C., Putyera, K., and Schwarz, J.A. *Carbon* 35: 83–94, 1997.
56. Bandosz, T.J., Jagiello, J., Contescu, C., and Schwarz, J.A. *Carbon* 31: 1193–1202, 1993.
57. Contescu, A., Vass, M., Contescu, C., Putyera, K., and Schwarz, J.A. *Carbon* 36: 247–258, 1998.
58. Jagiello, J. *Langmuir* 10: 2778–2785, 1994.

59. Benaddi, H., Bandosz, T.J., Jagiello, J., Schwarz, J.A., Rouzaud, J.N., Legras, D., and Beguin, F. *Carbon* 38: 669–674, 2000.
60. Radovic, L.R., Moreno-Castilla, C., and Rivera-Utrilla, *J. Chem. Phys. Carbon* 27: 227–405, 2001.
61. Papirer, E., Li, S., and Donnet, J.B. *Carbon* 25: 243–247, 1987.
62. Suarez, D., Menendez, J.A., Fuente, E., and Montes-Moran, M.A. *Langmuir* 15: 3897–3904, 1999.
63. Leon y Leon, C.A., Solar, J.M., Calemma, V., and Radovic, L.R. *Carbon* 30: 797–811, 1992.
64. Montes-Moran, M.A., Menendez, J.A., Fuente, E., and Suarez, D. *J. Phys. Chem. B* 102: 5595–5601, 1998.
65. Darmstadt, H. and Roy, C.H. *Carbon* 41, 2662–2665, 2003.
66. Menendez, J.A., Xia, B., Phillips, J., and Radovic, L.R. *Langmuir* 12: 4404–4410, 1996.
67. Menendez, J.A., Phillips, J., Xia, B., and Radovic, L.R. *Langmuir* 13: 3414–3421, 1997.
68. Menendez, J.A., Radovic, L.R., Xia, B., and Phillips, J. *J. Phys. Chem.* 100: 17243–17248, 1996.
69. Menendez, J.A., Illan-Gomez, M.J., Leon y Leon, C.A., and Radovic, L.R. *Carbon* 33: 1655–1659, 1995.
70. Noh, J.S. and Schwarz, J.A. *J. Colloid Interface Sci.* 130: 157–164, 1989.
71. Noh, J.S. and Schwarz, J.A. *Carbon* 28: 675–682, 1990.
72. Lopez-Ramon, M.V., Stoeckli, F., Moreno-Castilla, C., and Carrasco-Marin, F. *Carbon* 37: 1215–1221, 1999.
73. Stoeckli, F., Moreno-Castilla, C., Carrasco-Marin, F., and Lopez-Ramon, M.V. *Carbon* 39: 2231–2237, 2001.
74. Haydar, S., Moreno-Castilla, C., Ferro-Garcia, M.A., Carrasco-Marin, F., Rivera-Utrilla, J., Perrard, A., and Joly, J.P. *Carbon* 38: 1297–1308, 2000
75. Otake, Y. and Jenkins, R.G. *Carbon* 31: 109–121, 1993.
76. De la Puente, G., Pis, J.J., Menendez, J.A., and Grange, P. *J. Anal. Appl. Pyrol.* 41: 125–138, 1997.
77. Wu, X. and Radovic, L.R. *Carbon* 44: 141–151, 2006.
78. Menendez, J.A. *Thermochim. Acta* 312: 79–86, 1998.
79. Ehrburger, P., Louys, F., and Lahaye, J. *Carbon* 27: 389–393, 1989.
80. Ehrburger, P., Pusset, N., and Dziedzinl, P. *Carbon* 30: 1105–1109, 1992.
81. Radovic, L.R., Walker, P.L., Jr., and Jenkins, R.G. *Fuel* 62: 849–856, 1983.
82. Radovic, L.R., Steczko, K., Walker, P.L., Jr., and Jenkins, R.G. *Fuel Process. Technol.* 10: 311–326, 1985.
83. Laine, N.R., Vastola, F.J., and Walker, P.L., Jr. *J. Phys. Chem.* 67: 2030–2034, 1963.
84. Radovic, L.R., Walker, P.L., Jr., and Jenkins, R.G. *J. Catal.* 82: 382–394, 1983.
85. Lizzio, A.A., Piotrowski, A., and Radovic, L.R. *Fuel* 67: 1691–1695, 1988.
86. Arenillas, A., Rubiera, F., Barra, J.B., and Pils, J.J. *Carbon* 40: 1381–1383, 2002.
87. Radovic, L.R. *Carbon* 29: 809–811, 1991.
88. Kapteijn, F., Meijer, R., Moulijn, J.A., and Cazorla-Amoros, D. *Carbon* 32: 1223–1231, 1994.
89. Groszek, A.J. *Proc. R. Soc. Lond. A* 314: 473–498, 1970.
90. Groszek, A.J. *Carbon* 25: 717–722, 1987.
91. Groszek, A.J. *Carbon* 27: 33–39, 1989.
92. Groszek, A.J. and Parkyta, S. *Langmuir* 9: 2721–2725, 1993.
93. Groszek, A.J. *Carbon,* 35: 1399–1405, 1997.
94. Groszek, A.J. *Thermochim. Acta* 312: 133–143, 1998.
95. Groszek, A.J. and Aharoni, C. *Langmuir* 15, 5956–5960, 1999.

96. Domingo-Garcia, M., Groszek, A.J., Lopez-Garzon, F.J., and Perez-Mendoza, M. *Appl. Catal. A-Gen* 233: 141–150, 2002.
97. Beck, N.V., Meech, S.E., Norman, P.R., and Pears, L.A. *Carbon* 40: 531–540, 2002.
98. Groszek, A.J. and Templer, C.E. *Fuel* 67: 1658–1662, 1988.
99. Peña, J.M., Allen, N.A., Edge, M., Liauw, C.M., and Valange, B. *Polym. Degrad. Stabil.* 74: 1–24, 2001.
100. Szymanski, G.S., Biniak, S., and Rychlicki, G. *Fuel Process. Technol.* 79: 217–223, 2002.
101. Lopez-Ramon, M.V., Stoeckli, F., Moreno-Castilla, C., and Carraco-Marin, F. *Carbon* 38: 825–829, 2000.
102. Bradley, R.H., Daley, R., and Le Goff, F. *Carbon* 40: 1173–1179, 2002.
103. Phillips, J., Xia, B., and Menendez, J.A. *Thermochim. Acta* 312: 87–93, 1998.
104. Xie, F., Phillips, J., Silva, I.F., Palma, M.C., and Menendez, J.A. *Carbon* 38: 691–700, 2000.
105. Painter, P.C., Snyder, R.W., Starsinic, M., Coleman, M.C., Kuehn, D.W., and Davis, A. *Appl. Spectrosc.* 35: 475–485, 1981.
106. Thomasson, J., Coin, C., Kahraman, H., and Fredericks, P.M. *Fuel* 79: 685–691, 2000.
107. Zawadzki, J. *Chem. Phys. Carbon* 21: 147–380, 1989.
108. Salame, I.I. and Bandosz, T.J. *J. Colloid Interface Sci.* 210: 367–374, 1999.
109. Krzton, A., Cagniant, D., Gruber, R., Pajak, J., Fortin, F., and Rouzaud, J.N. *Fuel* 74: 217–225, 1995.
110. Nosyrev, I.E., Gruber, R., Cagniant, D., Krzton, A., Pajak, J., Stefanova, M.D., and Grishchuk, S. *Fuel* 75: 1549–1556, 1996.
111. Burg, P., Fydrych, P., Cagniant, D., Nansé, G., Bimer, J., and Jankowska, A. *Carbon* 40: 1521–1531, 2002.
112. Meldrum, B.J. and Rochester, C.H. *Fuel* 70: 57–63, 1991.
113. Koch, A., Krzton, A., Finqueneisel, G., Heintz, O., Weber, J.V., and Zimny, T. *Fuel* 77: 563–569, 1998.
114. Weber, J.V., Koch, A., and Robert, D. *J. Therm. Anal. Calorimetry* 53: 11–17, 1998.
115. Heintz, O., Petryniak, J., Kich, A., Krzton, A., Machnikowski, J., Weber, J.V., and Zimny, T. *The European Conference "Carbon 96,"* Newcastle, U.K., July 1996, Extended Abstracts, Vol. I, pp. 40–41.
116. Mastalerz, M. and Bustin, R.M. *Int. J. Coal Geol.* 33: 43–59, 1997.
117. Landais, P. and Rochdi, A. *Fuel* 72: 1393–1401, 1993.
118. Shin, S., Jang, J., Yoon, S.H., and Mochida, I. *Carbon* 35: 1739–1743, 1997.
119. Do, T.T., Celina, M., and Fredericks, P.M. *Polym. Degrad. Stabil.* 77: 417–422, 2002.
120. Proctor, A. and Sherwood, P.M.A. *Anal. Chem.* 54: 13–19, 1982.
121. Proctor, A. and Sherwood, P.M.A. *J. Electron Spectrosc. Relat. Phenom.* 27: 39–56, 1982.
122. Proctor, A. and Sherwood, P.M.A. *Surf. Interface Anal.* 4: 212–219, 1982.
123. Kozlowski, C. and Sherwood, P.M.A. *Carbon* 24: 357–363, 1986.
124. Watt, M., Allen, W., and Fletcher, T.H. In *Coal Science*, Pajares, J.A. and Tascon, J.M.D. Eds. Elsevier Science, 1995, pp. 1685–1688.
125. Zhu, Q., Thomas, K.M., and Russell, A.E. *The European Carbon Conference "Carbon 96,"* Newcastle U.K., 1996, pp. 14–15.
126. Zhu, Q., Money, S.L., Russell, A.E., and Thomas, K.M. *Langmuir* 13: 2149–2157, 1997.
127. Pels, J.R., Kapteijn, F., Moulijn, J.A., Zhu, Q., and Thomas, K.M. *Carbon* 33: 1641–1653, 1995.
128. Kelemen, S.R., Gorbaty, M.L., and Kwiatek, P.J. *Energy Fuels* 8: 896–906, 1994.

129. Kelemen, S.R., Gorbaty, M.L., Kwiatek, P.J., Fletcher, T.H., Watt, M., Solum, M.S., and Pugmire, R.J. *Energy Fuel* 12: 159–173, 1998.
130. Turner, J.A. and Thomas, K.M. *Langmuir* 15: 6416–6422, 1999.
131. Kambara, S., Takarada, T., Toyashima, M., and Kato, K. *Fuel* 74: 1247–1253, 1995.
132. Kapteijn, F., Moulijn, J.A., Matzner, S., and Boehm, H.P. *Carbon* 37: 1143–1150, 1999.
133. Wojtowicz, M.A., Pels, J.R., and Moulijn, J.A. *Fuel* 74: 507–516, 1995.
134. Kelemen, S.R., George, G.N., and Gorbaty, M.L. *Fuel* 69: 939–944, 1990.
135. Gorbaty, M.L., George, G.N., and Kelemen, S.R. *Fuel* 69: 945–949, 1990.
136. Kozlowski, M. *Fuel* 83: 259–265, 2004.
137. Olivella, M.A., del Rio, J.C., Palacios, J., Vairavamurthy, M.A., and de las Heras, F.X.C. *J. Anal. Appl. Pyrol.* 63: 59–68, 2002.
138. Olivella, M.A., Palacios, J.M., Vairavamurthy, A., del Rio, J.C., and de las Heras, F.X.C. *Fuel* 81: 405–411, 2002.
139. Kasrai, M., Brown, J.R., Bancroft, G.M., Yin, Z., and Tan, K.H. *Int. J. Coal Geol.* 32: 107–135, 1996.
140. Kelemen, S.R., Afeworki, M., Gorbaty, M.L., and Cohen, A.D. *Energy Fuels* 16: 1450–1462, 2002.
141. Takahagi, T. and Ishitani, A. *Carbon* 22: 43–46, 1984.
142. Hontoria-Lucas, C., Lopez-Peinado, A.J., De D Lopez-Gonzalez, J., Rojas-Cervantes, M.L., and Martin-Aranda, R.M. *Carbon* 33: 1585–1592, 1995.
143. Grzybek, T., Pietrzak, R., and Wachowska, H. *Fuel Process. Technol.* 77–78: 1–7, 2002.
144. Desimoni, E., Salvi, A.M., Langerame, F., and Watts, J.F. *J. Electron Spectrosc.* 85: 179–181, 1997.
145. Gong, B., Pigram, P.J., and Lamb, R.W. *Fuel* 77: 1081–1087, 1998.
146. Lee, W.H., Kim, J.Y., Ko, Y.K., Reucroft, P.J., and Zondlo, J.W. *Appl. Surf. Sci.* 141: 107–113, 1999.
147. Snape, C.E., McGhee, B.J., Martin, S.C., and Andresen, J.M. *Catal. Today* 37: 285–293, 1997.
148. Andresen, J.M., Martin, Y., Moinelo, S.R., Maroto-Valer, M.M., and Snape, C.E. *Carbon* 36: 1043–1050, 1998.
149. Nagano, Y., Gouali, M., Monjushiro, H., Eguchi, T., Ueda, T., Nakamura, N., Fukumoto, T., Kimura, T., and Achiba, Y. *Carbon* 37: 1509–1515, 1999.
150. Schmiers, H., Friebel, J., Streubel, P., Hesse, R., and Köpsel, R. *Carbon* 37: 1965–1978, 1999.
151. Murata, S., Hosokawa, M., Kidena, K., and Nomura, M. *Fuel Process. Technol.* 67: 231–243, 2000.
152. Knicker, H., Hatcher, P.G., and Scaroni, A.W. *Energy Fuels* 9: 999–1002, 1995.
153. Solum, M.S., Pugmire, R.J., Grant, D.M., Kelemen, S.R., Gorbaty, M.L., and Wind, R.A. *Energy Fuels* 11: 491–494, 1997.
154. Kelemen, S.R., Afeworki, M., Gorbaty, M.L., Kwiatek, P.J., Solum, M.S., Hu, J.Z., and Pugmire, R.J. *Energy Fuels* 16: 1507–1515, 2002.
155. Auer, E., Freund, A., Pietsch, J., and Tacke, T. *Appl. Catal. A.* 173: 259–271, 1998.
156. Jansen, R.J.J. and Van Bekkum, H. *Carbon* 33: 1021–1027, 1995.
157. Burg, P., Cagniant, D., Fydrych, P., Magri, P., Gruber, R., Bimer, J., Nansé, G., and Jankowska, A. *Fuel Process. Technol.* 79: 233–237, 2002.
158. Montes-Moran, M.A., Martinez-Alonso, A., and Tascon, J.M.D. *Fuel Process. Technol.* 77–78: 359–364, 2002.
159. Simon, F., Jacobash, H.J., Pleul, D., and Uhlmann, P. *Progr. Colloid Polym. Sci.* 101: 184–188, 1996.

160. Vickers, E., Watts, J.F., Perruchot, C.H., and Chemini, M.M. *Carbon* 38: 675–689, 2000.
161. Burg, P. and Cagniant, D. *Carbon* 41: 1031–1035, 2003.
162. Hüttinger, K.J., Höhmann-Wien, S., and Krekel, G. *Carbon* 29: 1281–1286, 1991.
163. Fowkes, F.M. *J. Adhes. Sci. Technol.* 1: 7–27, 1987.
164. Fowkes, F.M. *Ind. Eng. Chem.* 56: 41–52, 1964.
165. Zielke, U., Hüttinger, K.J., and Hoffman, W.P. *Carbon* 34: 1007–1013, 1996.
166. Zielke, U., Hüttinger, K.J., and Hoffman, W.P. *Carbon* 34: 999–005, 1996.
167. Zielke, U., Hüttinger, K.J., and Hoffman, W.P. *Carbon* 34: 1015–1026, 1996.
168. Conder, J.R. and Young, C.L. Eds. *Physicochemical Measurements by Gas Chromatography*. Wiley, New York, 1979. pp. 430–530.
169. Lloyd, D.R., Ward, T.C., and Shreiber, H.P. Eds. *Inverse Gas Chromatography, Characterization of Polymers and Other Materials*. American Chemical Society Symposium Series, 391, Washington, D.C., 1989, pp. 248-261.
170. Burg, P., Abraham, M.H., and Cagniant, D. *Carbon* 41: 867–879, 2003.
171. Dong, S., Brendlé, M., and Donnet, J.B. *Chromatographia* 28: 469–472, 1989.
172. Dorris, G.M. and Gray, D.G. *J. Colloid Interface Sci.* 77: 353–362, 1980.
173. Derouane, E.G., Andre, J.M., and Lucas, A.A. *J. Catal.* 110: 58–73, 1988.
174. St-Flour, C. and Papirer, E. *Ind. Eng. Chem. Prod. Res. Dev.* 21: 666–669, 1982.
175. Schultz, J., Lavielle, L., and Martin, C.J. *J. Chim. Phys.* 84: 231–237, 1987.
176. Lloyd, D.R., Ward, T.C., and Shreiber, H.P. Eds. *Inverse Gas Chromatography, Characterization of Polymers and Other Materials*. American Chemical Society Symposium Series, 391, Washington, D.C., 1989, pp. 185–202.
177. Brendlé, E. "Etude des propriétés de surface d'oxydes de fer". Ph.D. thesis. Université de Haute-Alsace, France, 1997.
178. Papirer, E. and Brendlé, E. *J. Chem. Phys.* 95: 122–149, 1998.
179. Wiener, H. *J. Am. Chem. Soc.* 69: 2636–2638, 1947.
180. Barysz, M., Jashari, G., Lall, R.S., Srivastava, V.K., and Trinajstic, N. *Chem. Appl. Topol. Graph Theory, Stud. Phys. Theor. Chem.* 28: 222–230, 1983.
181. St-Flour, C. and Papirer, E. *J. Colloid Interface Sci.* 91: 69–75, 1983.
182. Gutmann, V. The *Donor-Acceptor Approach to Molecular Interactions*, Plenum Press, New York, 1978.
183. Sidqi, M., Ligner, G., Jagiello, J., Balard, H., and Papirer, E. *Chromatographia* 8: 588–592, 1989.
184. Kovats, E. *Helv. Chim. Acta* 41: 1915–1932, 1958.
185. Guttierez, M.C., Rubio, J., Rubio, F., and Oteo, J.L. *J. Chromatogr.* 845: 53–66, 1999.
186. McReynolds, W.O. *J. Chromatogr. Sci.* 8: 685–691, 1970.
187. Larionov, O.G., Petrenko, V.V., and Platonova, N.P. *J. Chromatogr.* 537: 295–303, 1991.
188. Abraham, M.H., Doherty, R.M., Kamlet, M.J., and Taft, R.W. *Chem. Brit.* 551–553, 1986.
189. Poole, C.F. and Poole, S.K. *J. Chromatogr.* 965: 263–299, 2002.
190. Abraham, M.H. and Platts, J.A. *J. Org. Chem.* 66: 3484–3491, 2001.
191. Abraham, M.H., Grenn, C.J., and Acree, W.E., Jr. *J. Chem. Soc. Perkin. Trans. 2*, 281–286, 2000.
192. Abraham, M.H., Whiting, G.S., Doherty, R.M., and Shuely, W.J. *J. Chem. Soc. Perkin Trans. 2*, 1451–1460, 1990.
193. Abraham, M.H., Whiting, G.S., Doherty, R.M., and Shuely, W.J. *J. Chromatogr.* 587: 213–228, 1991.
194. Abraham, M.H. and Whiting, G.S. *J. Chromatogr.* 594: 229–241, 1992.
195. Abraham, M.H., Grellier, P.L., and McGill, R.A. *J. Chem. Soc. Perkin Trans. 2* 797–803, 1987.

196. Abraham, M.H., Chadha, H.S., Martins, F., Mitchell, R.C., Bradbury, M.W., and Gratton, J.A. *Pest. Sci.* 55: 78–88, 1999.
197. Abraham, M.H., Poole, C.F., and Poole, S.K. *J. Chromatogr.* 842: 79–114, 1999.
198. Vagner, C., Finqueneisel, G., Zimny, T., Burg, P., Grzyb, B., Machnikowski, J., and Weber, J.V. *Carbon* 41: 2847–2853, 2003.
199. Starck, J., Burg, P., Cagniant, D., Tascon, J.M.D., and Martinez-Alonso, A. *Fuel* 83: 845–850, 2004.
200. Burg, P., Fydrych, P., Abraham, M.H., Matt, M., and Gruber, R. *Fuel* 79: 1041–1045, 2000.
201. Boehm, H.P. *Carbon* 40: 145–149, 2002.

4 Sorption of Heavy Oils into Carbon Materials

Masahiro Toyoda, Norio Iwashita, and Michio Inagaki

Keywords: Exfoliated graphite, heavy oil, sorption, recovery, recycling

CONTENTS

I. INTRODUCTION

There have been a number of oil spill accidents in the world. In March 1989, an oil tanker accident occurred in Alaska and about 3.6×10^4 tons of heavy oil was spilled, which caused serious damage to the environment [1], and the situation was reported to be serious even 10 years later [2]. A number of tanker accidents can be pointed out, such as the oil spill of about 26×10^4 tons off Angora in May 1991, the wide spread of about 9×10^4 tons of spilled oil in the North Sea in January 1993, the collision of two tankers and the resulting serious disturbance of ship transport with about 2×10^4 tons of oil spilled in the Strait of Malacca in October 1997, the serious disaster on the coast of western France in December 1999, the contamination by 70×10^4 L of heavy and diesel oils that caused anxiety about the survival of ancient species in the Galapagos Islands in January 2001, and the stranding of a tanker on the coast of Spain, with about 4×10^4 tons of oil spilled in November 2002. In addition to these accidents, which were mainly due to nasty weather, the demolition of storage tanks in Kuwait during the 1991 Gulf War spilled a large amount of heavy oil into the sea. Another possibility for oil pollution of the ocean occurs during oil rig drilling.

In January 1997, the oil tanker *Nakhodka* spilled almost 4.5×10^3 tons of oil near the coast of Japan Sea. This accident had a strong impact because the spilled oil moved along the coastline, extending over some 250 km, and it seriously contaminated the shoreline [3]. Actually, this particular tanker accident motivated the present series of investigations, that is, sorption, recovery, and recycling of spilled heavy oils using macroporous carbon materials.

Although such disastrous accidents resulted in massive oil spills, it has been reported [4] that tankers and rig accidents accounted for only 12.5 and 1.5% of the total amount of spilled oil, respectively. The principal loss of oil actually occurs during its transportation and storage [4]. A continuous leaking of oil through pipe joints, for example, may produce serious contamination of soil, river water, and, sometimes, even subterranean water, and this has detrimental effects on human life, as well as on various plants, fishes, and water [5]. These oil spills, although not as massive as tanker accidents, have occurred frequently and resulted in not only a great deal of damage to the environment, including its ecological cycles, but also in a great loss of energy resources, particularly heavy oils. The occurrence of these oil spill accidents has been discussed from different points of view, mainly with a view to protecting the environment [6–10].

A common treatment technique used in oil spill accidents on the sea is containment of the oil with large floating barriers (so-called oil fences), followed by skimming with specialized ships that either vacuum the oil off the sea or soak it up with sorbent materials [11,12]. In most cases, skimming has been done manually, as shown in Figure 4.1a. The primary cleanup method on the shore has also been manual—dipping up of the water containing the spilled oil and burning it out in specialized furnaces, or removing the sand and rocks contaminated by the spilled oil. In the case of the *Nakhodka* accident on the Japan Sea coast, different treatment techniques were applied: manual skimming of spilled oil confined

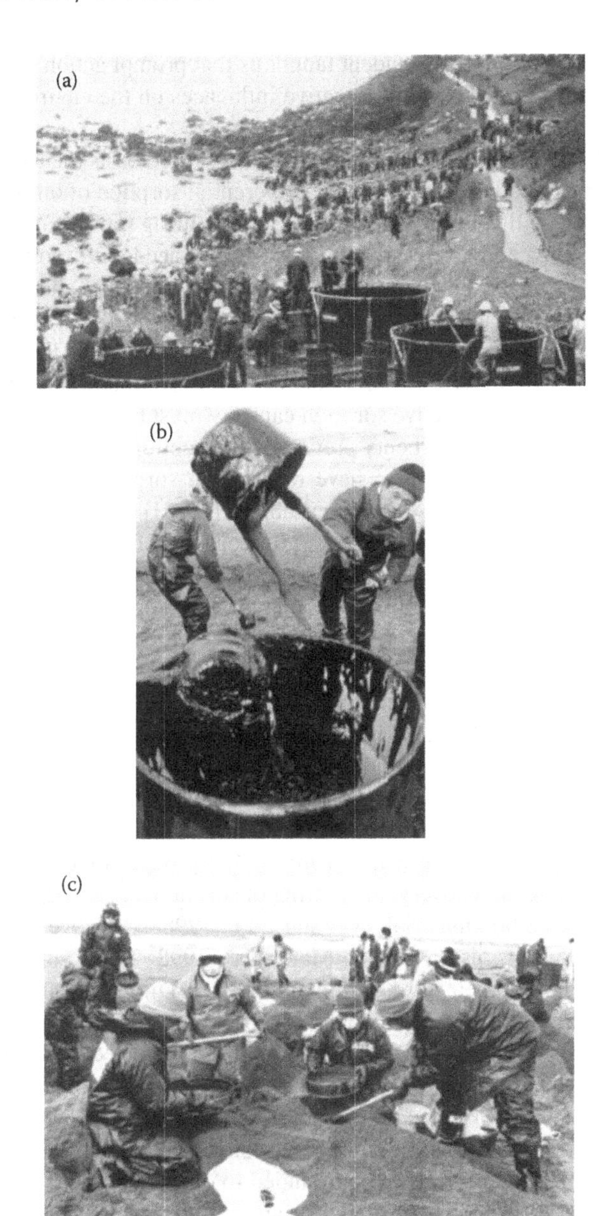

FIGURE 4.1 Manual skimming of spilled heavy oil (*Nakhodka* accident in January 1997).

in large floating barriers; dipping up the spilled oil by using ladles and containers, and burning it out (Figure 4.1b); and removing the contaminated sands and rocks (Figure 4.1c). These on-site treatment activities were carried out by more than 200,000 people—most of them volunteers living near the coast, and some even coming from different places far from the coast—treatment continued for

3 months. Lessons from this accident taught us that prompt action is crucial and, also, that we have to expect some negative influences on the environment over a long period.

One of the techniques for skimming massive spilled oil is the use of sorbents with high sorption capacity and rate, the preferential sorption of oil being desired in the case of oil spills in water. For leakage of oil during transportation, sorbents with large capacity, and also easy manipulation, are needed to avoid serious pollution. So far, some porous polymers such as poly(propylene) and poly(ethylene terephthalate) have been used for the sorption of spilled oil. Their sorption capacity is in the range of 10 to 30 kg of heavy oil per kg of polymer [13,14]. However, most polymers sorb water as well as heavy oil, with no special selectivity for heavy oils, and so their effective sorption capacity must be lower than the figures mentioned. Some natural sorbents prepared from cotton fiber, milkweed floss, and kenaf plant were reported to have rather high sorption capacity and have potential for oil recovery and sorbent reusability [15–21].

As early as in 1979, a patent proposal for sorption of heavy oil by exfoliated graphite was presented in Japan [22]. However, it did not attract attention, and more detailed studies were not carried out. A paper on the sorption of heavy oil into exfoliated graphite was first presented at a carbon conference by a Chinese group in 1996 [23]. The reported capacity of the exfoliated graphite used was about 25 kg/kg. Unfortunately, this paper was not subsequently published in any international journal, and so it did not attract greater worldwide attention. In 1997, just after the *Nakhodka* accident, we followed up on the previous studies of heavy oil sorption into exfoliated graphite and found that this material was able to sorb a large amount of the heavy oil floating on water at room temperature very quickly—more than 80 kg/kg within 1 min [24,25]. Based on our results, a new research project on the recovery and cycling of sorption and desorption of spilled heavy oils using carbon materials was started in 1998—"Recovery and Recycle of Spilled Heavy Oils Using Carbon Materials" (1998–2000), supported by the New Energy Development Organization (NEDO). Different carbon materials were studied through the determination of sorption capacity, sorption kinetics, and repeated sorption–desorption cycles. The principal objectives of this project were the following: (1) recovery of spilled heavy oils by using carbon materials and (2) recycling both the heavy oils and carbon materials. Promising results were obtained with exfoliated graphite, carbonized fir fibers, and carbon fiber felt [24–51]. This research was then expanded to other oils, such as cooking and engine lubrication oils. (joint research program between Japan Society for the Promotion of Science [JSPS] and National Natural Science Foundation of China [NSFC] [2000–2003] and also a frontier research project at the Aichi Institute of Technology supported by the Japanese Ministry of Education.) The Chinese group worked on the sorption of not only heavy oils but also biological fluids [52]. Recently, these studies were reviewed and discussed in relation to our work [53].

The results of the very high sorption capacity of exfoliated graphites promoted detailed studies on the characterization of their pore structure using newly developed techniques such as image analysis [54–62]. Previously in this series, we reviewed exfoliation phenomena of graphite flakes and the applications of exfoliated graphite [63].

In the present review we summarize the experimental results of studies on the sorption capacity and kinetics for heavy oils, mainly A-grade and viscous C-grade heavy oils, and the cycling performance of sorption and recovery on various carbon materials: exfoliated graphite, carbonized fir fibers, and carbon fiber felt. Macroporous (in other words, bulky) carbon materials showed efficient sorption capacity. However, because the bulkiness of carbon materials caused difficulties in handling, storage, and also reuse, it was necessary to study the recycling of not only the heavy oils recovered but also of the carbon sorbents. The recycling performance of only heavy oil sorption was studied, although there may be other possibilities for the recycling of carbon materials. Based on these fundamental studies, some preliminary experiments for the practical recovery of spilled heavy oils were carried out on a small scale in the laboratory.

II. CARBON MATERIALS AND HEAVY OILS USED

Exfoliated graphite has been industrially produced in huge amounts and is formed into flexible graphite sheets for various industrial applications [63–65]. It is prepared by rapid heating to a high temperature, about 1000°C, of residue compounds of natural graphite flakes with sulfuric acid, this treatment being usually called *exfoliation*. Recently, the exfoliation of graphite was also efficiently performed by microwave irradiation of the residue compounds at room temperature [66–68]. Because this exfoliation was caused by rapid decomposition of residual sulfuric acid molecules—mostly located in the gallery between graphite basal planes—into gaseous species, the original graphite flakes were converted to wormlike particles owing to preferential exfoliation perpendicular to the basal planes. A typical scanning electron microscopy (SEM) image of these particles is shown in Figure 4.2a. In exfoliated graphite, three kinds of pores are differentiated: large pores (or void spaces) between the wormlike particles (Figure 4.2a), crevicelike pores on the surface of the wormlike particles (Figure 4.2b), and small pores inside the particles, which are usually ellipsoidal (Figure 4.2c). Large void spaces formed in exfoliated graphite were studied by mercury porosimetry using a special dilatometer [54] and also by impregnation of paraffin oil [58]. Crevice-like pores on particle surfaces were evaluated by measuring the distance between the edges [55]. Small pores inside the particles were analyzed with the aid of image processing, determining the distribution of cross-sectional areas, lengths along the major and minor axes, and aspect ratio [56,57,59]. Based on these studies, the exfoliation process was discussed [60–62].

Fibrous components of the fir tree (*Abis sachalinensis Fr. Schm*), which are waste material in forestry, were separated from the lignin components by treating with saturated water vapor (about 9 kgf/cm^2 at 179°C) for 1 min [69]. The fibers thus separated were carbonized at either 380 or 900°C for 1 h in a flow of high-purity argon gas, the solid yield being about 31 or 18 mass%, respectively. The appearance of fibers carbonized at 900°C is shown in Figure 4.3. Three kinds of pores

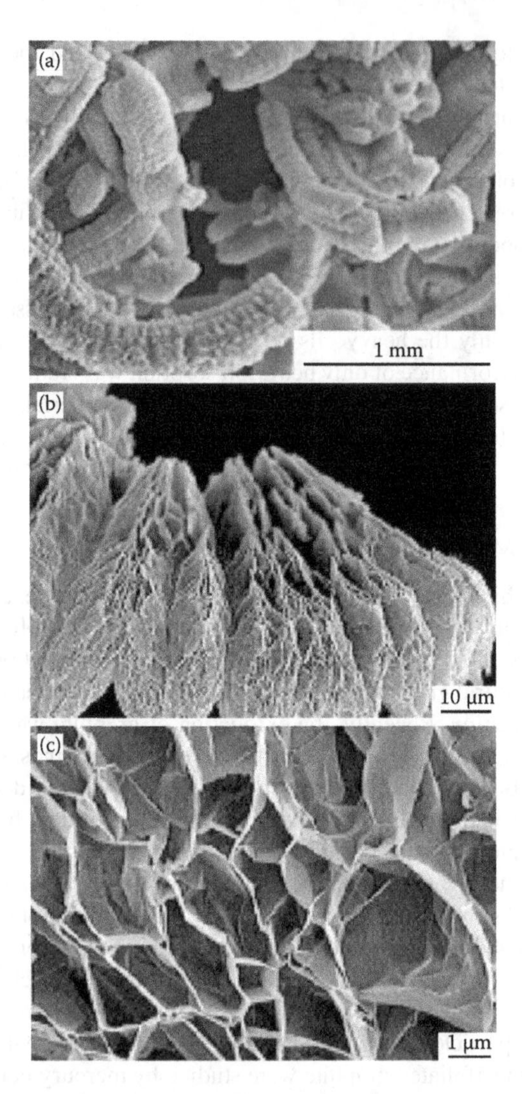

FIGURE 4.2 SEM images of exfoliated graphite. (a) Wormlike particles and large spaces among them, (b) crevice-like pores on the surface of wormlike particles, and (c) pores inside of the particle (cleaved cross section).

are also observed here: macropores (void spaces) between the fibers (Figure 4.3a); pores along the fiber axes with rectangular cross section (Figure 4.3b); and small round pores on the walls of fibers that seem to be connected to the rectangular pores (Figure 4.3c). The bulk density of the fir fibers was controlled by compressing the original fibers in an alumina crucible before carbonization, and its value was determined after carbonization from volume and mass measurements.

Four grades of commercially available carbon fiber felt were used, consisting of either PAN-based or pitch-based carbon fibers. Their bulk density was

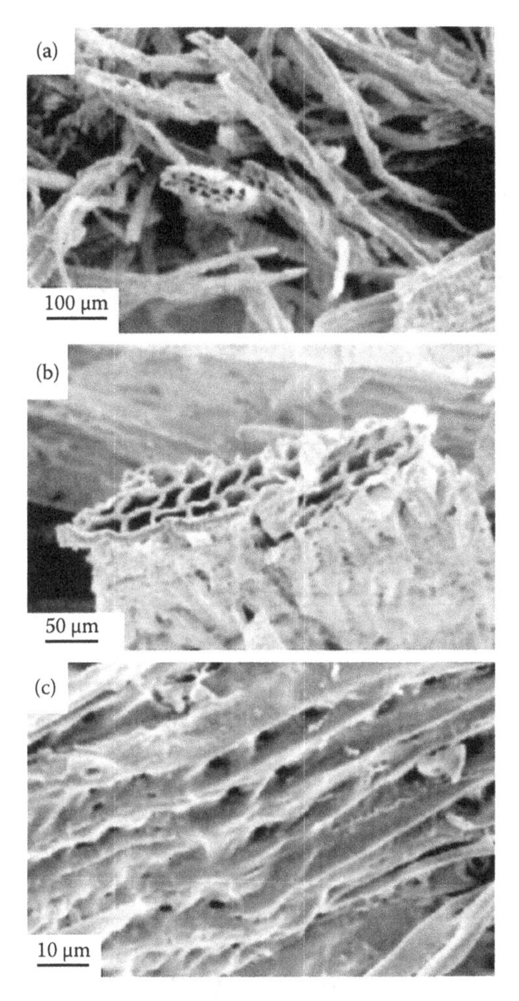

FIGURE 4.3 SEM images of fir fibers carbonized at 900°C. (a) Fibrous particles and large spaces between them, (b) rectangular pores along the fiber axis, and (c) small pores on the wall of the fibers.

determined from volume and mass before heavy oil sorption. A SEM image of one carbon fiber felt is shown in Figure 4.4. Here, only one kind of pore exists, that is, macropores (large void spaces) between fibers, because each fiber has a rather smooth surface and is essentially nonporous because of the low BET (Brunauer Emmett Teller) surface area (a few m^2/g).

In addition to these three carbon materials, different charcoals and fibrous carbon materials, including activated carbon fibers, were used for comparison.

The viscosity and the density of heavy oils used in these studies are listed in Table 4.1, with the code employed in Japanese Industrial Standard (JIS-K2204-1980). The less viscous A-grade and the highly viscous C-grade were used in most experiments.

FIGURE 4.4 SEM image of carbon fiber felt showing large void spaces among carbon fibers.

TABLE 4.1
Heavy oils used

Heavy Oil	Specific Gravity (kg/m^3)	Viscosity at Room Temperature (Pa·s)
A-grade	864	0.004
B-grade	890	0.27
C-grade	945	0.35
Crude	826	0.004

III. SORPTION CAPACITY FOR HEAVY OILS

A small piece of carbon material is added either onto heavy oil floating on water or heavy oil alone in a 500-mL beaker.

Typical changes in appearance are shown in Figure 4.5 of the heavy oil floating on water by the action of exfoliated graphite. A-grade heavy oil floating on water (Figure 4.5a) was completely sorbed into the exfoliated graphite added (Figure 4.5b). The characteristic brown color of A-grade heavy oil disappeared within 1 min after the addition of exfoliated graphite when the amount of heavy oil was less than the sorption capacity of the exfoliated graphite added. The exfoliated graphite with sorbed oil lost its luster and appeared deep black (Figure 4.5b). Even after sorption of heavy oil, the exfoliated graphite remained floating on water. After removing the exfoliated graphite, no contamination appears in the water and also no contamination is observed even if it is transferred onto a white filter paper, as shown in Figure 4.5d. When the amount of heavy oil was a little

FIGURE 4.5 Appearance of sorption and recovery of heavy oil on water using exfoliated graphite (EG). (a) A-grade heavy oil floating on water, (b) 2 min after the addition of EG with less amount of oil than the sorption capacity of EG, (c) heavy oil with more than the sorption capacity of EG added, and (d) after transfer of EG after sorbing heavy oil onto a white filter paper.

larger than the sorption capacity of the exfoliated graphite added, the periphery of the graphite was trimmed by oil. When a large quantity of excess oil is present, the entire exfoliated graphite appears wetted, and the brown color of oil remained on water, as shown in Figure 4.5c. The exfoliated graphite with the sorbed heavy oil could easily be separated from water by conventional filtration.

A piece of exfoliated graphite was placed on the surface of A-grade heavy oil directly; it was picked up by using a stainless steel mesh after 2 h of immersion and then kept for 1 h to drain off the excess oil. In the case of viscous C-grade heavy oil, immersion for 15 h and draining-off for 3 h was employed.

The sorption capacity was calculated by measuring the mass of exfoliated graphite before and after sorption; it is expressed as kilograms of sorbed heavy oil per kilogram of carbon. When a stainless steel mesh was used for picking up the sorbing material after sorption of oil, the mass of the oil that adhered onto the mesh was subtracted from the observed mass increase, though it was negligibly small in the case of A-grade oil.

A. Exfoliated Graphite

In Figure 4.6, sorption capacities of two exfoliated graphite samples with slightly different bulk densities (EG-1 and EG-2) are compared on four grades of heavy oil (Table 4.1). In the case of A-grade heavy oil, up to 83 kg was sorbed into 1 kg of the exfoliated graphite EG-1, whose bulk density was 6 kg/m^3. This sorption capacity is much higher than that of poly(propylene) nonwoven web, which has

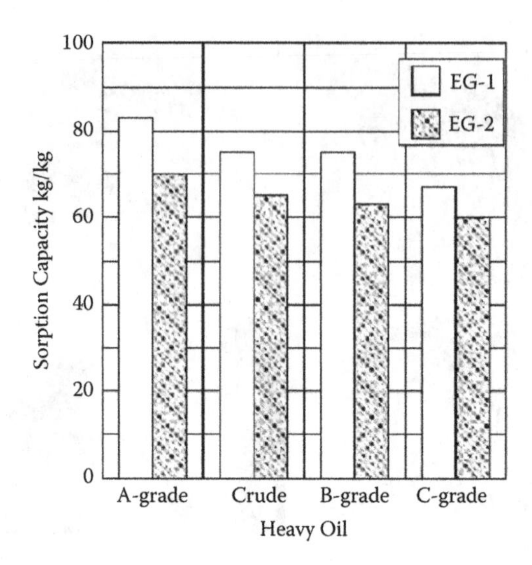

FIGURE 4.6 Sorption capacity of two exfoliated graphite samples with slightly different bulk densities for four heavy oils.

also been used as a sorbent for heavy oils [17]. The sorption rate of A-grade heavy oil was high: it was essentially complete within 1 min. The exfoliated graphite with a bulk density of 10 kg/m^3 had a slightly lower sorption capacity (about 70 kg/kg). The sorption capacities of these two samples were similar for crude oil, being 75 and 65 kg/kg, respectively. The sorption rate was also high, reaching saturation within 2 min. In the case of C-grade heavy oil, which has a much higher viscosity, the sorption capacity was only somewhat less, 67 and 60 kg/kg of EG-1 and EG-2, respectively; however, sorption proceeded very slowly—about 8 h being needed for saturation. The B-grade heavy oil showed a capacity similar to that of crude oil, but its sorption rate was similar to that of C-grade oil. These results suggested a dependence of sorption capacity of exfoliated graphite on the viscosity of the sorbate heavy oils, and also a strong dependence on the bulk density of the sorbent. This is illustrated in Figure 4.7. Thus, for example, viscous C-grade heavy oil was not sorbed at all when the bulk density of the sorbent was more than 40 kg/m^3. In the whole range of bulk density, sorption capacity for viscous heavy oils (C- and B-grade) was smaller than for the less viscous oils (A-grade and crude). A bulk density of 100 kg/m^3 was achieved by a strong compression, and the product lost the key characteristics of exfoliated graphite, that is, bulky association of wormlike particles.

In Figure 4.8 it is seen that sorption capacity is linearly related to total pore volume for pores with a 1- to 600-μm radius, measured by mercury porosimetry using a special dilatometer [54]. Therefore, it can be concluded that effective pore sizes for sorption of heavy oil are in the range of 1 to 600 μm, most of which are located among the wormlike particles; in other words, most of the heavy oil is sorbed into the void spaces formed by entanglements of the wormlike particles of

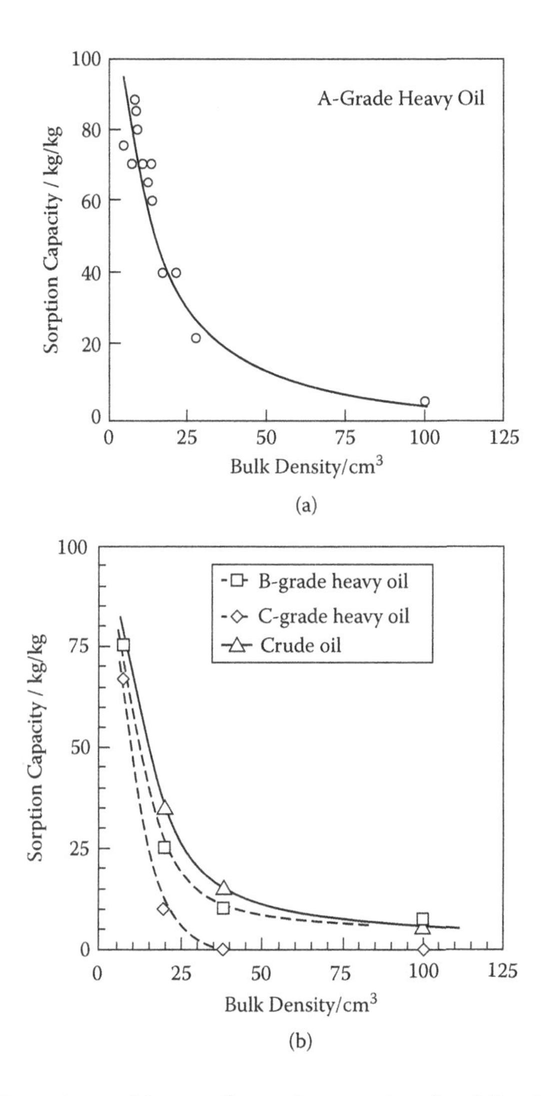

FIGURE 4.7 Dependence of heavy oil sorption capacity of exfoliated graphite on its bulk density. (a) A-grade heavy oil, (b) B- and C-grade heavy oils and crude oil.

exfoliated graphite, as shown in Figure 4.2a. Experimental points for total pore volume above 0.1 m^3/kg tend to deviate from the linear relationship, which may suggest that the larger void spaces cannot be completely filled with heavy oil.

Large void spaces among wormlike particles were quantitatively evaluated by image analysis using thin slices prepared from an exfoliated graphite after impregnation with paraffin oil. To compare the volume of these large spaces with sorptivity, they were found to be responsible for about 70% of total heavy oil sorption capacity [58]. However, crevicelike pores on the surface of particles and the ellipsoidal pores inside the particles also have important roles in heavy oil

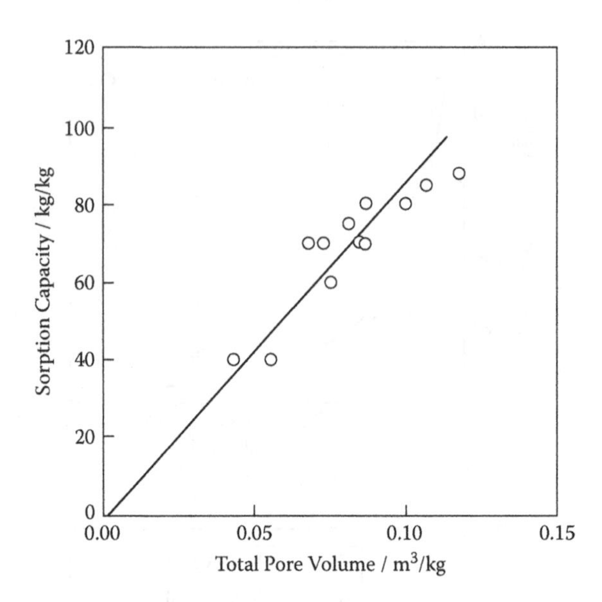

FIGURE 4.8 Sorption capacity versus total pore volume of exfoliated graphites with different bulk densities.

sorption. Observation under an optical microscope showed that oil was coming up at the edges of crevices (Figure 4.2b) formed on the particle surface in the beginning of sorption, to occupy whole crevicelike pores and then fill large void spaces quickly [36]. Such complicated pore structure is thought to result in rather strong holding of sorbed heavy oil that did not segregate on the filter paper during filtration to recover the sorbent from the water surface (Figure 4.5d). The oleophilic (hydrophobic) nature of the carbon surface is also advantageous for such sorption and occlusion of oils.

In Figure 4.9a, it is seen that sorption capacity tends to decrease with decreasing temperature. In the case of viscous C-grade heavy oil, this temperature dependence is so pronounced that no sorption is detected below 15°C. At the other extreme, in the case of less viscous A-grade heavy oil, only a slight decrease in capacity is observed. This temperature dependence was reasonably supposed to be due to changes in viscosity of the oil with temperature. In Figure 4.9b, therefore, sorption capacity observed at different temperatures (Figure 4.9a) was replotted against viscosity, and a pronounced dependence was observed. From such data, this dependence may be divided into a low-viscosity range, <0.01 Pa·s, and a high-viscosity range, >0.2 Pa·s.

The sorption capacity of exfoliated graphite was also measured for other oils, including kerosene, different cooking oils, and motor oils, most of which had a viscosity between 0.01 and 0.2 Pa·s [51]. When these data are included, the correlation is found to be continuous, as shown in Figure 4.9c, that is, the dependence of sorption capacity on oil viscosity shows a plateau in an intermediate viscosity range. Such a relationship may be explained as follows: oils having a viscosity

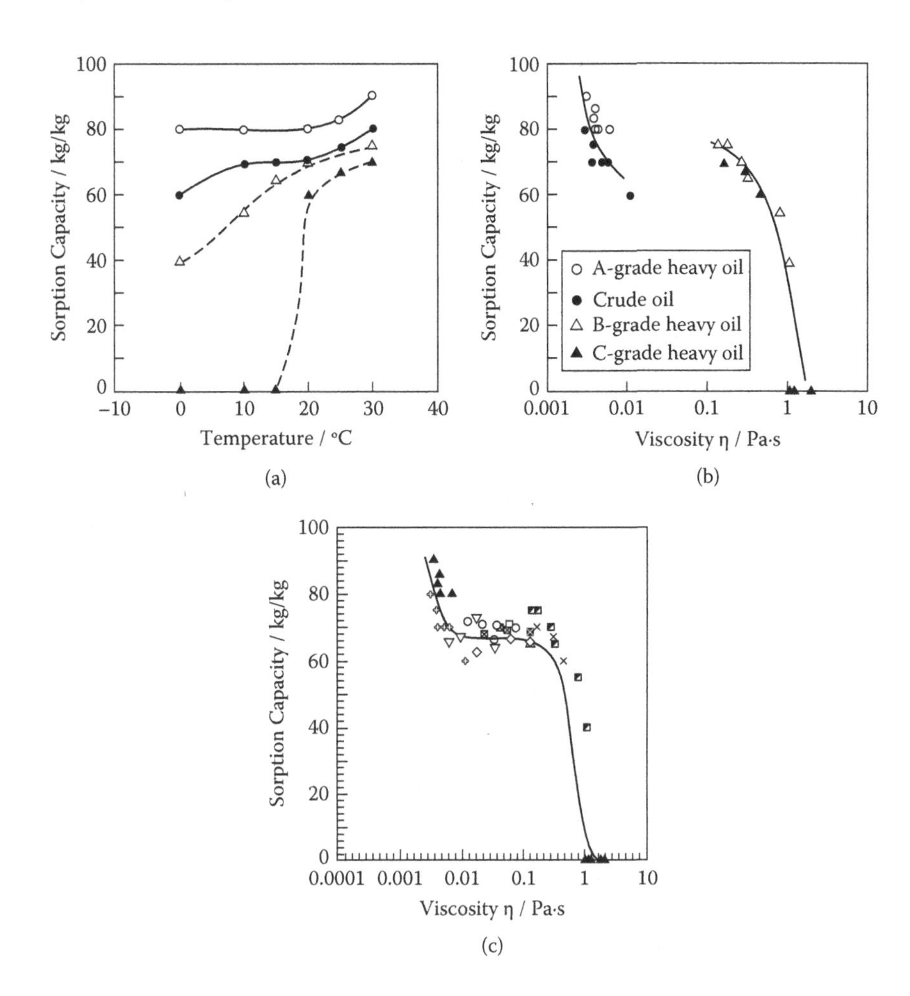

FIGURE 4.9 Dependence of sorption capacity of exfoliated graphite on oil viscosity. (a) Effect of temperature, (b) effect of viscosity for the four heavy oils, and (c) effect of viscosity for different oils, including the heavy oils studied here. (From Inagaki, M., Toyoda, M., and Nishi, Y., *Kagaku Kougaku* 64, 179, 2001 [in Japanese]. With permission.)

in the range of 0.01 to 0.2 Pa·s can be sorbed at about 60 to 80 kg/kg into exfoliated graphite with a bulk density of about 6 kg/m^3, where almost all pores seem to be filled with oil. In contrast, oils with a viscosity above 0.2 Pa·s cannot gain access to smaller pores, such as those pores inside the wormlike particles and tips of crevicelike pores on their surface. Oils with viscosity higher than 1 Pa·s cannot even pass through the smaller pores and channels that are the entrances to the inside void spaces of exfoliated graphite. When viscosity is less than 0.01 Pa·s, a sorption capacity of more than 80 kg/kg was observed; this may suggest more efficient filling of pores, particularly the larger ones. More experimental and theoretical studies are required on the dependence of sorption capacity of carbon materials on the viscosity of sorbate liquids, by taking into consideration the experimental fact that sorption capacity for water is only a few kg/kg, as will be shown in Section IV.

The sorption behavior of exfoliated graphite for several biomedical molecules, including ovalbumin, serum albumin, bovine serum albumin (BSA), lysine, and herring sperm DNA, was also studied [53]. The sorption capacity and behavior of these molecules are very similar to those of heavy oils.

B. Carbonized Fir Fibers

In Figures 4.10a and 4.10b, the sorption capacity of fir fibers carbonized at 380 and 900°C for A-grade and C-grade heavy oils, respectively, is plotted against their bulk density. Some of the experimental points for exfoliated graphite (Figure 4.7) are also shown for comparison. A strong bulk density dependence is seen for both A-grade and C-grade heavy oils. This is similar to the case of exfoliated graphite, even though the experimental points for the fibers are scattered over a rather wide range. For A-grade heavy oil, the capacity seems to be slightly lower than that of exfoliated graphite, particularly in the range of low bulk density, but it is still rather high, 60 to 80 kg per kg of the fibers having bulk density of about 6 kg/m^3. Above 40 kg/m^3, the sorption capacity is comparable to that of exfoliated graphite, 10 to 20 kg/kg. For viscous C-grade heavy oil, however, sorption capacity of the fibers is much higher than that of exfoliated graphite, particularly those having bulk density above 10 kg/m^3. Although exfoliated graphite with bulk density of more than 40 kg/m^3 could not sorb C-grade heavy oil, carbonized fir fibers with the same bulk density could sorb about 15 kg/kg (Figure 4.10b).

This strong dependence of sorption capacity on the bulk density of carbonized fir fibers suggests that the spaces formed between entangled fibrous particles with irregular surfaces, as shown in Figure 4.3, are primarily responsible for heavy oil sorption.

As shown in Figure 4.10, carbonization temperature does not seem to have a marked effect on sorption capacity for both A-grade and C-grade oils. A wide range of carbonization temperatures, 380 to 1200°C, was employed, but no marked effect on sorption capacity for A-grade heavy oil was found [38]. This experimental result is consistent with the foregoing discussion; that is, the large spaces between entangled fibrous particles are responsible for heavy oil sorption. However, the original as-separated fir fibers with low bulk density showed relatively low capacity for A-grade oil, in the range of 17 to 24 kg/kg. Carbonization of the fibers was thus essential to achieve capacities as high as 80 kg/kg.

For the relatively high sorption capacity (about 30 kg/kg) of natural sorbents, such as milkweed, the presence of wax was pointed out to be important [17]. However, the present results clearly show that fibers carbonized to 900°C, in which no organic materials such as wax remained, could have a high sorption capacity, even higher than the original fibers. Therefore, the hydrophobic or oleophilic nature of the carbon surface is an important factor for achieving high sorption capacity, in addition to large void spaces between sorbent particles.

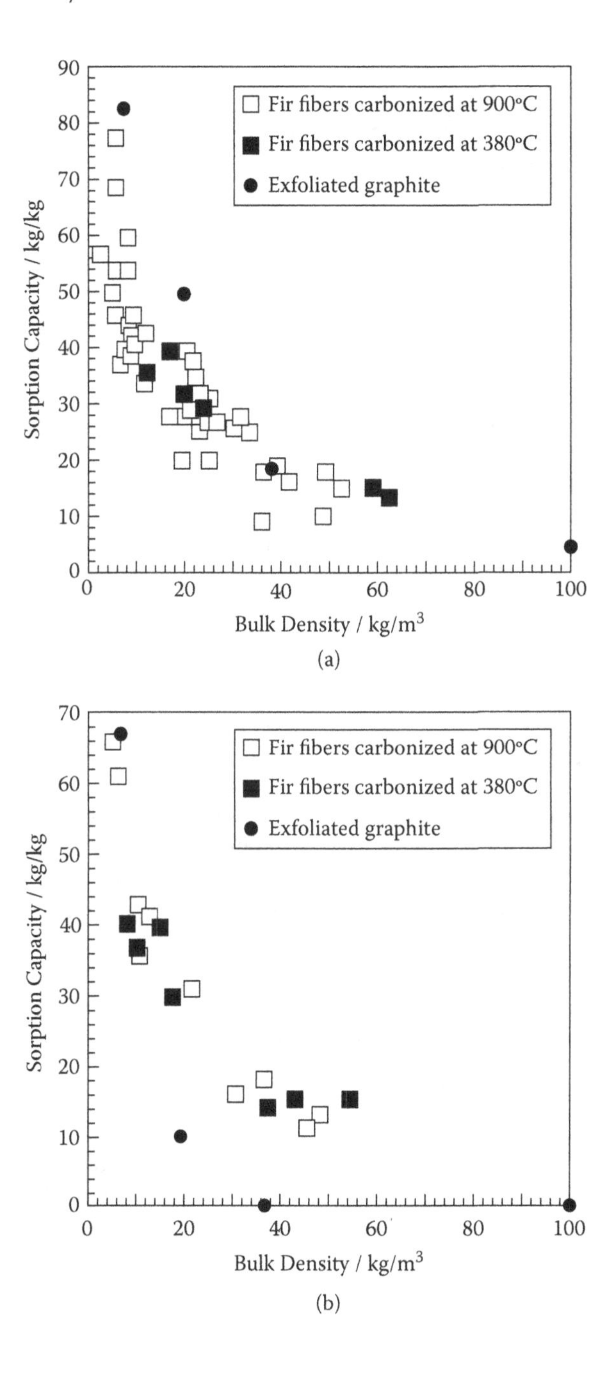

FIGURE 4.10 Dependence of sorption capacity of carbonized fir fibers on their bulk density. (a) A-grade heavy oil and (b) C-grade heavy oil.

TABLE 4.2
Sorption capacity of different sorbents for less viscous A-grade and more viscous C-grade heavy oil

Sorbent Material	Average Bulk Density (kg/m³)	Sorption Capacity for A-Grade (kg/kg)	Sorption Capacity for C-Grade (kg/kg)
Felt A of PAN-based carbon fibers	53.6	17	22
Felt B of PAN-based carbon fibers	65.3	14	16
Felt C of PAN-based carbon fibers	77.0	13	17
Felt D of pitch-based carbon fibers	74.6	11	12
Activated carbon fibers A (720 m²/g)	76.5	10	—
Activated carbon fibers B (920 m²/g)	36.7	20	—
Activated carbon fibers C (1220 m²/g)	42.7	19	—
Carbon fiber fabric A	190	6.1	—
Carbon fiber fabric B	109	7.4	—
Exfoliated graphite	7.2	83	67
Fir fibers carbonized at 900°C	5.6	78	66

C. Carbon Fibers

As listed in Table 4.2, the various carbon fibers tested do not show high sorption capacities in comparison with the carbonized fir fibers and the exfoliated graphite (capacities for low-bulk-density materials are listed in the table). For example, activated carbon fibers C, having a high surface area, do not have a high sorption capacity for the low-viscosity A-grade heavy oil. The surface area of the other samples was reasonably supposed to be negligibly small (few m^2/g). Different granular activated carbons with high surface area (>1000 m^2/g), such as those derived from coconut shells and from coals, were also used for heavy oil sorption, but their capacity was also very low, about 1 kg/kg or less.

Although certain scatter was observed, a correlation between sorption capacity and average bulk density of these carbon fibers was found to be the same as for exfoliated graphite and carbonized fir fibers for A- and C-grade heavy oils (Figures 4.7 and 4.10, respectively). Therefore, it is concluded that the sorption capacity of carbon materials depends predominantly on their bulk density, supporting the assumption that the spaces between fibrous particles of carbonized fir fibers and wormlike particles of exfoliated graphite, which are entangled in a complicated fashion with each other and thus produce void spaces with irregular shapes and a wide range of sizes, are responsible for oil sorption.

The carbon fiber felt used also does not have a high sorption capacity for A- and C-grade heavy oils (Table 4.2), but it showed excellent cycling performance, as will be discussed in Section VI.C.

D. Charcoals

Fibrous components of several plants, such as milkweed, kenaf, cotton, and sugi (don), have been applied for recovery of spilled heavy oils [15–21,70]. In Section III.B, carbonized fir fibers were described as useful sorbents for heavy oils. The hydrophobic (oleophilic) nature of most carbon materials is expected to be advantageous for oil sorption. It seemed interesting from the viewpoint of global environment to apply as oil sorbents porous charcoals with a low bulk density prepared from quickly growing woods and plants.

Charcoals were prepared from three plants, balsa from Ecuador, giant ipil-ipil from the Philippines, and bamboo from Japan, with various conditions of heating rate—8, 10, and 50°C/min—and final heat treatment temperature (HTT) ranging from 300 to 700°C.

Pore size distributions of charcoals, evaluated by mercury porosimetry, are shown in Figure 4.11a, together with those of the parent plants. Dependences of the total pore volume are shown in Figure 4.11b. As expected, the pore volume of charcoals is much larger than that of the parent plants. Balsa charcoals have a pore volume of 20 to 30 mL/g and an average pore radius of about 10 μm—the largest values among the present three samples. It is interesting to note that the giant ipil-ipil charcoals show a very wide pore size distribution, but their pore volume is relatively small, as shown in Figure 4.11.

Sorption capacity of the charcoals for A-grade heavy oil measured by direct soaking is plotted against HTT in Figure 4.12. The giant ipil-ipil and bamboo charcoals prepared at the high heating rate of 50°C/min in vacuo show relatively large sorption capacities of 1.5 kg/kg, but these are much smaller than those of the balsa charcoals. Charcoals prepared at the higher heating rate also showed the broader pore size distribution, as measured by mercury porosimetry. This is thought to be the reason why these charcoals have a relatively large sorption capacity.

In Figure 4.13, the efficiency of utilization of pores for oil sorption, that is, the ratio of the volume of heavy oil sorbed to total pore volume is plotted for the three charcoals against their average pore radius. It is concluded that heavy oil sorption into pores under 1 μm in radius is difficult. However, sorption efficiency is also strongly dependent on pore structure, charcoals prepared at higher heating rate showing a relatively high sorption efficiency.

E. Summary

The capacity of carbon sorbents for heavy oils was found to depend mainly on their bulk density, rapidly decreasing with increasing bulk density, because the macropores in the sorbent in the range of 1 to 600 μm are primarily responsible for heavy oil sorption. As both exfoliated graphite and carbonized fir fibers can have very low bulk density, for example, 6 to 7 kg/m^3, they have a very high sorption capacity of about 80 kg/kg. The sorption capacity of exfoliated graphite was also found to depend on the viscosity of the oil. Carbon fiber felts, however, do not have such low bulk density, and so their sorption capacity is correspondingly low.

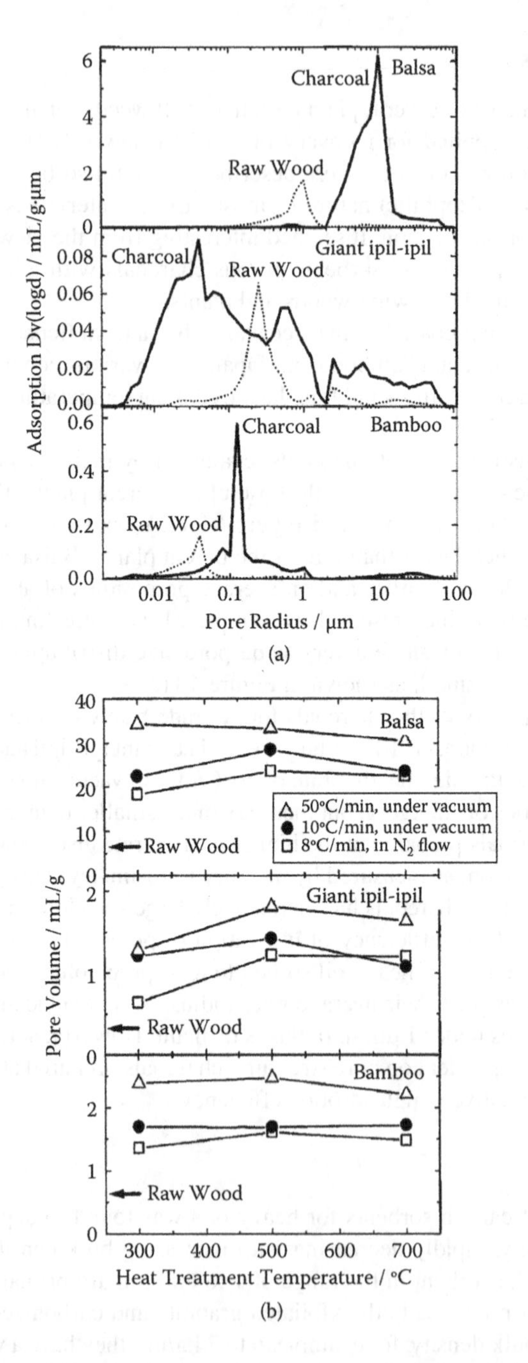

FIGURE 4.11 Pore size distribution of charcoals prepared from balsa, grant ipil-ipil, and bamboo. (a) Pore size distribution of raw wood and charcoal carbonized at 700°C for 1 h at a heating rate of 50°C/min and (b) dependence of pore volume on heat treatment temperature and rate.

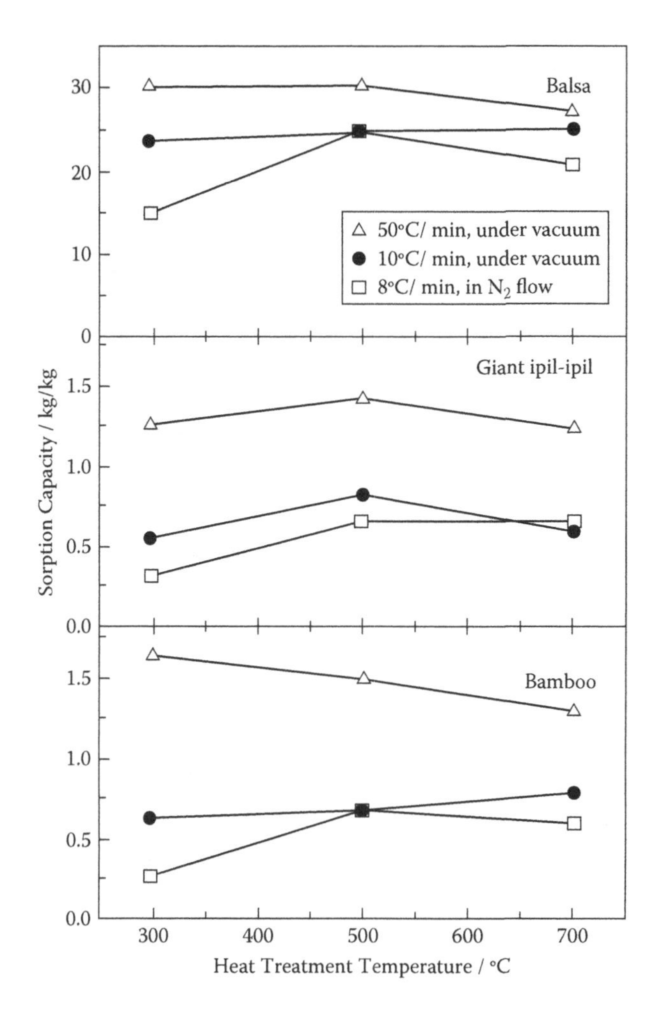

FIGURE 4.12 Sorption capacity of charcoals as a function of heat treatment conditions.

The pore volume, and consequently the sorption capacity of charcoals, depend strongly on carbonization conditions. Charcoals prepared from balsa exhibited a relatively high sorption capacity of 30 kg/kg.

IV. SELECTIVITY OF SORPTION

Exfoliated graphite could sorb a small amount of water; for example, the capacity of our exfoliated graphite with 6 kg/m^3 bulk density was about 1.8 kg/kg, and no difference in sorption capacity was detected among distilled water, tap water, and seawater, as shown in Table 4.3. Even after saturation with water, the exfoliated graphite floated on the water surface.

When A-grade heavy oil was dropped onto one end of this water-saturated exfoliated graphite, water was observed to come out from another end of the

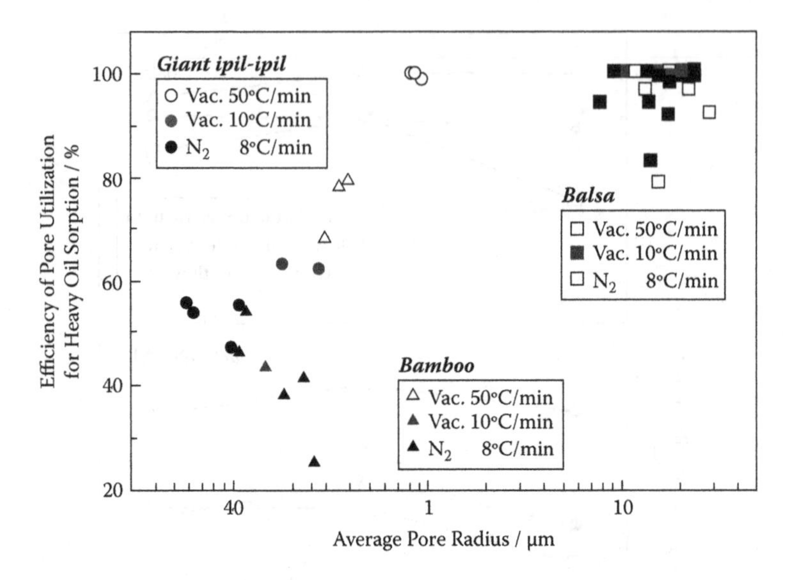

FIGURE 4.13 Efficiency of utilization of pores for heavy oil sorption, i.e., ratio of volume of heavy oil sorbed to total pore volume, against average pore radius of charcoals.

TABLE 4.3
Water adsorption (g/g) by exfoliated graphite and carbonized fir fibers

Sample	Distilled Water	Tap Water	Seawater
Exfoliated graphite	1.8	1.8	1.8
Fir fibers carbonized at 380°C	5.3	2.8	0.25
Fir fibers carbonized at 900°C	4.6	2.0	0.23

lump, as shown in Figure 4.14. By continuing the dropping of heavy oil, water came out continuously and, finally, heavy oil appeared to come out. This experimental fact reveals that water sorbed into exfoliated graphite can be replaced by heavy oil; in other words, exfoliated graphite sorbs heavy oil preferentially rather than water. The same experiment was carried out using commercially available poly(propylene) (PP) mats. PP mats could sorb about 4 kg/kg of water, but no water was seen to come out from a PP mat by dropping A-grade heavy oil.

The sorption capacity of the exfoliated graphite for A-grade heavy oil, after saturation either with seawater or tap water, was measured to be 70 kg/kg, and the time for achieving this value was about 4 h; this is slightly lower and much longer than for exfoliated graphite without prior saturation by water (83 kg/kg and within 1 min, respectively). This experimental fact reveals that heavy oil displaces water sorbed into exfoliated graphite, but not completely. The reason for the lowering of capacity and increasing time for sorption is postulated to be the

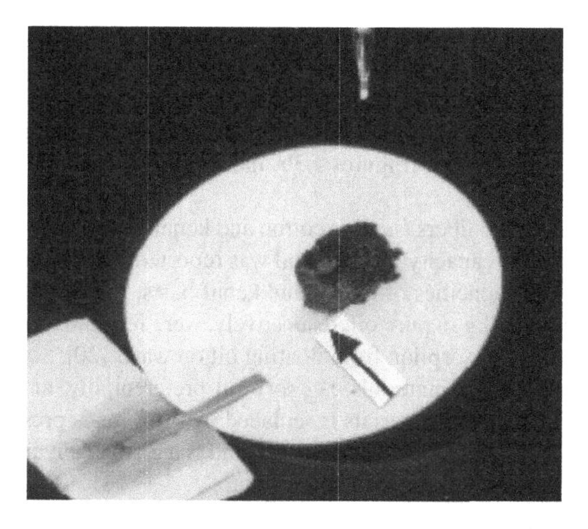

FIGURE 4.14 Water coming out from water-saturated exfoliated graphite by dropping A-grade heavy oil on its top.

following: water preadsorbed in the small pores could not be replaced completely by heavy oil—particularly the water adsorbed in small pores inside the wormlike particles and the tips of crevice-like pores on their surface.

In the case of carbonized fir fibers, water adsorption depends strongly on the type of water, as shown in Table 4.3: the fibers adsorb a relatively large amount of distilled water, but only a small amount of seawater; adsorption is much more for distilled water and much less for seawater than for exfoliated graphite. Fir fibers carbonized at 900°C adsorbed a little less water than those carbonized at 380°C. After the fir fibers were saturated with distilled water, heavy oil was dropped onto it, but drainage of water, as in the case of exfoliated graphite (Figure 4.14), was not observed. The sorption capacity of water-saturated 900°C-carbonized fir fibers for A-grade heavy oil was very similar to that for the dry fibers. These experimental results suggest that adsorbed water was trapped in small pores inside the fibrous particles (Figures 4.3b and 4.3c) and therefore could not be replaced by heavy oil.

For lignocellulosic fibers (such as cotton and kenaf fibers), a more pronounced reduction in sorption capacity for diesel oil was reported when the oil was floating on water; sorption capacities of cotton and kenaf basts, which were measured to be about 30 and 7 kg/kg in pure oil, respectively, were reduced to about 8 and 1.5 kg/kg, respectively, by sorption from floating oil on water [20].

In summary, carbon materials can sorb oil preferentially, and a part of the water adsorbed in carbon materials is replaced by oil. This is presumably due to the hydrophobic nature of the carbon surface and is a distinct advantage of carbon materials, particularly for oils spilled on water.

V. SORPTION KINETICS

The sorption rate of heavy oils, A- and C-grade, into carbon materials was evaluated by applying the so-called wicking method [71]. The system for the measurement is schematically shown in Figures 4.15a and 4.15b: mass increase by capillary suction of heavy oil from the bottom into carbon sorbents (either a column of exfoliated graphite and fir fibers carbonized at 900°C, which are packed into a glass tube with a cross-sectional area of 314 mm^2 with different densities, or a piece of carbon fiber felt) was measured at room temperature as a function of time.

Some of the sorption curves for exfoliated graphites with different bulk densities are shown in Figures 4.16a and 4.16b. The initial slope depends strongly on both the bulk density of exfoliated graphite and the viscosity of heavy oil. For A-grade heavy oil, very rapid suction and saturation within 10 s were observed for bulk densities in excess of 20 kg/m^3 (Figure 4.16a). A more gradual suction of heavy oil is observed for a bulk density of 7 kg/m^3, and saturation is not reached even after 50 s. The initial slope increases with increasing bulk density, but the saturation mass increase shows a maximum at a bulk density of around 12 kg/m^3. Very similar dependence on the bulk density of exfoliated graphite is also observed for the much more viscous C-grade heavy oil, but saturation requires much longer times (Figure 4.16b).

The time needed to reach 50% of saturation mass increase, half-time $t_{1/2}$, was employed as a measure of sorption rate. In Figures 4.17a and 4.17b, $t_{1/2}$ is plotted against bulk density of exfoliated graphite. Marked dependence of the sorption rate on bulk density of the sorbent with a bulk density of 6 kg/m^3 is seen for both A- and C-grade heavy oils. A large difference in $t_{1/2}$ is clearly shown between A- and C-grade heavy oils, the former being faster by more than two orders of magnitude. Such a difference is expected to result from the difference in the viscous properties of the oils. By densification of the exfoliated graphite from 7 to

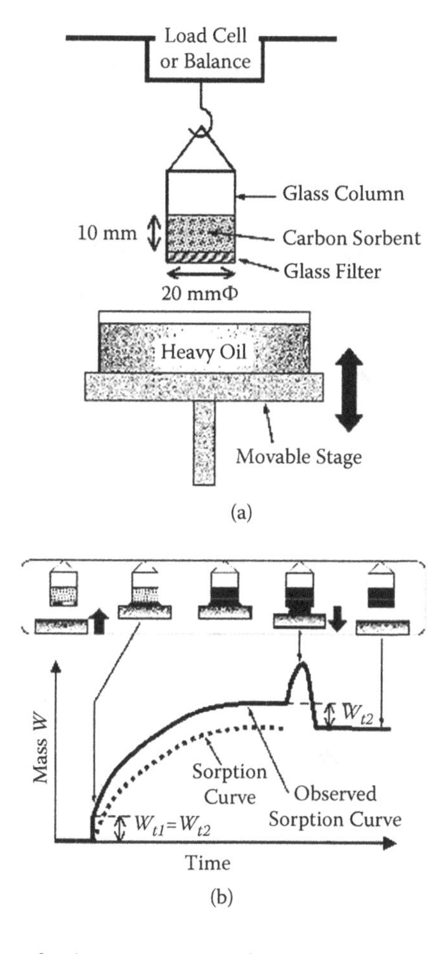

FIGURE 4.15 Scheme for the measurement of sorption kinetics. (a) System for the measurement and (b) mass increase due to heavy oil pumping into exfoliated graphite.

20 kg/m^3, the value of $t_{1/2}$ decreases to one-fourth for both grades of oil. However, densification to levels higher than 20 kg/m^3 is seen to have almost no further effect on the sorption rate.

A more detailed quantitative analysis of sorption kinetics was carried out, employing the conservation of momentum, Poiseuille's law, and capillary phenomena [72]. The sorption curve before saturation is well approximated by the equation

$$m_s = K_s\, t_{1/2} + \mathrm{B}, \tag{4.1}$$

where m_s is mass increase per cross-sectional area of the sorbent column, t is time, K_s is the sorptivity or liquid sorption coefficient, and B is a constant. Therefore, K_s (kg/m^{2}s$^{1/2}$) is also a measure of sorption kinetics. This equation was theoretically derived for the penetration of a liquid into the cylindrical capillary of a

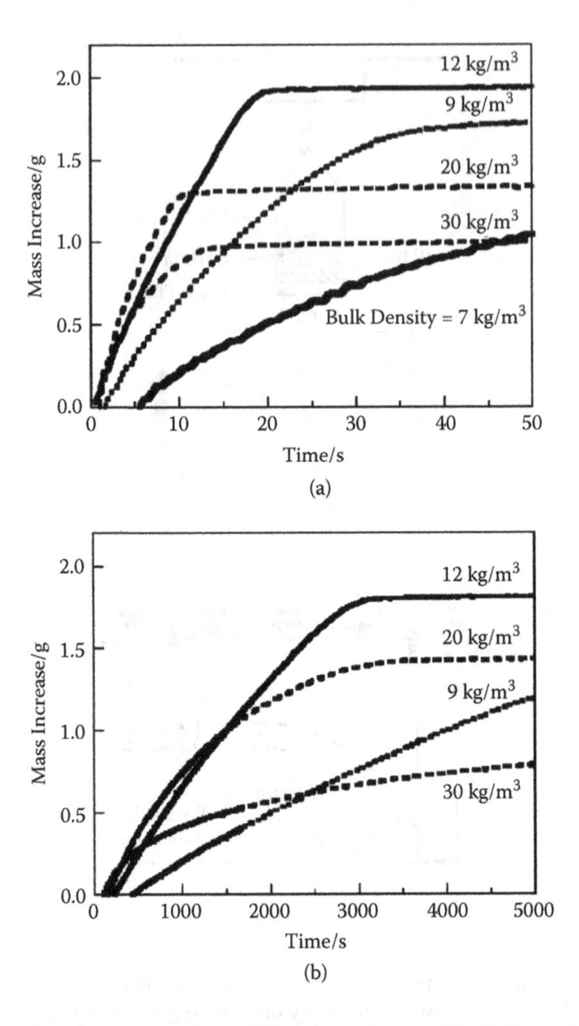

FIGURE 4.16 Sorption curves for exfoliated graphites with different bulk densities. (a) A-grade heavy oil and (b) C-grade heavy oil.

porous body [72]. It was successfully applied to porous ceramics [73], and used to experimentally study the influences of the nature of sorbate (density, viscosity, and surface tension of liquids) and also of the pore structure of ceramic sorbents (porosity and surface nature) on the sorptivity K_s [73–76].

On a column of exfoliated graphite packed to a bulk density of 14 kg/m³ with different column heights in a glass tube, m_s is plotted against $t_{1/2}$ in Figure 4.18. The initial slope of m_s versus $t_{1/2}$ plots (sorptivity, K_s) is independent of the specimen height, though the saturated values of m_s are different.

In Figures 4.19a through 4.19c, the plots of m_s versus $t_{1/2}$ for A-grade heavy oil are shown for three carbon sorbents with different bulk densities: exfoliated graphite, carbonized fir fibers, and carbon fiber felt. The sorptivity depends

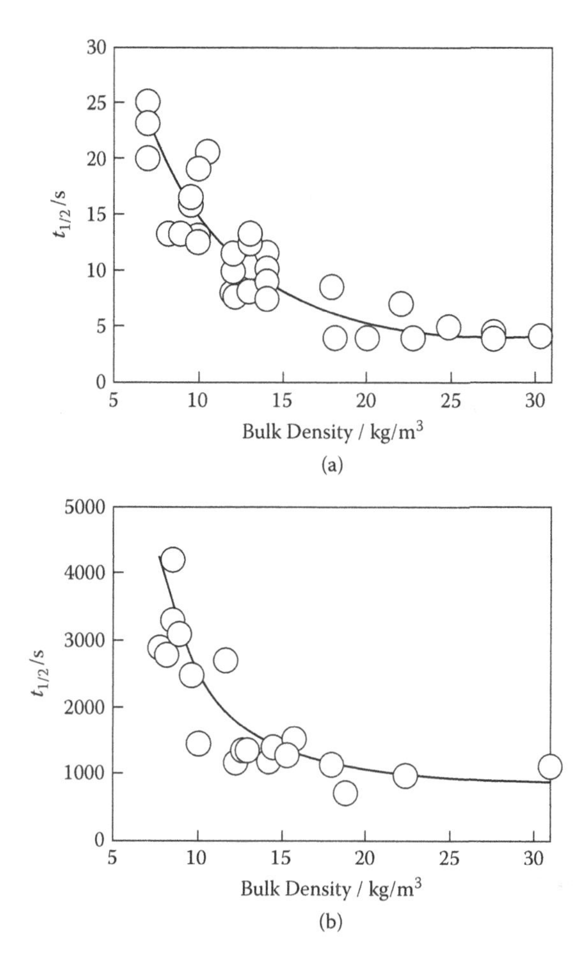

FIGURE 4.17 Time to reach 50% of mass increase saturation, $t_{1/2}$, versus bulk density of exfoliated graphite. (a) A-grade heavy oil and (b) C-grade heavy oil.

strongly on the carbon sorbent and its bulk density. This is shown in Figure 4.20. For A-grade heavy oil, K_s for carbonized fir fibers drastically increases with increasing bulk density in the region from 8 to 20 kg/m^3 and reaches saturation at about 5.5 kg/m^{2}s$^{1/2}$ when the fibers are densified above 30 kg/m^3. In the case of carbon fiber felt, which has high bulk density, the value of K_s is approximately constant at about 5.5 kg/m^{2}s$^{1/2}$, even though there is some scatter of experimental points. For the exfoliated graphite, the maximum K_s is about half that for the others, achieved at 16 kg/m^3.

For viscous C-grade heavy oil, the value of sorptivity was only about 0.2 kg/m^{2}s$^{1/2}$. A slight dependence on bulk density of two carbon sorbents, carbonized fir fibers and carbon fiber felts, was observed, but it is difficult to discuss this in detail at present.

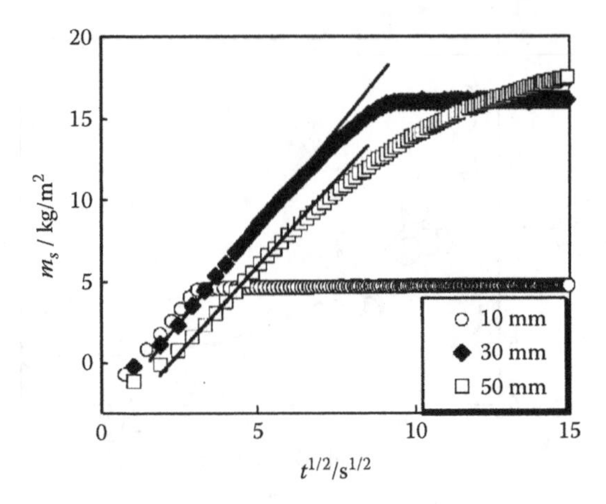

FIGURE 4.18 Mass increase per cross-sectional area of the sorbent m_s (exfoliated graphite with bulk density of 14 kg/m^3) against $t_{1/2}$ as a function of sorbent height.

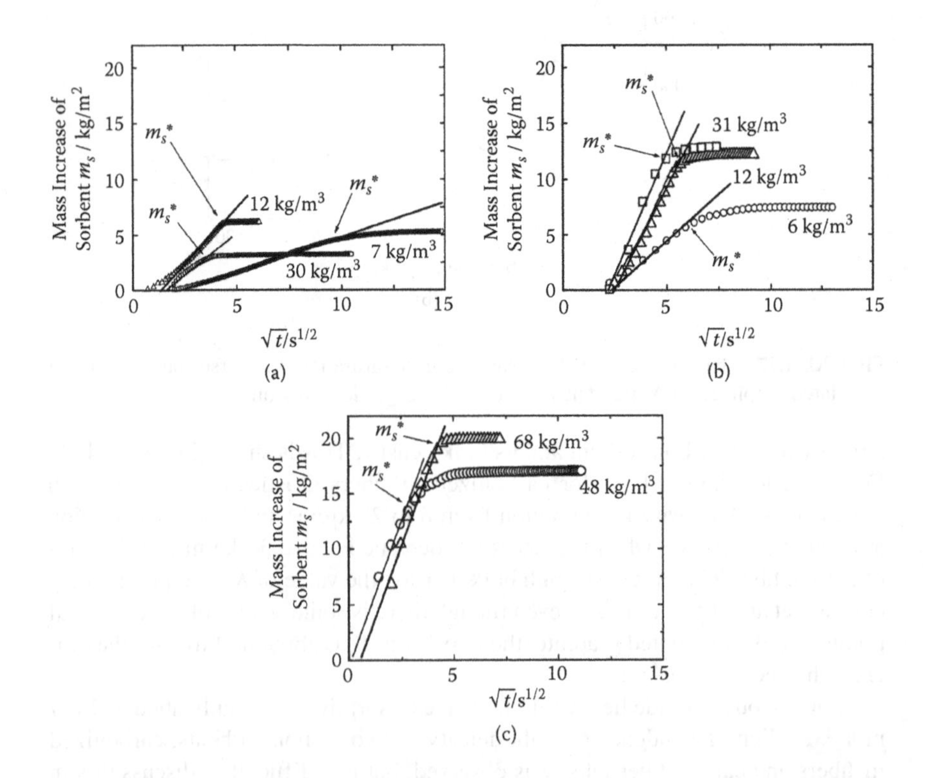

FIGURE 4.19 Plots of m_s versus $t_{1/2}$ on carbon sorbents with different bulk densities. (a) Exfoliated graphite, (b) carbonized fir fibers, and (c) carbon fiber felt.

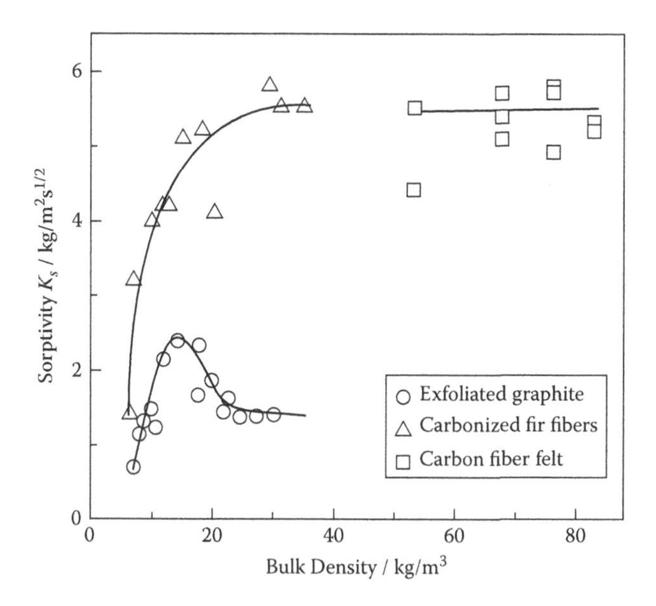

FIGURE 4.20 Dependence of sorptivity K_s on bulk density of the three carbon sorbents.

The sorptivity K_s has been expressed theoretically as follows [73]:

$$K_s = \left[d_l \cdot \sqrt{\frac{\gamma}{\mu}} \right] \cdot \left[\frac{\varepsilon^*}{\lambda} \cdot \sqrt{r_0} \right] \cdot \left[\sqrt{\frac{\cos\theta}{2}} \right] \tag{4.2}$$

where d_l is density; γ, surface tension; μ, viscosity of the sorbate; ε^*, effective sorbent porosity; λ, average tortuosity factor of the capillaries ($\lambda > 1$); r_0, average pore radius of the sorbent; and θ, the contact angle of the interface between liquid sorbate and pore wall in the solid sorbent. Thus, the first term in Equation 4.2 contains parameters of the sorbate, the second contains properties of the sorbent, and the third characterizes the sorbate–sorbent interface. To understand the dependence of sorptivity on sorbent bulk density for A-grade heavy oil, the first term in Equation 4.2 must be constant, and the third term can be approximated to be a constant because the contact angle θ does not change much among the carbon materials used. Therefore, sorptivity depends on three sorbent parameters: effective sorbent porosity ε^*, average tortuosity factor λ, and average pore radius r_0.

The effective sorbent porosity, ε^*, corresponds to the volume of the macropores directly involved in the sorption process. According to Beltran et al. [73], it is calculated from the value m_s^* at which the linear relation between m_s and $t_{1/2}$ ceases (see Figure 4.19), using the following equation:

$$\varepsilon^* = \frac{m_s^*}{L \cdot d_l}, \tag{4.3}$$

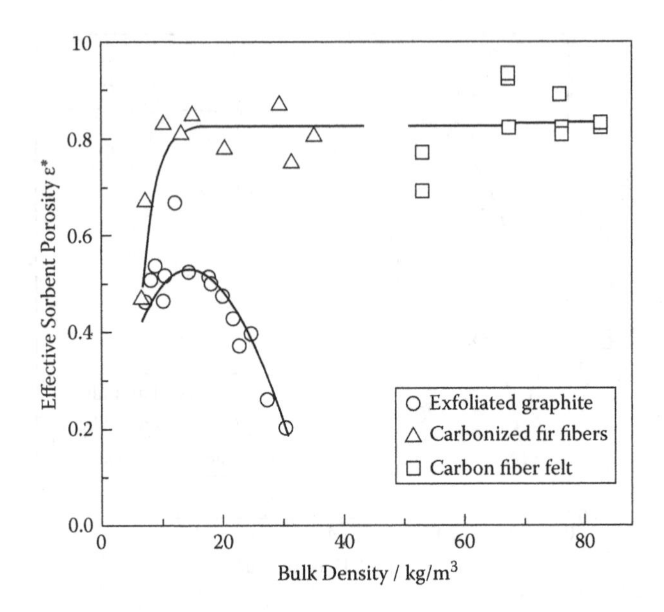

FIGURE 4.21 Dependence of effective sorbent porosity ε* on bulk density of carbon sorbents.

where L is the height of the sorbent column. In Figure 4.21, effective sorbent porosity is plotted against sorbent bulk density. The value of ε* for carbonized fir fibers increases with increasing bulk density from 7 to 15 kg/m^3, and its saturated value is about 0.85. In the case of carbon fiber felt, ε* can be considered constant at around 0.85. The ε* of exfoliated graphite is the smallest among these samples. The dependence of ε* on bulk density is very similar to that of sorptivity (see Figure 4.20).

The tortuosity factor of the capillaries is defined as $\lambda > 1$, because $\lambda = 1$ corresponds to a smooth surface of cylindrical pores. Quantitative evaluation of λ is difficult because of the complicated surface structure of the wormlike particles of exfoliated graphite and carbonized fir fibers and also the complicated entanglement of the fibrous particles, as shown in the SEM images in Figures 4.2, 4.3, and 4.4. From these microscopic observations, however, it can be assumed that exfoliated graphite has much higher λ values than the other two carbon sorbents because of the rough surface due to crevices on its wormlike particles; also, the λ value for carbonized fir fibers is reasonably expected to be larger than that for carbon fiber felt.

A quantitative comparison of values of the average pore radius r_0 is also difficult, for the same reason. As shown in Figure 4.22, the pore size distribution of carbon sorbents is quite complicated and broad, particularly for exfoliated graphite and carbonized fir fibers of low bulk density. With increasing bulk density, it tends to become simpler and shifts to smaller pore sizes.

For the carbon fiber felt, ε* is almost independent of bulk density and has a rather high value in comparison with the other two sorbents, and its λ is expected

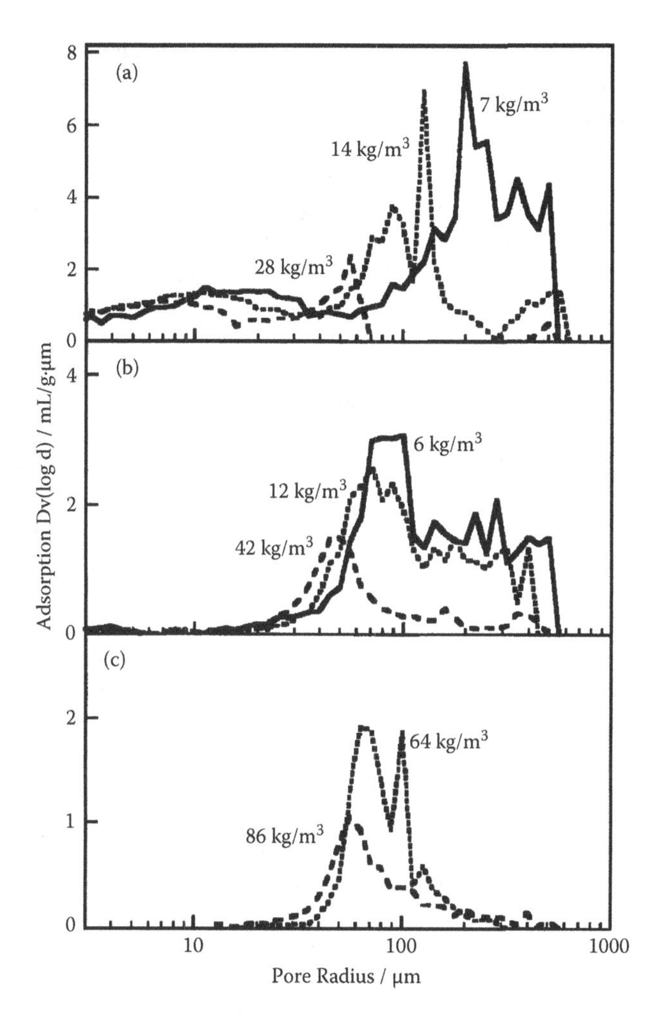

FIGURE 4.22 Pore size distributions of carbon sorbents measured by mercury porosimetry. (a) Exfoliated graphite, (b) carbonized fir fibers, and (c) carbon fiber felt.

to be the smallest among the three sorbents. Its pore size distribution is relatively simple (see Figure 4.22c), and r_0 is almost independent of bulk density. As a consequence, the value of K_s turns out to be high and is almost independent of bulk density, as shown in Figure 4.20.

In the case of carbonized fir fibers, however, the dependence of K_s on bulk density is mainly governed by ε^*, which also reveals a pronounced dependence on bulk density, even though the average pore radius seems to become smaller with increasing bulk density. The values of K_s and ε^* of the carbonized fir fibers both increase with increasing density from 6 to 20 kg/m^3. The pore size distribution becomes sharper at higher density. The peak at a density of 42 kg/m^3 is located at about 50 μm, and the pore sizes are extremely small when compared to specimens with low density (Figure 4.22b). Because λ might be reduced with increasing density, K_s increases with increasing density until 30 kg/m^3. For fibers with a low density, the pore size distribution is so broad that sorption of heavy oil may not be performed quickly.

It is reasonably assumed that the relatively small value of K_s of exfoliated graphite results from a small value of ε^* (Figures 4.20 and 4.21) and a large value of λ. From the pore size distribution, it was expected for exfoliated graphite with bulk density of 7 kg/m^3 to contain large spaces among its particles, but this could not be detected by mercury porosimetry [54]. Thus, the entire volume of these spaces does not contribute to the sorption of heavy oil. This is why exfoliated graphite has a small value of ε^* even though it has the largest sorption capacity among the carbons used. The exfoliated graphite, with a slightly larger bulk density, 14 kg/m^3, has a relatively narrow pore size distribution, with average pore radii from 50 to 100 μm (Figure 4.22a). As a consequence, the heavy oil was quickly sorbed. By further densifying to 28 kg/m^3, the volume of larger pores, that is, above 60 μm, was reduced, and small pores below 50 μm could be clearly observed. The reduction of effective sorbent porosity for bulk density above 16 kg/m^3 (Figure 4.21) is due to the disappearance of pores larger than 60 μm, which seems to be suitable for heavy oil sorption.

In summary, the sorption rate for A-grade heavy oil, evaluated by sorptivity in the present work, depends on three factors, ε^*, λ, and r_0 of the sorbent carbons—ε^* increasing rapidly but r_0 decreasing with increasing bulk density, particularly in the low-density region. λ is mainly governed by the smoothness of the surface of sorbent particles. To have a high sorptivity, therefore, a low value of λ, that is, a smooth particle surface, and a high bulk density, where changes in both ε^* and r_0 become small, are desired. These conditions are best fulfilled by the carbon fiber felt among the three carbon sorbents used. For exfoliated graphite, sorptivity is rather low, probably because of the high value of λ due to the complicated surface of its wormlike particles. Carbonized fir fibers show a wide range of sorption kinetics, from a low rate comparable to that of exfoliated graphite to a high rate comparable to that of carbon fiber felt.

The sorption kinetics for the other oils, including kerosene, light oil, various cooking oils, different motor oils, and diesel oil, into a column of exfoliated graphite with a bulk density of about 7 kg/m^3 was also studied, and similar results were obtained [51]. The dependence of sorptivity K_s for these oils on their viscosity accorded with the results for heavy oils.

VI. CYCLIC PERFORMANCE OF CARBON MATERIALS FOR SORPTION OF HEAVY OILS

Recovery of sorbed heavy oil from carbon materials was carried out either by filtering under suction (about 5 to 7 kPa pressure), washing with solvents, centrifuging at 3800 rpm, or by squeezing mechanically. The recovered sorbent was repeatedly subjected to sorption of the heavy oil. To determine its cycling performance, this cycle of sorption and recovery (desorption) was repeated up

FIGURE 4.23 Equipment used for the recovery of heavy oils sorbed into carbon sorbents by filtration under suction.

to ten times with each sample of the carbon sorbent. For filtration under suction, as shown in Figure 4.23, a simple equipment is used with a filter paper. In the process of washing out the sorbed heavy oil, *n*-hexane was used in the case of A-grade heavy oil, but for viscous C-grade heavy oil A-grade heavy oil was employed in addition to *n*-hexane.

A. Exfoliated Graphite

Both A-grade heavy oil and crude oil were recovered without marked changes in the texture of the exfoliated graphite, as shown in Figure 4.24. The recovery ratios for low-viscosity A-grade heavy oil and crude oil were about 60 and 42%, respectively, after suction for more than 90 min. The high-viscosity B- and C-grade heavy oils could not be recovered by such filtration, as reducing the suction pressure below 7 kPa was not effective in recovering the oils. These oils could be partially recovered by compression of exfoliated graphite, but this process was never employed in the present work, because exfoliated graphite lost its bulky

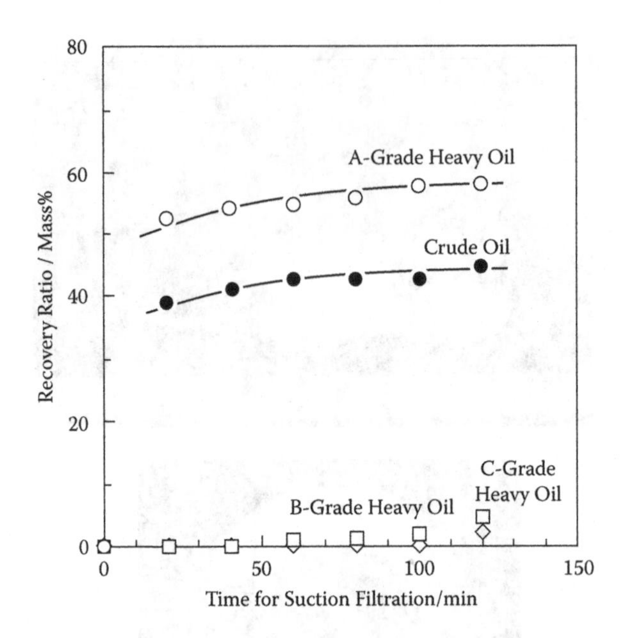

FIGURE 4.24 Dependence of recovery ratio of four heavy oils on filtration time under suction of about 7 kPa.

texture after compression and could not be used as sorbent again. These results show that heavy oils with relatively low viscosity are easily recovered from sorbed exfoliated graphite without marked disruption of its bulky texture. It has to be pointed out that heavy oils recovered from exfoliated graphite by filtration showed a somewhat lighter color; they were even transparent at the very beginning of filtration. This is attributed to the fact that a fraction of the large aromatics in heavy oil are trapped in the sorbent, as will be discussed in Section VII.

In Figure 4.25 it is seen that the sorption capacity is as high as 80 kg/kg, and about 50% of sorbed oil is recovered. Subsequently, sorbed and recovered amounts of oil decrease rather markedly with increasing cycling. The reason for such rapid decrease between the first and second cycles seems to be oil retention in the pores and on the surface of the exfoliated graphite after filtration. In Figure 4.26a an SEM image (so-called wet-SEM) is shown for the wormlike exfoliated graphite particles after sorption of A-grade heavy oil. The crevicelike pores on the surface are difficult to recognize, but they are clearly observed before sorption (Figure 4.2b) and also after oil recovery by filtration (Figure 4.26b). This suggests the retention of a certain amount of heavy oil in these pores. In Figures 4.26c and 4.26d SEM images of wormlike particles after sorption of C-grade heavy oil are shown, revealing that the crevicelike pores are filled with oil even after the filtration process.

In Figures 4.27a and 4.27b, the sorption capacity with cycling by filtration is seen to decrease to 50% after the fourth cycle. By insertion of either hexane washing or a heat treatment step, sorption capacity further dropped markedly.

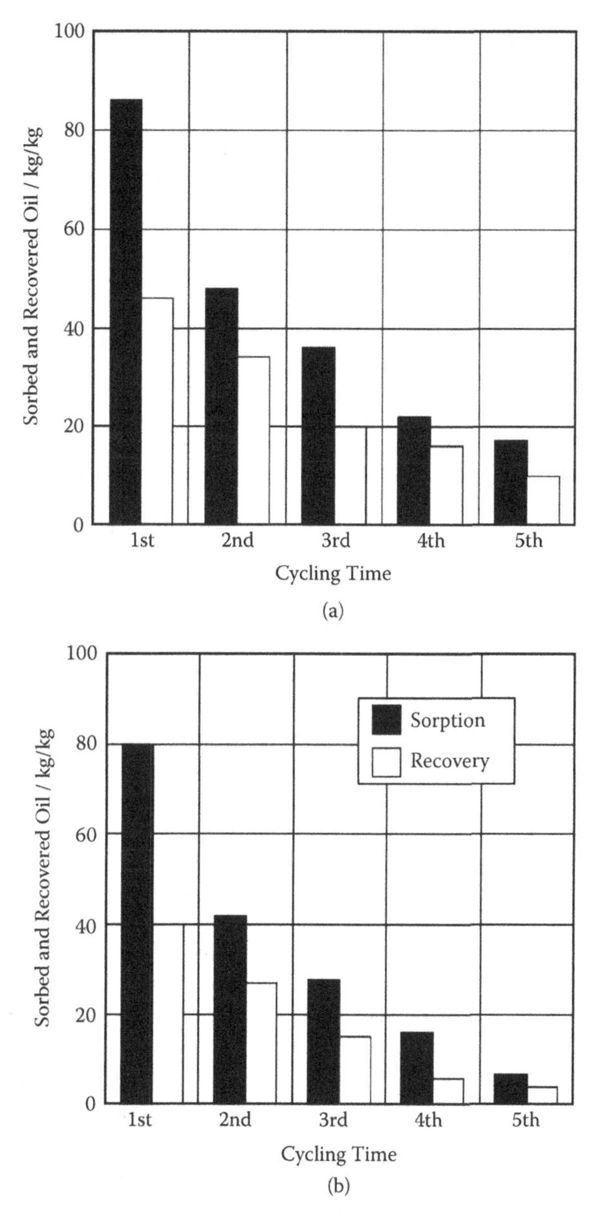

FIGURE 4.25 Changes with cycling time of the amount of heavy oil sorbed into exfoliated graphite with a bulk density of about 6 kg/m^3 and subsequently recovered by filtration under suction. (a) A-grade heavy oil and (b) crude oil.

Shrinkage of exfoliated graphite was observed after hexane washing, which is thought to account for such reduction in sorption capacity. Washing with kerosene or xylene and heating in a vacuum at 400°C for 7 h also resulted in marked reduction of sorption capacity.

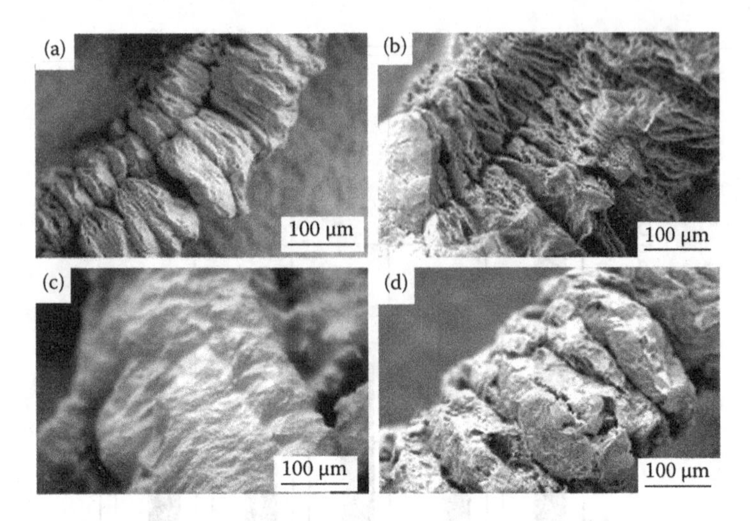

FIGURE 4.26 SEM images of the surface of wormlike particles of exfoliated graphite after sorption of heavy oil. (a) After sorption of A-grade heavy oil, (b) after recovering by suction filtration, (c) after sorption of C-grade heavy oil, and (d) after suction filtration.

By squeezing exfoliated graphite after sorption, nearly all the sorbed oil could be recovered, but this completely destroyed the characteristic bulky texture of the exfoliated graphite, and so there was no further sorption of oil. Therefore, oil recovery through mechanical squeezing or compression does not suit the purpose of the present study.

B. Carbonized Fir Fibers

A-grade heavy oil could be recovered from fir fibers carbonized at 900°C by simple filtration with mild suction (5 kPa) and the fibers could be reused, although the sorption capacity did decrease with cycling, as shown in Figure 4.28.

In the first cycle, the 5.5-kg/m^3 fibers (Figure 4.28a) sorbed about 46 kg/kg, and about 91% of the sorbed oil (42 kg/kg) was recovered. After the second cycle, sorption capacity decreased. The recovery ratio of about 90% was maintained up to the eighth cycle, although there was some scatter in the results for both sorbed and recovered amounts of oil. Figure 4.28b shows another series of cycling and recovery experiments with fibers having a high bulk density; sorption capacity decreases gradually, and is maintained at 26 kg/kg even after the eighth cycle (about 84% of the first cycle). In the first cycle, the recovery ratio is rather low, about 80%, but it becomes more than 90% upon further cycling. In comparison with exfoliated graphite, fir fibers carbonized at 900°C have better cycling performance for sorption and recovery of A-grade heavy oil.

As illustrated in Figures 4.28a and 4.28b, carbonized fir fibers with a lower bulk density have a higher sorption capacity but a poorer cycling performance because the fiber entanglement is more fragile. In Figure 4.28c the relative sorption capacity S/S_{1st} (ratio of sorption capacity in *n*th cycle to that in the first cycle)

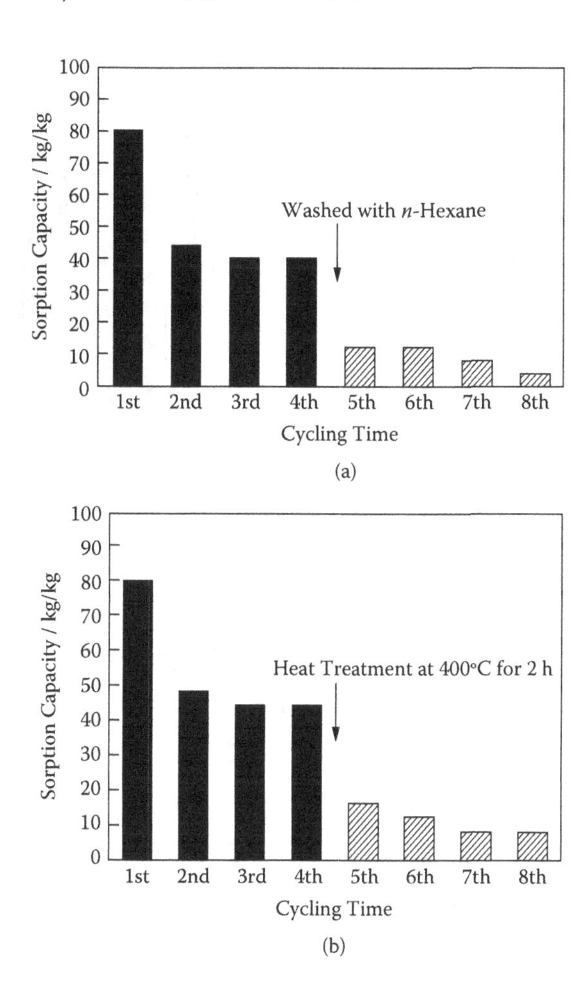

FIGURE 4.27 Changes in sorption capacity of exfoliated graphite with a bulk density of 6 kg/m^3 for A-grade heavy oil. (a) Interrupted by washing with *n*-hexane and (b) interrupted by heat treatment at 400°C for 2 h in vacuum.

and the recovery ratio R/S (ratio of recovered to sorbed amount) are plotted for each cycle against cycling times for fir fibers with different bulk densities. By simple filtration under suction, the capacity of the 6-kg/m^3 fibers after the eighth cycle was reduced to 60%, but for the 35-kg/m^3 fibers it remained more than 90%, although the absolute sorption capacity was relatively low. The recovery ratio was in a narrow range—90 to 95%—irrespective of bulk density of the fibers. It is noteworthy that carbonized fir fibers exhibit rather good cycling performance, gradual decrease in sorption capacity, and high recovery ratio upon cycling. For the fibers carbonized at 380°C, S/S_{1st} and R/S are also plotted in Figure 4.28c, their cycling performance being similar to that of the fibers carbonized at 900°C.

The difference in cycling performance between fir fibers and exfoliated graphite seems to be mainly due to the strength of the fibrous particles: fir fibers are

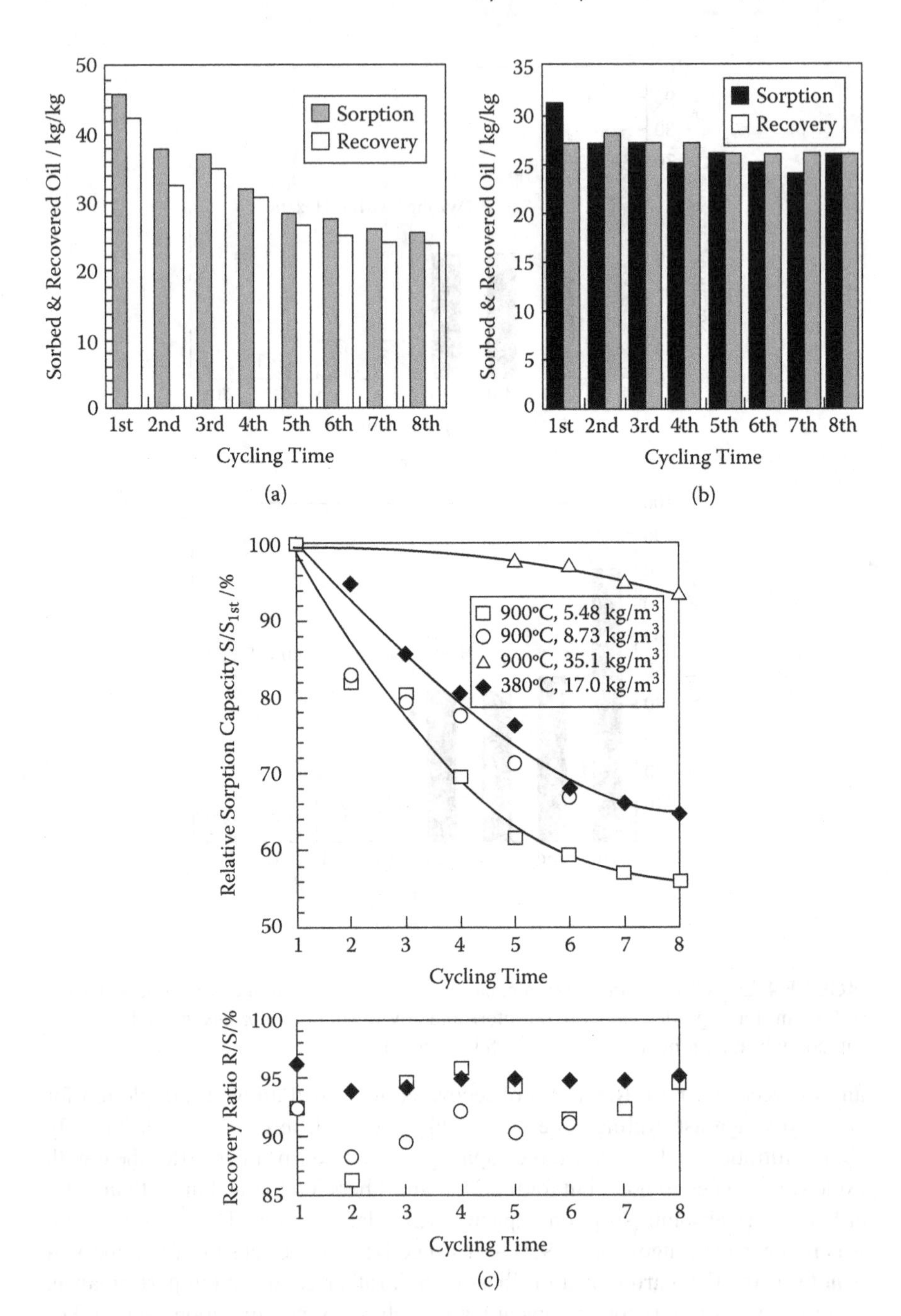

FIGURE 4.28 Changes in sorbed amount of A-grade heavy oil for carbonized fir fibers and in recovered amount by filtration under suction. (a) Fibers with a bulk density of 5.5 kg/m^3, (b) fibers with a bulk density of 20 kg/m^3, and (c) changes in relative sorption capacity S/S_{1st} and recovery ratio R/S with cycling time.

much stronger than wormlike particles of exfoliated graphite, and so the bulky network of fir fibers is somewhat stronger (more rigid). Hence, the deformation of the bulky network of fibrous particles is smaller in fir fibers than in exfoliated graphite upon cycling.

Heavy oil sorbed into carbonized fir fibers could be recovered by washing with *n*-hexane. In Figure 4.29a an example of cycling performance of fibers carbonized at 900°C is shown; more than 90% of the sorbed oil is recovered. Although C-grade heavy oil could not be recovered by filtration, it could be recovered with a high efficiency by washing with *n*-hexane, as shown in Figure 4.29b. After washing the fibers could be reused for sorption of C-grade heavy oil, but their sorption capacity decreased upon cycling, at a rate dependent on their bulk density.

In Figures 4.30a and 4.30b, the relative sorption capacity and recovery ratio are plotted against cycling time with hexane washing for A-grade and C-grade heavy oils as a function of bulk density. For both oils, fibers carbonized at 900°C with higher bulk density have a lower sorption capacity but a slower decrease in sorption capacity and a higher recovery ratio, that is, a better cycling performance. Using fibers carbonized at 380°C, C-grade heavy oil sorption is comparable to that with fibers carbonized at 900°C, and recovery ratio is also close to 100%. It has to be mentioned that the sorption capacity of fibers for viscous C-grade heavy oil is relatively low, but the recovery ratio can be close to 100% by washing with a solvent, mainly because of the relatively strong network of fibrous particles.

The C-grade heavy oil could be recovered by washing with the less viscous A-grade heavy oil, as shown in Figure 4.31. In this case, relative sorption capacity S/S_{1st} also decreases with recycling, but the recovery ratio R/S is more than 90%. Because B-grade heavy oil is a mixture of A-grade heavy oil with C-grade, the washing process of sorbed C-grade heavy oil with A-grade heavy oil corresponds to the preparation of B-grade heavy oil.

C. Carbon Fiber Felt

Using carbon fiber felt, an example of sorbed and recovered amounts of A-grade heavy oil with cycling time is shown in Figure 4.32. More than 90% of sorbed oil could be recovered by simple filtration under suction of 5 kPa, though the absolute capacity for sorption is rather low: 11 kg/kg. In the first cycle, the recovered amount is about 90% of the sorbed amount. Sorption capacity and recovery ratio R/S are seen not to change with cycling. This cycling performance was much better than that of exfoliated graphite and carbonized fir fibers, as described earlier. The same cycling performance was observed for all carbon fiber felts used. Recovery of the viscous C-grade heavy oil was impossible by filtration under suction, in agreement with the results for exfoliated graphite and carbonized fir fibers.

In Figure 4.33 representative cycling performance with *n*-hexane washing is shown for A- and C-grade heavy oils. The recovery ratio is almost 100% in each cycle for both heavy oils. It was also possible to wash out the viscous C-grade heavy oil with the less viscous A-grade heavy oil, similar to the behavior of carbonized fir fibers.

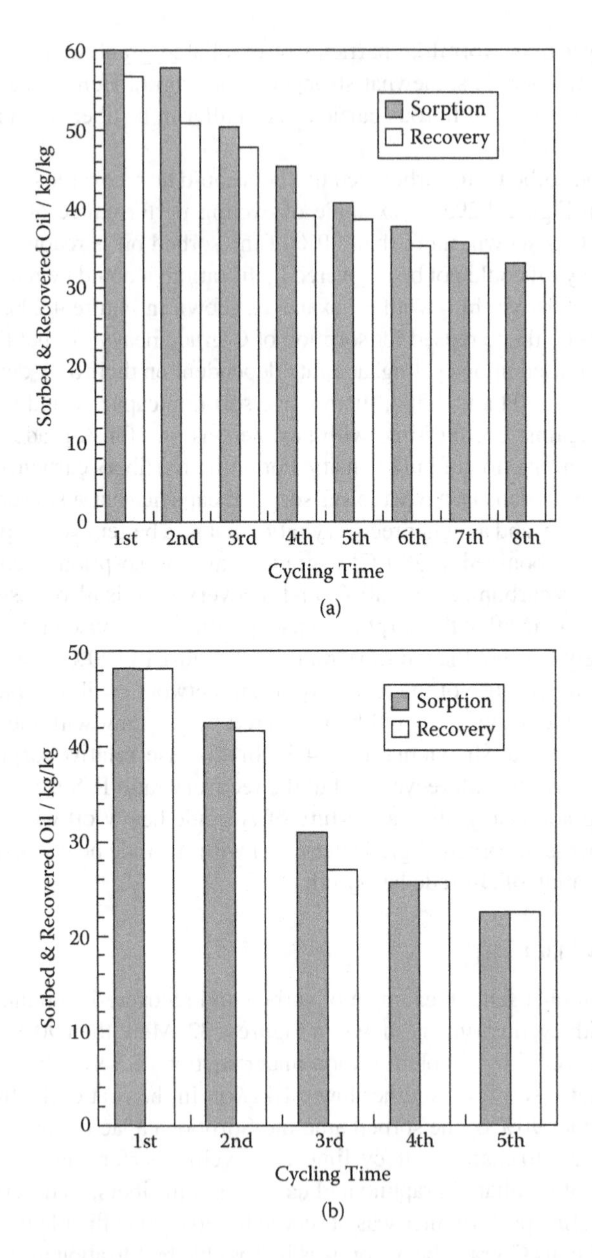

FIGURE 4.29 Changes in sorbed amount of heavy oil for fir fibers carbonized at 900°C and recovered amount by washing with *n*-hexane. (a) A-grade heavy oil for fibers with a bulk density of 8 kg/m^3 and (b) C-grade heavy oil for fibers with 6.4 kg/m^3.

As the felts used in the present work are expected to have good mechanical properties, the spent felts were exposed to centrifugation at 3800 rpm. As shown

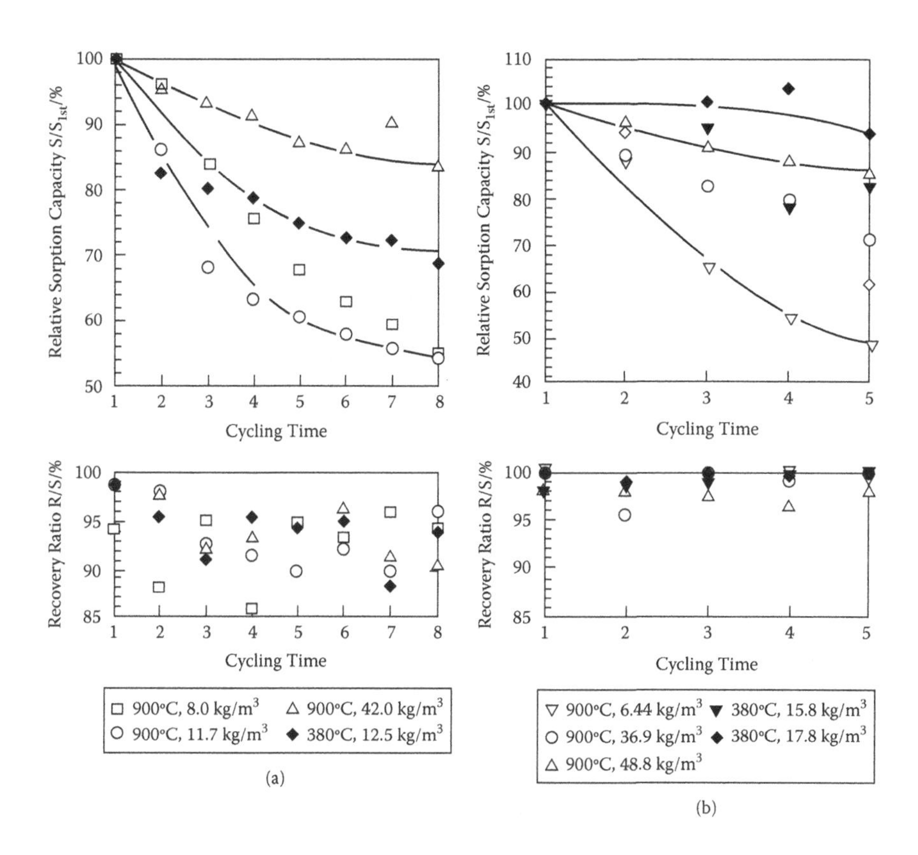

FIGURE 4.30 Changes in relative sorption capacity S/S_{1st} and recovery ratio R/S with cycling time on carbonized fir fibers with different bulk densities. (a) A-grade heavy oil and (b) C-grade heavy oil.

in Figure 4.34, the recovery ratio for both A- and C-grade oils is almost 100%, and the sorption capacity can be kept almost at the same level as in the first cycle even after eight cycles.

Recovery by squeezing and twisting was also performed for the felts sorbed with heavy oil. The cycling performance is shown in Figure 4.35 for felt C (PAN-based). The bulky network of carbon fibers in the felt appeared to be altered to some extent under squeezing and twisting, and so the sorption capacity decreased after the first cycle. In subsequent cycles, however, only a small additional change in sorption capacity occurs, but the recovery ratio is close to 100%. For felt B (also PAN-based), excellent recycling performance was also observed up to the fifth cycle, but it was fragmented during twisting. Felt D (pitch-based), on the other hand, could not be subjected to squeezing because of its low mechanical strength. For the other carbon materials studied, use of such extreme operations for oil recovery (centrifuging and squeezing) was not possible, because their bulky texture was completely destroyed.

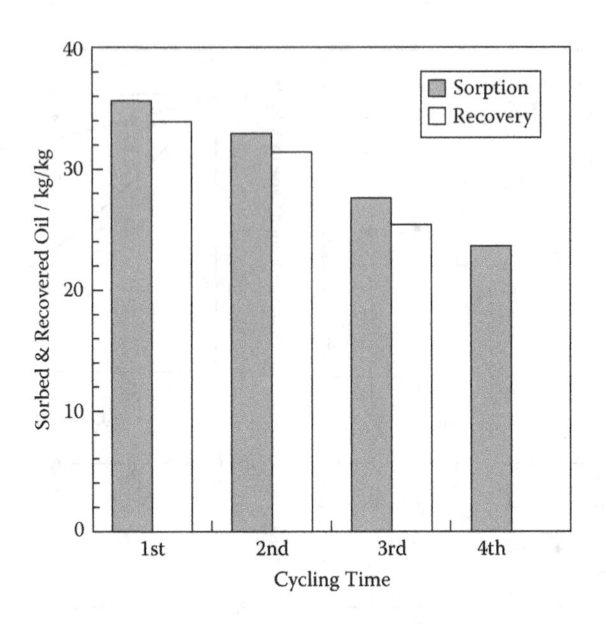

FIGURE 4.31 Changes in sorbed amount of C-grade heavy oil for fir fibers carbonized at 900°C with a bulk density of 11 kg/m^3 and recovered amount by washing with A-grade heavy oil.

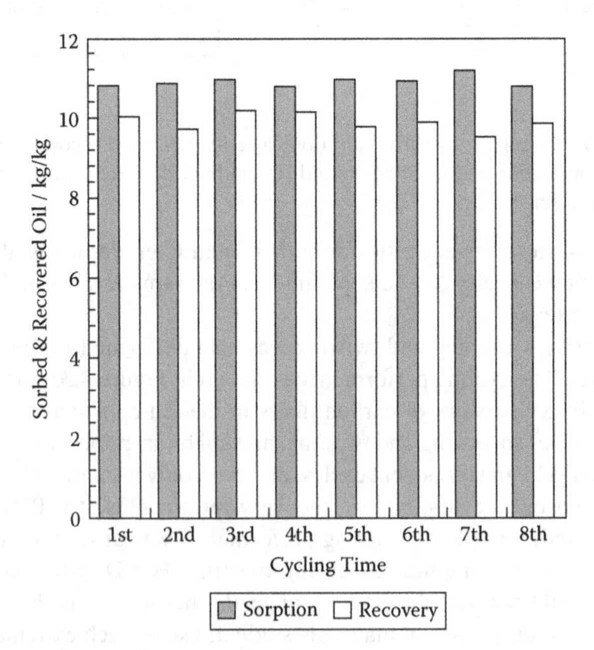

FIGURE 4.32 Changes in sorbed amount of A-grade heavy oil for carbon fiber felt with a bulk density of 72.6 kg/m^3 with cycling time and recovered amount by filtration under suction of 0.5 kPa pressure.

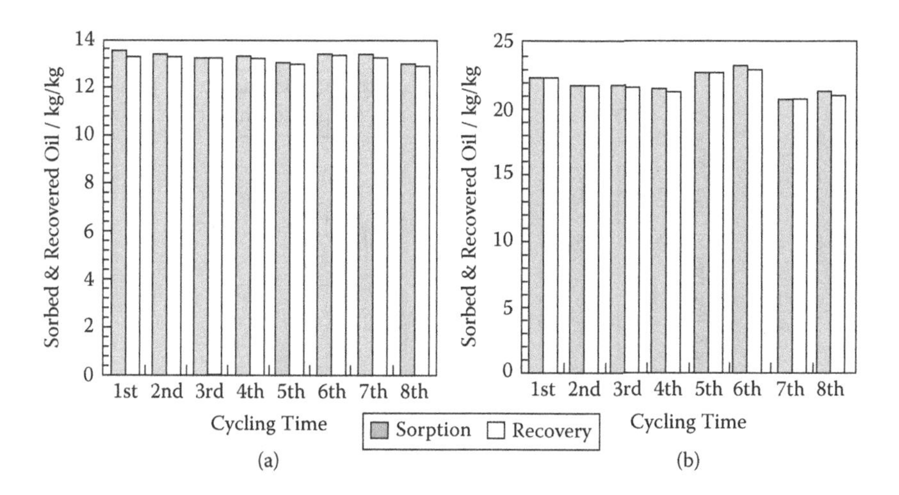

FIGURE 4.33 Changes in sorbed amounts of heavy oils for carbon fiber felt and recovered by *n*-hexane washing. (a) A-grade heavy oil for felt with a bulk density of 68.7 kg/m^3 and (b) C-grade heavy oil for felt with 51.7 kg/m^3.

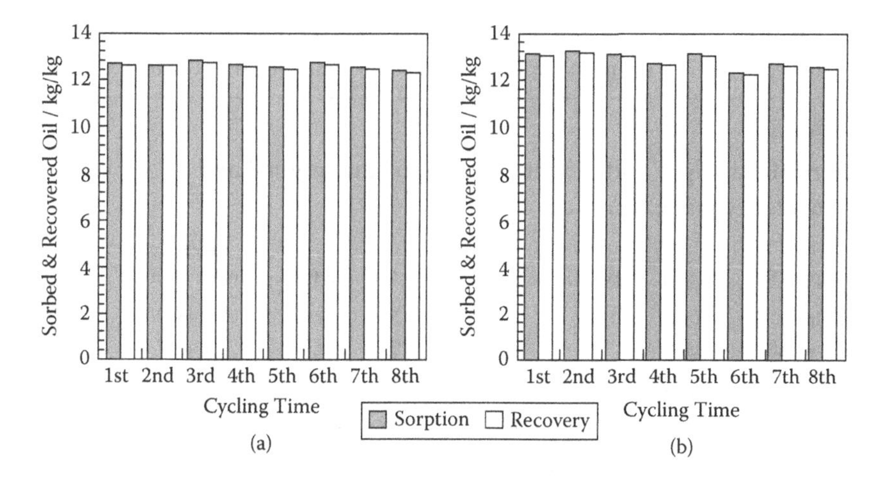

FIGURE 4.34 Changes in sorbed amount of A-grade heavy oil for carbon fiber felt and recovered amount with centrifugation at a rotation speed of 3800 rpm. (a) Felt with bulk density of 63.6 kg/m^3 and (b) felt with 85.5 kg/m^3.

D. Summary

Exfoliated graphite was able to be subjected only to the filtration under suction for recovering less viscous heavy oil sorbed (A-grade heavy oil and crude oil), which is a much milder operation than others, but its cycling performance is rather poor, even though its sorption capacity for the first cycle is very high. Carbonized fir fibers had better cycling performance for filtration than exfoliated graphite.

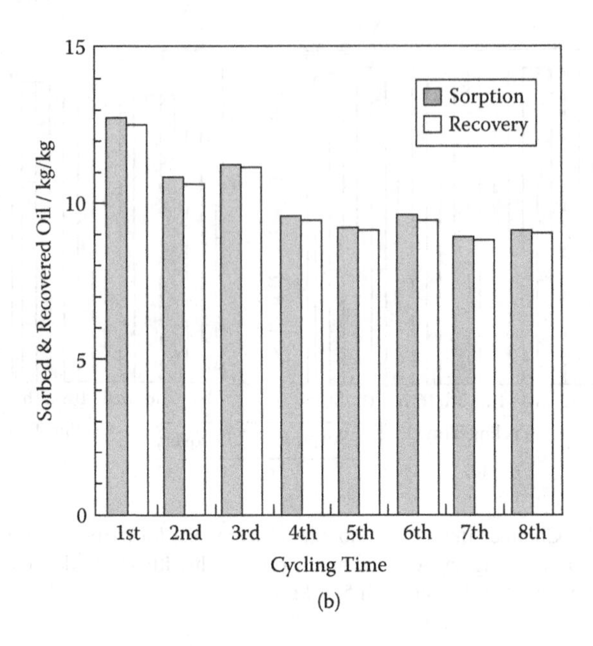

FIGURE 4.35 Changes in sorbed amount of A-grade heavy oil for carbon fiber felt with a bulk density of 77.0 kg/m^3 and recovered amount by squeezing with twisting.

TABLE 4.4
Elemental analysis and refractive index for the original and recovered A-grade heavy oil

Sample Oil	C (%)	H (%)	N (%)	O (%)	S (%)	Refractive Index at 20°C	F_{arom}
Original	85.3	13.9	< 0.1	0.3	0.81	1.4669	4.0
Recovered	85.2	13.9	< 0.1	0.3	0.81	1.4667	4.2

Note: F_{arom} = fraction of aromatic hydrocarbons (%).

Recovery operations by washing with organic solvents (such as *n*-hexane) were able to be applied to carbonized fir fibers with rather good performance. Carbon fiber felt, particularly from PAN-based carbon fibers, had excellent cycling performance for filtration under suction, washing with solvents, centrifuging, and even squeezing, although its sorption capacity is not very high.

VII. ANALYSIS OF RECOVERED OILS

Composition and structure analyses were performed by different techniques on the original and recovered A-grade heavy oils. Elemental analysis data, refractive index, and fraction of aromatic hydrocarbons determined by ^{1}H-NMR are shown in Table 4.4. Number-averaged (Mn) and weight-averaged (Mw) molecular weights and their ratios were measured by FI-MS (flow injection mass spectroscopy) and GPC (gel permeation chromatography), and are summarized in Table 4.5. In the latter measurements, three different detectors, RI (refractive index), IR at 2940 cm^{-1}, and UV at 254 nm, were used. By FI-MS, molecular types of hydrocarbons, classified by the number z in the molecular formula C_nH_{2n+z}, can be estimated. The results for the original and recovered A-grade heavy oils are summarized in Table 4.6. The fractions of saturated, aromatic, and resinous hydrocarbons, as well as asphaltenes, were determined from chromatographic analysis using silica/alumina gels. The results are shown in Table 4.7. Between the original and recovered A-grade heavy oil, no difference was detected in elemental composition, refractive index, molecular weight values, and molecular types of hydrocarbons. The FT-IR spectra of the original and recovered A-grade heavy oil also showed no difference.

In Table 4.8 the results of the analysis of hydrocarbon composition in the number of aromatic rings by HPLC (high-performance liquid chromatography) are summarized. The content of aromatics with more than three rings decreased to 1.0 vol% from 2.1% in the original oil, suggesting that these aromatics remain in the exfoliated graphite. This result is consistent with the experimental fact that the oils recovered by filtration under suction had a slightly lighter color than the original, as mentioned in Section VI.A.

Distillation curves measured by gas chromatography could not differentiate between the original and the recovered A-grade heavy oils, but there was a small difference in the high-boiling-point components, namely, a little less content in the recovered oil, as shown in Table 4.9. This result does agree with that of HPLC analysis (Table 4.8) and also with the fact that the recovered oil has a slightly lighter color than the original.

For crude oil and C-grade heavy oil, the fractions of aromatic hydrocarbons measured by ^{1}H-NMR and average molecular weight values measured by FD-MS (field desorption mass spectrometry) analysis are listed in Table 4.10. The fraction of aromatic hydrocarbons, F_{arom}, tends to decrease a little for the two grades of heavy oil recovered from exfoliated graphite, more markedly for the C-grade, even though a detectable difference was not found for the A-grade heavy oil by FI-MS analysis (Table 4.5). From the FD-MS analysis, no pronounced differences between the original and the recovered oil are detected in both Mn and Mw. Only a slight decrease in Mw/Mn is observed for C-grade heavy oil, suggesting a slight narrowing of the molecular weight distribution.

All these analyses of the recovered oils did not show notable differences with respect to the original oils. The only exception is with the high-molecular-weight hydrocarbons containing more than three rings, a part of which tends to remain in the sorbent carbons. Therefore, it can be concluded that the recovered oils can be used in the same manner as the original ones.

In Table 4.11 are listed the water content in A-grade heavy oils that have been stirred in water for different periods and then recovered by using exfoliated graphite. As the limit of this determination is supposed to be 0.05 vol%, the water content in the recovered oils even after stirring in water for 20 h cannot be differentiated from that of the original.

TABLE 4.5
Average molecular weight values of the original and recovered A-grade heavy oil measured by FI-MS and GPC with different detectors

Sample A-Grade Heavy Oil	FI-MS			GPC: RI			GPC: IR at 2940 cm^{-1}			GPC: UV at 254 nm		
	Mn	Mw	Mw/Mn	Mn	Mw	Mw/Mn	Mn	Mw	Mw/Mn	Mn	Mw	Mw/Mn
Original	258.1	273.5	1.06	154	175	1.1	276	327	1.2	141	166	1.2
Recovered	258.1	274.4	1.06	154	174	1.1	284	331	1.2	142	166	1.2

Note: Mn = number-average molecular weight; Mw = weight-average molecular weight.

TABLE 4.6
Analysis of hydrocarbon types in the original and recovered A-grade heavy oil by FI-MS

z*	Type of Hydrocarbon**	Original (%)	Recovered (%)
+2 or −12	**Paraffin**, naphthalene, etc.	48.4	47.9
0 or −14	**Naphthenes with 1 ring**, biphenyl, etc.	17.7	17.9
−2 or −16	**Naphthenes with 2 rings**, fuluprene, etc.	9.2	9.4
−4 or −18	**Naphthenes with 3 rings**, anthracene, etc.	4.5	4.5
−6	**Alkylbenzenes**, naphthenes with four rings, etc.	10.8	10.9
−8	**Indan, tetralin**, naphthenes with five rings, etc.	6.3	6.4
−10	Indene, naphthenes with six rings, etc.	3.1	3.0
Total		100.0	100.0
Weight-average molecular weight		273.5	274.4

* C_nH_{2n+z}.
** Bold letters show main component.

TABLE 4.7
Chromatographic analysis of the original and recovered A-grade heavy oil

Sample Oil	Saturated Hydrocarbons (%)	Aromatic Hydrocarbons (%)	Resinous Hydrocarbons (%)	Asphaltenes (%)
Original	87.8	11.7	0.4	0.1
Recovered	88.0	11.5	0.4	0.1

TABLE 4.8
Composition of hydrocarbon components of the original and recovered A-grade heavy oil

Sample Oil	Saturated Hydrocarbons (vol%)	Olefins (vol%)	Aromatics (vol%)		
			One Ring	Two Rings	Three or More Rings
Original	76.3	0.0	15.5	6.1	2.1
Recovered	77.4	0.0	15.6	6.0	1.0

TABLE 4.9
Distillation temperature (°C) versus distillates from the original and recovered A-grade heavy oil

	Distillates (vol%)								
Sample Oil	**IBP**	**5**	**10**	**15**	**20~80**	**85**	**90**	**95**	**FBP**
Original	128.7	159.9	186.3	209.4	—	346.6	359.3	376.5	458.9
Recovered	132.4	164.0	189.4	212.1	—	344.8	356.9	373.0	433.1

Note: IBP = initial boiling point; FBP = final boiling point.

TABLE 4.10
Average molecular weight values measured by FD-MS analysis and fractions of aromatic hydrocarbons in original and recovered crude oil and C-grade heavy oil

			FD-MS Analysis				
Sample Heavy Oil		**F_{arom}**	**Mn (%)**	**Mw (%)**	**Mz (%)**	**Mw/Mn**	**Mz/Mw**
Crude oil	Original	4.9	645	869	1102	1.35	1.27
	Recovered	4.5	672	915	1147	1.36	1.25
C-grade heavy oil	Original	5.4	1071	1768	2428	1.65	1.37
	Recovered	4.6	1207	1839	2393	1.52	1.30

Note: F_{arom} = fraction of aromatic hydrocarbons; Mn = number-average molecular weight; Mw = weight-average molecular weight.

TABLE 4.11
Content of water in A-grade heavy oils

	Water Content (vol%)
Original	0.00
Oil sorbed onto exfoliated graphite from its floating on water surface	0.05
Oil stirred in water for 20 h and then sorbed into exfoliated graphite	0.05

VIII. PRELIMINARY EXPERIMENTS FOR PRACTICAL RECOVERY OF SPILLED HEAVY OILS

Carbon materials have a large sorption capacity for heavy oils, as explained in Section III. Also, the sorbed heavy oils can be recovered by a rather simple recycling process, as explained in Section VI. It can be said that carbon materials

TABLE 4.12
Sorption of A-grade heavy oil into plastic bags containing exfoliated graphite

Bag Material	Mesh_ Opening (kg/m²)	Thickness of Heavy Oil Layer on Water		
		0.9 mm (100 mL)	4.3 mm (500 mL)	9.0 mm (1000 mL)
Poly(ethylene)	0.010	×	○	○
	0.015	×	○	○
	0.030	×	○	○
	0.050	×	○	○
Poly(propylene)	0.050	×	○	○

Note: × = poor contact between exfoliated graphite and heavy oil; ○ = good contact between exfoliated graphite and heavy oil.

have good potential for practical application in the recovery of spilled heavy oil. Among the materials used, exfoliated graphite had the highest sorption capacity, but its fragility and bulkiness are expected to cause certain difficulties in handling and storage for practical applications. Therefore, some preliminary experiments (trials) were carried out with exfoliated graphite on a small scale, with the sorbent packed into various plastic bags and formed into various bulk densities. Also, the recovery of heavy oil mixed into sand was studied by using exfoliated graphite. In the heavy oil spill accidents in the sea, heavy oil was changed to a mousse by mixing with water during floating; therefore, preliminary sorption experiments of heavy oil mousse with exfoliated graphite and carbonized fir fibers were also carried out.

A. Exfoliated Graphite Packed into Plastic Bags

Exfoliated graphite was packed into bags of poly(ethylene) and poly(propylene) with different mesh openings (characterized by the unit of kg/m^2), as shown in Table 4.12. The bag packed with exfoliated graphite ($200 \times 200 \times 50$ mm^3) was placed on the heavy oil floating on water at different layer thicknesses (different amounts of A-grade heavy oil on 500 mL of water in a $373 \times 309 \times 43$ mm^3 tray). The appearance of the bag placed on the heavy oil layer is shown in Figure 4.36.

When the oil layer is as thin as 0.9 mm (100 mL heavy oil used), the contact of exfoliated graphite inside the bag with the heavy oil was not sufficient, as shown in Figure 4.36a. A thin oil layer seems to be pushed away from exfoliated graphite, as schematically illustrated in Figure 4.36c. When the oil layer is thick enough, that is, thicker than about 4 mm (more than 500 mL of heavy oil used), exfoliated graphite was wetted completely, as shown in Figure 4.36b.

In Table 4.13 the sorption capacity of exfoliated graphite packed at different bulk densities in plastic bags with or without waterproof treatment is summarized. Each bag was immersed into enough A-grade heavy oil for 1 h and dipped up using stainless steel mesh; excess oil was drained off for 5 min, and then the mass

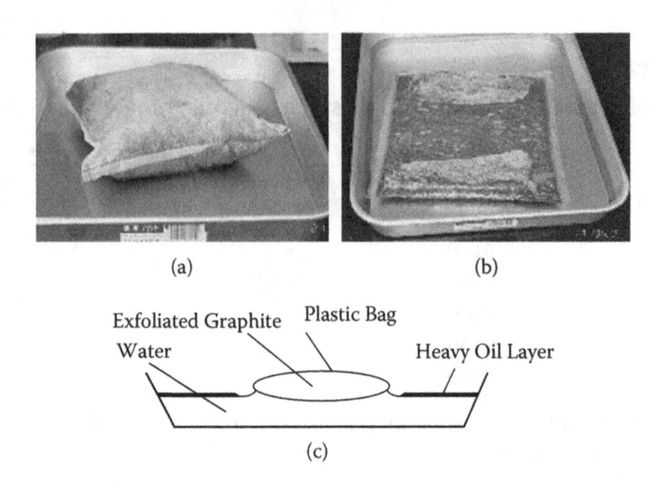

FIGURE 4.36 Appearance of dipping a plastic bag packed with exfoliated graphite into heavy oil. (a) Thin heavy oil layer on water, (b) thick heavy oil layer, and (c) schematic illustration of the relation between exfoliated graphite and heavy oil in the case of a thin layer.

TABLE 4.13
Sorption capacity of exfoliated graphite packed into plastic bags

Sample Code	Bag Material	Waterproof Treatment	Exfoliated Graphite Packed (kg)	Bulk Density of EG Packed (kg/m³)	Sorption Capacity (kg/kg)
A-1	Poly (ethylene)	Yes	0.018	9	36.1
A-2		Yes	0.022	11	31.9
A-3		No	0.018	9	36.3
A-4		No	0.022	11	30.6
B-1	Poly (propylene)	Yes	0.018	9	33.7
B-2		Yes	0.022	11	28.9
B-3		No	0.018	9	36.7
B-4		No	0.022	11	33.4

increase was measured. In Figure 4.37 the sorption capacity of exfoliated graphite packed into plastic bags is plotted against its bulk density and compared with the results for exfoliated graphite itself. Irrespective of bag material and waterproofing treatment, the bulk density is shown to govern the sorption capacity.

Bags of poly(ethylene terephthalate) containing exfoliated graphite with different bulk densities were dipped into A-grade heavy oil. After sorption, the bags were separated from the oil by two methods: hung up using a clamp and dipped up using a stainless steel mesh. In Figure 4.38 the sorption capacity of exfoliated graphite in the bag separated from the oil by these two methods is plotted as a

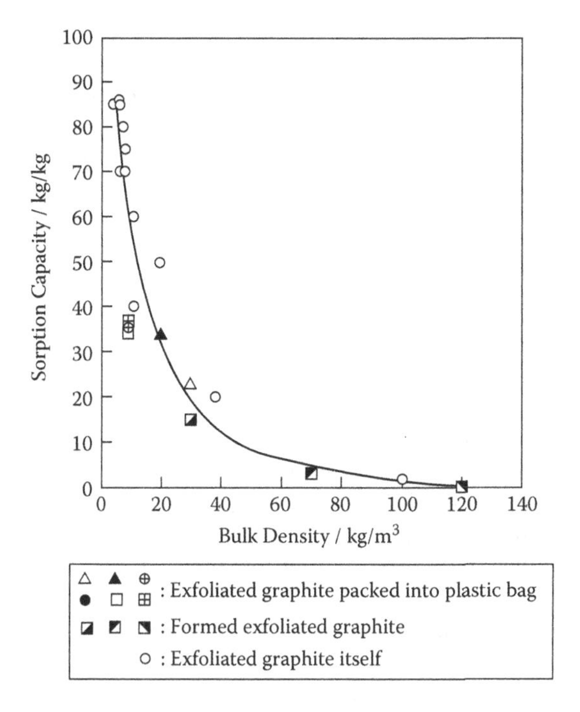

FIGURE 4.37 Bulk density dependence of sorption capacity of exfoliated graphite packed into plastic bag and of formed exfoliated graphite.

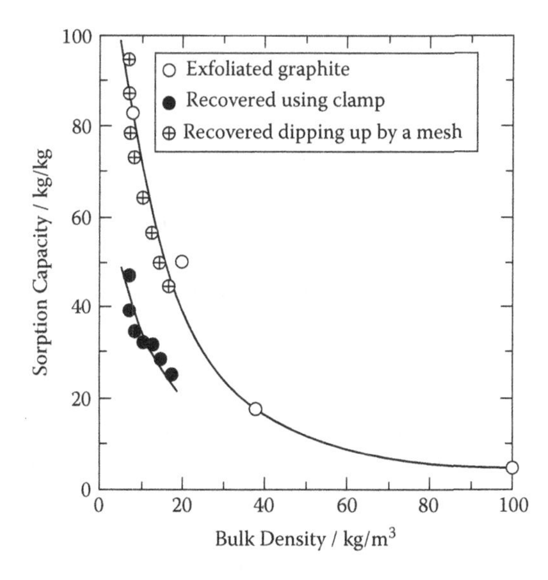

FIGURE 4.38 Bulk density dependence of sorption capacity of exfoliated graphite packed into plastic bag when separated from heavy oil by two different methods.

TABLE 4.14
Sorption capacity of exfoliated graphite compacts

Sample Code	Bulk Density (kg/m³)	Sample Weight (g)	Sorbed Amount of Heavy Oil (g)	Sorption Capacity (kg/kg)
G-1	30	1.168	17.02	14.6
G-2	30	1.147	17.36	15.1
G-3	30	0.878	17.47	19.9
G-4	70	1.253	4.273	3.4
G-5	120	1.027	1.949	1.9

function of bulk density and compared with the behavior of the exfoliated graphite itself. When carefully dipped up using a metal mesh, almost the same sorption capacity is obtained as for the exfoliated graphite itself. When hung up using a clamp, however, sorption capacity is reduced almost by half, probably because of squeezing of the bag by itself during this procedure. This result shows that the recovery process is also important if the high sorption capacity of exfoliated graphite is to be achieved.

B. Formed Exfoliated Graphite

Instead of packing into plastic bags, exfoliated graphite was formed into compacts with a size of $40 \times 40 \times 10\ mm^3$ by compression in a mold without using any binding material. The compacts thus prepared could be easily handled, but their bulk density became more than 30 kg/m^3. By changing the amount of exfoliated graphite in the mold, compacts with bulk densities of 30, 70, and 120 kg/m^3 were prepared. The results of sorption of A-grade heavy oil into these formed exfoliated graphites are summarized in Table 4.14. Their sorption capacity value for A-grade heavy oil is plotted against their bulk density in Figure 4.37.

The sorption capacity of formed exfoliated graphites is comparable to that of exfoliated graphite itself, when their high bulk density is taken into consideration.

C. Heavy Oil Sorption from Contaminated Sand

The experimental setup was as follows: in a glass cylinder, a mixture of sand with A-grade heavy oil (contaminated sand) was placed in the lower part, and exfoliated graphite with different packed densities was placed on top of the contaminated sand. As heavy oil ascended, the exfoliated graphite appeared wet and changed its color to much darker black than its original color. This change of the height with darker black color was measured as a function of time.

To understand the effect of particle size of the sand, α-alumina powders with different particle sizes were selected as model sand; their clear white color helped visualize the contamination by heavy oil. The alumina powders used are listed in Table 4.15. Sand sampled from the seashore where the accident of the tanker

TABLE 4.15
Alumina powders used as model sands

Grade of alumina powder	#30	#36	#46	#60	#90
Average particle size (μm)	713	603	425	303	175

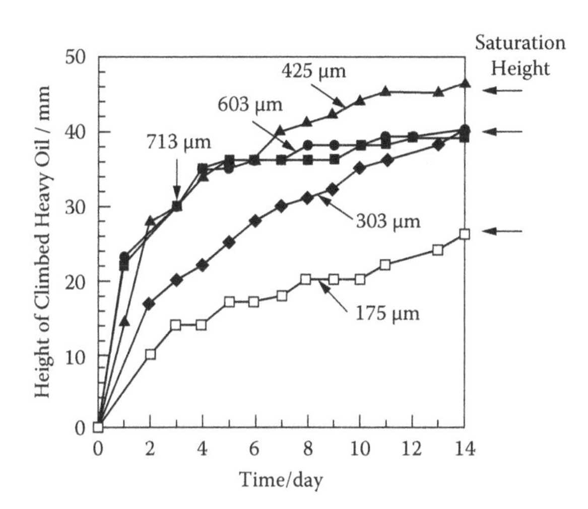

FIGURE 4.39 Time changes in the height of heavy oil climbed in exfoliated graphite for contaminated sand with different particle sizes.

Nakhodka happened in 1997 was also used. It had a brown color and also a distribution of particle sizes.

In Figure 4.39 are shown the changes in height of heavy oil with time for model sands with different particle sizes; the bulk density of packed exfoliated graphite was kept at about 10 kg/m^3. With increasing time, the height increases gradually and appears to reach saturation. After 14 days, saturation is reached, except for the smallest particles. When the sand particle size is about 425 μm, heavy oil could climb up to the highest position in exfoliated graphite with a bulk density of about 10 kg/m^3; in other words, the largest amount of heavy oil could be pumped into exfoliated graphite, although the weight of the sorbed heavy oil could not be determined. With a smaller particle size of 175 μm, sorption into the exfoliated graphite is very slow, so the climbed height is rather low, increasing gradually even after 14 days. With a larger particle size of 713 μm, the climbing rate is almost the same as for 425-μm particles, but it seems to reach saturation in a shorter time. The saturation height is determined after 14 days for each sand particle size (Figure 4.39).

The saturation height was seen to depend strongly on the bulk density of exfoliated graphite and also on the particle size of the sand, as shown in Figure 4.40, suggesting that there is an appropriate combination for achieving a high saturation height between particle size of sand contaminated by heavy oil and bulk density of exfoliated graphite; in other words, a balance in pore size between sand and

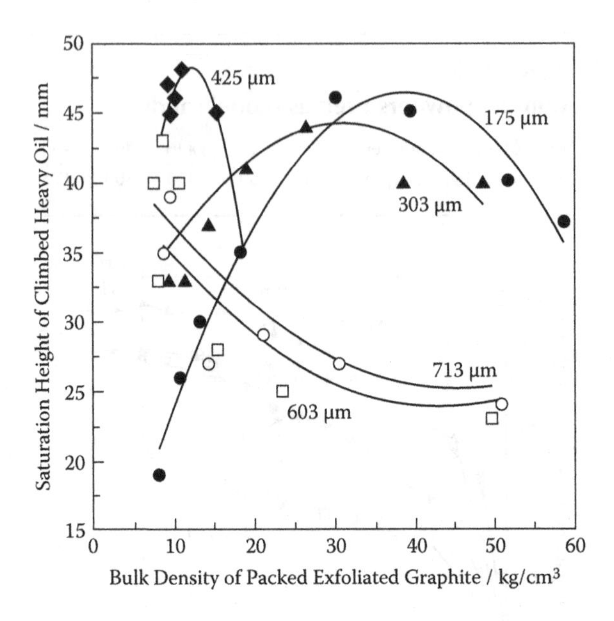

FIGURE 4.40 Dependence of saturation height of climbed heavy oil into exfoliated graphite on its bulk density for contaminated sand with different particle sizes.

exfoliated graphite. When the sand with very fine size, for example, about 175 μm, is contaminated by A-grade heavy oil, the exfoliated graphite with a bulk density of about 35 kg/m^3 gives the highest saturation, that is, the highest efficiency of pumping. When the sand particle size is large, for example, 425 μm, we need exfoliated graphite with about 10 kg/m^3, that with 40 kg/m^3 being unable to pump heavy oil. For larger-size sand, lower bulk density of exfoliated graphite is needed.

For real sea sand, whose average particle size was determined to be 236 μm (from its distribution measured by sieving), the same experiment was performed using exfoliated graphite with a bulk density of 10 kg/m^3. The results are shown in Figure 4.41, where saturation height is plotted against sand particle size, in comparison with alumina powders. The result on real sea sand is consistent with the relation obtained by model sand alumina having well-defined particle size. Because the sea sand used has relatively small particle size, the exfoliated graphite with slightly higher bulk density than 10 kg/m^3 (Figure 4.41) can be used for effective pumping.

D. Sorption of Heavy Oil Mousse

When heavy oil is mixed with water, it is known to change to a mousse-like material. Sorption of such heavy oil mousse into exfoliated graphite and carbonized fir fibers was studied. The heavy oil mousse was prepared by mixing 50 mL of C-grade heavy oil with 25 mL of water. In Figure 4.42 the amount of mousse sorbed is plotted against soaking time for three carbon materials: exfoliated graphite with a bulk density of about 6 kg/m^3 and two carbonized fir fibers (380 and 900°C-carbonized) with bulk density of about 7.2 kg/m^3. Sorption capacity

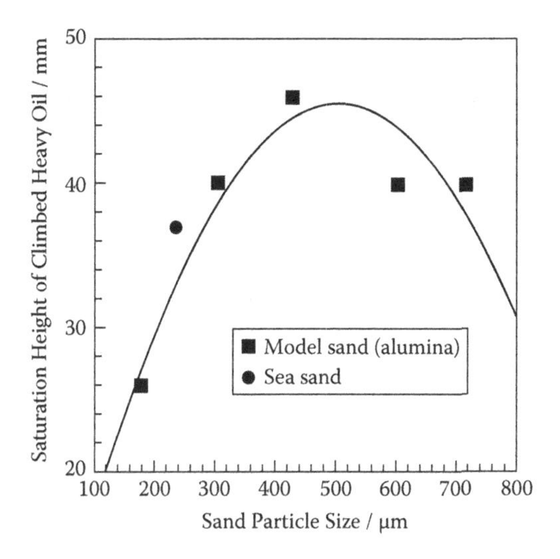

FIGURE 4.41 Dependence of saturation height of climbed heavy oil into exfoliated graphite with a bulk density of 10 kg/m^3 on sand particle size.

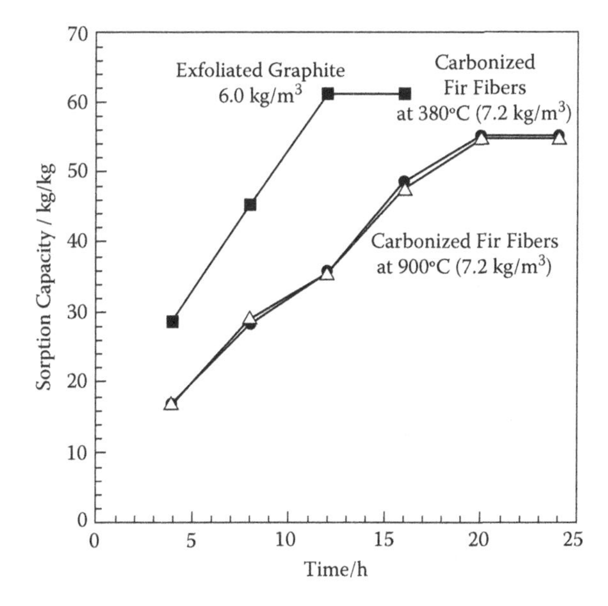

FIGURE 4.42 Changes in amount of heavy oil mousse sorbed into three carbon sorbents with time.

was saturated at about 61 kg/kg after 10 h for exfoliated graphite, at a lower capacity and for a longer time than for pure C-grade heavy oil (83 kg/kg and 8 h, respectively). For carbonized fir fibers, the sorption capacity for the mousse was slightly lower and saturation was achieved after about 20 h.

TABLE 4.16
Comparison of heavy oil sorption performance among various materials

Sorbent Materials	Oil	Sorption Capacity (kg/kg)	Sorptivity K_s ($kg/m^2s^{1/2}$)	Recovering Process	Refs.
Exfoliated graphite	A	87	0.7[a]	Only filtration	29,30
	C	67	—	No recovery	
Carbonized fir fibers	A	78	1.4[a]	Filtration and washing	38
	C	66	0.1[a]	Washing	
Carbon fiber felt	A	20	5.5[a]	Filtration, washing, centrifuging, and squeezing	42
	C	22	0.2[a]	Washing, centrifuging, and squeezing	
Poly(propylene) fiber nonwoven web	Crude	20	—	Squeezing	17
		11	—	—	
Cotton fibers	Crude	33	—	Squeezing	17
Cotton grass fiber	Diesel	~18			79
Milkweed fibers	Crude	~50	—	Squeezing	17
Kenaf heated to 400°C	Salad	33	—	—	20
Silky floss fiber	Crude	85	—	—	80
Wheat straw after acetylation	Machine	12~29	—	—	84

[a] Approximate value.

Even though the sorption performance (both capacity and rate) for heavy oil mousse was a bit inferior to that of pure oil, it was experimentally demonstrated that heavy oil mousse can be recovered by using carbon materials. However, the separation of water from heavy oil mousse was not observed throughout the sorption process.

IX. CONCLUDING REMARKS

A. Comparison among Carbon Materials

In Table 4.16 we summarize the heavy oil sorption performance of the three carbon sorbents used by listing their sorption capacity, sorptivity as a measure of sorption rate, and the cycling processes that can be applied. Both sorption capacity and sorptivity K_s depend strongly on sorbent bulk density, as shown in Figure 4.43.

Exfoliated graphite has very high sorption capacity for heavy oil, but its sorption rate is rather low. By increasing its bulk density, the sorption rate can be improved slightly, but sorption capacity decreases at the same time. Fir fibers have a sorption capacity and rate that are comparable with exfoliated graphite for

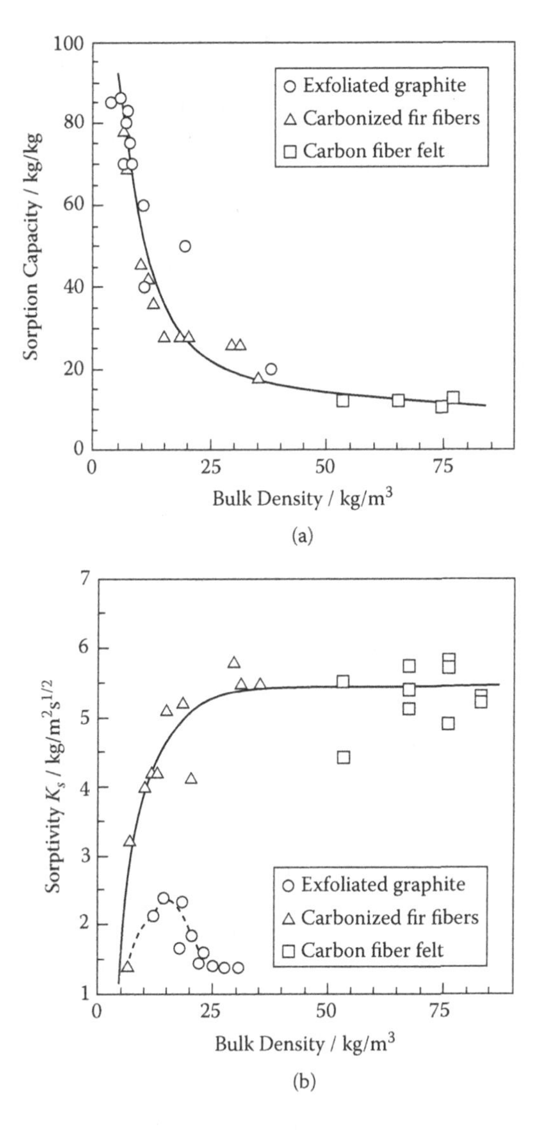

FIGURE 4.43 Dependence of heavy oil sorption parameters on bulk density of three carbon sorbents. (a) Sorption capacity and (b) sorptivity K_s.

both A- and C-grade heavy oils. By densification of carbonized fir fibers, sorption rate increases rather rapidly, but sorption capacity decreases. On the other hand, carbon fiber felts have relatively low capacity but high rate, because of their high bulk density; sorptivity seems to be at its upper limit for carbon sorbents.

For practical application in heavy oil spill accidents, prompt action is so crucial that both high sorption capacity and high rate of sorption are required. In currently available carbon sorbents, however, these two requirements cannot be met in the same material, as illustrated in Figure 4.43. Therefore, a judicious balance between capacity and rate has to be attained by carefully selecting the bulk density of the sorbent. From Figure 4.43, a bulk density of about 20 kg/m^3 seems to be the most appropriate, with a capacity of about 40 kg/kg and sorptivity close to the highest value of 5.5 $kg/m^2s^{1/2}$.

Sorbent recovery is also an important point for practical applications. Thus, bulky exfoliated graphite and carbonized fir fibers have some associated problems and, also, there may be problems related to handling and storage of large amounts. To solve such problems, exfoliated graphite was packed into a plastic bag and examined for heavy oil sorption and recovery. If sufficient attention is paid to maintain contact with the heavy oil and to recover the bag after sorption by dipping, a high capacity can be achieved. No problems related to handling and storage are expected for carbon felt, but its sorption capacity is not as high.

Recycling of recovered heavy oils is also necessary from the viewpoint of energy resources. The recovery process for exfoliated graphite is very limited: only filtration under suction with less viscous A-grade and crude oil is possible, but cycling efficiency is poor. Carbonized fir fibers have somewhat better cycling performance than exfoliated graphite. The fibers were able to withstand washing using organic solvents, with a rather high recovery ratio. For the felt, not only filtration and washing but also centrifugation and even squeezing processes could be applied with high cycling efficiency.

When an oil spill accident occurs at sea, or in a lake or a river, the spilled oil is usually dispersed over a large water area. To remove the spilled oil, its area must be limited by using floating barriers to avoid further oil spread, and it should then be skimmed by a pump or ladle. Sometimes a sorbent is added into this spilled oil to make the skimming process easier and more efficient, the sorbent with spilled oil being dipped up by sections. When a sorbent in the form of small particles, such as chopped straw and cane, is used, this sorbent–oil system is usually unstable, and a certain amount of oil, taken up by sorbents adjacent to the zone where the sorbent–oil mixture was dipped up by using a scoop or shovel, moves gradually to that zone; in other words, the contaminated water surface cannot be made small until the entire oil spill is removed. When finely dispersed carbon is added into an oil layer and kept for several hours, however, the mixture of oil with carbon sorbent becomes so intimate that a part of it is easily dipped up, but no oil oozes from neighboring carbon sorbent–oil mixtures [77]. This is another advantage of carbon materials when used as oil sorbents. It is often referred to as *consolidated bed* and consists of carbon sorbent and oil. The carbon used in these experiments was produced from waste wood and had a bulk density of 170 kg/m^3, and its sorption capacity for light and diesel oils was about 4 kg/kg [77].

The formation of a consolidated bed of carbon sorbent with oil is attributed to the hydrophobic and oleophilic nature of the carbon surface. The same effect is expected for the carbon materials used in our research: exfoliated graphite, carbonized fir fibers, and carbon fiber felt, as suggested by the experimental evidence presented in Figures 4.5b and 4.5d.

B. Comparison with Other Materials

For oil spill cleanup, three types of methods have been applied: physical methods such as skimming and use of sorbents; chemical methods such as dispersion, in-situ burning, and the use of solidifiers; and biological methods or bioremediation [78]. The use of a sorbent is effective and practical, although a combination of various methods has usually been employed to achieve the most effective cleanup. Oil sorbents used and studied so far can be classified into three groups: inorganic minerals such as perlite and vermiculite, organic synthetic materials such as poly(propylene) and poly(urethane), and biomass such as peat moss, kenaf, straw, and wood fibers.

Organic synthetic materials are already commercialized as sorbents in the form of either nonwoven web or foam. Not only the size but also the shape of pores formed in these materials was shown to be important—rectangular rather than circular pores are preferred—by using poly(propylene) filaments [13]. These organic synthetic sorbents have been used as reference materials for other sorbents in most research papers. Research on the development of biomass-derived sorbents is very active now, mainly because most of them are low-cost waste materials, and many of them are biodegradable, even though the sorption capacity of these materials is not so high, typically about 10 kg/kg [13,15,17,78]. This is reasonable because an environment-friendly process has to be applied to remediate environmental contamination such as a heavy oil spill, most biomass materials being biodegradable and safe for disposal. Many kinds of biomass have been tried; for example, milkweed (*Asclepias*) fibers [17], cotton fibers [17], kenaf (*Hibiscus cannabinus*) bast and core fibers [20], bagasse (a by-product of sugar extraction from sugarcane) [78], rice hull [78], cotton grass fiber as a by-product of peat excavation [79], and fibers from various plants such as sisal (*Agave sisalana*), coir (*Cocos nucifera*), sponge-gourd (*Luffa cylindrica*), and floss-silk (*Chorisia speciosa*) [80].

The sorption capacity of most of these biomass sorbents was not so high, about 10 kg/kg, much less than that of carbon materials and even less than that of commercially available synthetic sorbents. For milkweed and cotton fibers, however, relatively high sorption capacities were reported, about 33 and 50 kg/kg, respectively, for light oils (a diesel oil and a crude oil) [17]; similar values were reported for unstructured cotton fibers [17]. The silk-floss fibers showed high sorption capacity (85 kg/kg) for crude oil, an uptake comparable to that of carbon materials [80]. For biomass sorbents, water uptake has to be taken into account as well. After soaking the sorbent in water, its capacity for oil was reported to decrease slightly [17]. Therefore, the surface of many of such sorbents must be changed from hydrophilic to hydrophobic. For this purpose, acetylation of free hydroxyl groups was carried out on various biomass materials, for example, rice straw [81], raw cotton [82], sugarcane bagasse [83], and wheat straw [84]; sorption capacity for oil was reported to double, from about 10 to 20 kg/kg.

In Table 4.16 the sorption performance of some of these materials is compared to that of carbon materials. Because no quantitative rate measurements have been reported on these sorbents, a full comparison cannot be made. Although the heavy oils used in these experiments are not exactly the same, the two carbon sorbents, exfoliated graphite and carbonized fir fibers, are seen to have a markedly higher sorption capacity even for viscous C-grade heavy oil, more than twice that of conventional poly(propylene) foam and most natural fibrous sorbents.

For the process recovery of heavy oil from these sorbents, only squeezing was reported, and most of them showed very high performance. The experimental result that oils taken up by sorbents can be recovered by squeezing is reasonably understood to suggest that milder processes for recovery, that is, filtration under suction and washing with organic solvents, can easily be applied. From the viewpoint of recovery performance, therefore, exfoliated graphite and carbonized fir fibers are inferior to poly(propylene) nonwoven webs and natural fibrous sorbents.

ACKNOWLEDGMENTS

The present work was supported mainly by the NEDO project on energy and environmental technology (No. 98Ec-12-002) and partly by a joint research program between Japan (Japanese Society for the Promotion of Science) and China (National Natural Science Foundation of China) and a grant of Frontier Research Project from the Ministry of Education, Culture, Sports, Science, and Technology, Japan. The authors would like to express their sincere thanks to Prof. H. Konno of Hokkaido University, Prof. Y. Sawada of Osaka City University, Dr. Y. Nishi of Toyo Tanso Co., Ltd., Prof. G. Dai of South West Jiaotong University for their cooperation and encouragement, and also to Drs. R. Aizawa and J. Fujita of Hitachi Chemicals Co., Ltd. and Drs. M. Harakawa and T. Kihara of Nippon Oil Co., Ltd. for their cooperation in the NEDO project. The authors also thank Prof. L. R. Radovic, editor of this book, for his kind advice in writing this chapter.

REFERENCES

1. Westermeyer, W.E., *Environ. Sci. Technol.*, 25, 196, 1991.
2. Mitchell, J.G., *National Geographic*, March, 124, 1999.
3. Oil Spill Accident of Tanker Nakhodka, Home page of Fukui Prefecture. http://www.erc.pref.fukui.jp/news/oil.html (1197. 4. 22) [in Japanese]
4. Goldberg, R., *Academy of Natural Sciences*, USA, May 1–3, 1991.
5. Owens, E.H., *Proceedings of the 14th World Petroleum Congress*, 1994, 375.
6. Gaaseidnes, K. and Turbeville, J., *Pure Appl. Chem.*, 71, 96, 1999.
7. Owens, E.H., *Pure Appl. Chem.*, 71, 83, 1999.
8. Fiocco, R.J. and Lewis, A., *Pure Appl. Chem.*, 71, 27, 1999.
9. Ogawa, T., Hirao, K., and Osawa, S., *Sekyu Gakkaishi*, 43, 207, 2000 [in Japanese].
10. Bucas, G. and Saliot, A., *Marine Pollut. Bull.*, 44, 1388, 2002.
11. Wolfe, D.A., Hameedi, M.J., and Galt, J.A., *Environ. Sci. Technol.*, 28, 561, 1994.
12. Oaine, R.T., *Annu. Rev. Ecol. Syst.*, 27, 197, 1996.
13. Zahid, M.A., Halligan, J.E., and Johnson, R.F., *Ind. Eng. Chem. Process Des. Dev.*, 11, 550, 1972.
14. Fukuoka, T., *Fushokufu-Jyoho* 1999 [No.11], 6, 1999 [in Japanese].
15. Johnson, R.F., Manjrekar, T.G., and Halligan, J.E., *Environ. Sci. Technol.*, 7, 439, 1973.

16. Drelich, J., Hupka, J., and Gutkowski, B., *Chemistry for Protection of the Environment 1987* (Studies in Environmental Science, Vol. 34) Pawlowski, L., Mentasti, E., Lacy, W.J., and Sarzanini, C., Eds., Elsevier, 207, 1988.
17. Chol, H.M. and Cloud, R.M., *Environ. Sci. Technol.*, 26, 772, 1992.
18. Yamamoto, H., *Cellulose Commun.*, 5, 148, 1998.
19. Miyata, N., *Sen'i Gakkaishi*, 55, 576, 1999 [in Japanese].
20. Lee, B.G., Han, J.S., and Rowell, R.M., *Kenaf Properties, Processing and Products*, Sellers, T. and Reichert, N.A., Eds., Mississippi State Univ. 423, 1999.
21. Saito, M., Ishii, N., Ogura, S., Maemura, S., and Suzuki, H., *Nihon Zousen-Gakkai Ronbunsyu*, No. 190, 287, 2001 [in Japanese].
22. Fujiraito Ind. Co., Japanese Patent Application (No. 95333), 1979 [in Japanese].
23. Shen, W.C., Cao, N.Z., Wen, S.Z., Gu, J.Z., and Wang, Z.D., *Extended Abstracts of the European Conference, Carbon '96,* Newcastle upon Tyne, U.K., 256, 1996.
24. Toyoda, M., Aizawa, J., and Inagaki, M., *Desalination,* 115, 199, 1998.
25. Toyoda, M., Aizawa, J., and Inagaki, M., *Nihon Kagaku Kaishi,* 1998, 563, 1998 [in Japanese].
26. Toyoda, M., Moriya, K., Aizawa, J., and Inagaki, M., *Nihon Kagaku Kaishi,* 1999, 193, 1999 [in Japanese].
27. Toyoda, M., Moriya, K., and Inagaki, M., *TANSO,* 1999 [No. 187], 96, 999 [in Japanese].
28. Toyoda, M., Moriya, K., and Inagaki, M., *Nihon Kagaku Kaishi,* 2000 [No. 3], 217, 2000 [in Japanese].
29. Toyoda, M., Moriya, K., Aizawa, J., Konno, H., and Inagaki, M., *Desalination,* 128, 205, 2000.
30. Inagaki, M., Konno, H., Toyoda, M., Moriya, K., and Kihara, T., *Desalination,* 128, 213, 2000.
31. Inagaki, M., Shibata, K., Setou, S., Toyoda, M., and Aizawa, J., *Desalination,* 128, 219, 2000.
32. Tryba, B., Kalenczuk, R.J., Kang, F., Inagaki, M., and Morawski, A.W., *Mol. Cryst. Liq. Cryst.*, 340, 113, 2000.
33. Toyoda, M. and Inagaki, M., *Carbon,* 38, 199, 2000.
34. Toyoda, M., Moriya, K., and Inagaki, M., *Sekiyu Gakkaishi,* 44, 169, 2001 [in Japanese].
35. Inagaki, M., Toyoda, M., and Nishi, Y., *Kagaku Kougaku,* 64, 179, 2001 [in Japanese].
36. Inagaki, M., Toyoda, M., Iwashita, N., Nishi, Y., and Konno, H., *Carbon Sci. (Korea),* 2, 1, 2001.
37. Inagaki, M., Kawahara, A., and Hayashi, T., *Res. Reports Aichi Inst. Tech.,* No. 36, 69, 2001 [in Japanese].
38. Inagaki, M., Kawahara, A., and Konno, H., *Carbon*, 40, 105, 2002.
39. Toyoda, M., Dogawa, N., Seki, T., Fujita, A., and Inagaki, M., *TANSO,* 2001 [No.199], 166, 2001 [in Japanese].
40. Tsumura, T., Kojitani, N., Umemura, H., Toyoda, M., and Inagaki, M., *Appl. Surf. Sci.,* 196, 429, 2002.
41. Inagaki, M., Toyoda, M., Iwashita, N., Nishi, Y., Konno, H., Fujita, A., and Kihara, T., *TANSO,* 2002 [No. 201], 16, 2002 [in Japanese].
42. Inagaki, M., Kawahara, A., Iwashita, N., Nishi, Y., and Konno, H., *Carbon*, 40, 1487, 2002.
43. Toyoda, M., Nishi, Y., Iwashita, N., and Inagaki, M., *Desalination,* 151, 139, 2002.
44. Nishi, Y., Dai, G., Iwashita, N., Sawada, Y., and Inagaki, M., *Mater. Sci. Res. Intl.*, 8, 243, 2002.
45. Nishi, Y., Iwashita, N., Sawada, Y., and Inagaki, M., *Water Res.*, 36, 5029, 2002.

46. Inagaki, M., Kawahara, A., and Konno, H., *Desalination*, 17, 77, 2004.
47. Tryba, B., Morawski, A.W., and Inagaki, M., *Spill Sci. Tech. Bull.*, 8, 569, 2003.
48. Toyoda, M. and Inagaki, M., *Spill Sci. Tech. Bull.*, 8, 467, 2003.
49. Inagaki, M., Toyoda, M., Iwashita, N., and Kang, F., *Adsorption by Carbons*, Tascon, J.M.D., Ed., Elsevier, 2008 (in press).
50. Inagaki, M., Nishi, Y., Iwashita, N., and Toyoda, M., *Fresenius Environ. Bull.*, 13, 183, 2004.
51. Inagaki, M., Nagata, T., Suwa, T., and Toyoda, M., *New Carbon Mater.*, 21, 97–101, 2006.
52. Shen, W., Wen, S., Cao, N., Zheng, L., Zhou, W., Liu, Y., and Gu, J., *Carbon*, 37, 351, 1999.
53. Kang, F., Zheng, Y.-P., Zhao, H., Wang, H.-N., Wang, L.-N., Shen, W., and Inagaki, M., *New Carbon Mater.*, 18, 161, 2003.
54. Nishi, Y., Iwashita, N., and Inagaki, M., *TANSO* 2002 [No. 201], 31, 2002 [in Japanese].
55. Kang, F., Zheng, Y.P., Wang, H.N., Nishi, Y., and Inagaki, M., *Carbon*, 40, 1575, 2002.
56. Inagaki, M. and Suwa, T., *Carbon*, 39, 915, 2001.
57. Inagaki, M., Tashiro, R., and Suwa, T., *Res. Reports Aichi Inst. Tech.*, No. 37, 53, 2002.
58. Zheng, Y.P., Wang, H.-N., Kang, F.-Y., Wang, L.-N., and Inagaki, M., *Carbon*, 42, 2603, 2004.
59. Inagaki, M., Tashiro, R., Toyoda, M., Zheng, Y.-P., and Kang, F., *J. Ceram. Soc. Japan Suppl.*, 112, S1513, 2004.
60. Inagaki, M., Tashiro, R., Washino, Y., and Toyoda, M., *J. Phys. Chem. Solids*, 65, 133, 2004.
61. Inagaki, M., Toyoda, M., Kang, F., Zheng, Y. P., and Shen, W., *New Carbon Mater.*, 18, 241, 2003.
62. Inagaki, M., Saji, N., Zheng, Y.-P., Kang, F., and Toyoda, M., *TANSO* 2004 [No. 215], 258, 2004.
63. Inagaki, M., Kang, F.Y., and Toyoda, M., *Chemistry and Physics of Carbon*, Vol. 29, L. Radovic, Ed., Marcel Dekker, 1, 2004.
64. Chung, D.D.L., *J. Mater. Sci.*, 22, 4190, 1987.
65. Furdin, G., *Fuel*, 77, 479, 1998.
66. Kwon, O.Y., Choi, S.-W., Park, K.-W., and Kwon, Y.-B., *J. Ind. Eng. Chem. (Korea)*, 9, 743, 2003.
67. Tryba, B., Morawski, A.W., and Inagaki, M., *Carbon*, 43, 2417, 2005.
68. Inagaki, M., Kobayashi, S., and Tryba, B., *TANSO*, 2004 [No. 215], 249, 2004.
69. Umehara, K., Nakamura, S., and Saito, M., *27th Symposium on Chemical Treatment of Woods, Proceedings*, 49, 1997 [in Japanese].
70. Iwashita, N., Nishi, Y., Sawada, Y., and Inagaki, M., *Zairyou*, 53, 818, 2004 [in Japanese].
71. Aggarwal, R., *Carbon*, 15, 291, 1977.
72. Washburn, E.W., *Phys. Rev.*, 17, 273, 1921.
73. Beltran, V., Escardino, A., Feliu, C., and Rodrigo, M.D., *Br. Ceram. Trans. J.*, 87, 64, 1988.
74. Beltran, V., Barba, A., Rodrigo, M.D., and Escardino, A., *Br. Ceram. Trans. J.*, 88, 219, 1989.
75. Beltran, V., Barba, A., Jaque, J.C., and Escardino, A., *Br. Ceram. Trans. J.*, 90, 77, 1991.
76. Escardino, A., Beltran, V., Barba, A., and Sanchez, E., *Br. Ceram. Trans. J.*, 98, 225, 1999.

77. Samoilov, N.A., Khlestkin, R.N., Osipov, M.I., and Chichirko, O.P., *Russ. J. Appl. Chem.*, 77, 327, 2004.
78. Bayat, A., Aghmiri, S.F., Moheb, A., and Vakili-Nezhaad, G.R., *Chem. Eng. Technol.*, 28, 1525, 2005.
79. Suni, S., Kosunen, A.-L., Hautala, M., Pasila, A., and Romantschuk, M., *Marine Pollut. Bull.*, 49, 916, 2004.
80. Annunciado, T.R., Sydenstricker, T.H.D., and Amico, S.C., *Marine Pollut. Bull.*, 50, 1340, 2005.
81. Sun, X.F., Sun, R., and Sun, J.X., *J. Agric. Food Chem.*, 50, 6428, 2002.
82. Adebajo, M.O. and Frost, R.L., *Spectrochim. Acta Part A*, 60, 2315, 2004.
83. Sun, X.F., Sun, R.C., and Sun, J.X., *Biores. Technol.*, 95, 343, 2004.
84. Sun, R.C., Sun, X.F., Sun, J.X., and Zhu, Q.K., *Comptes Rendus, Chim.*, 7, 125, 2004.

Index

D

E

F

G

H

I

K

L

M

N

O

P

V

X

Z

Printed in the United States
by Baker & Taylor Publisher Services